W9-BCS-884

Chaos in Wonderland

Also by Clifford A. Pickover and available from St. Martin's Press

Computers, Pattern, Chaos, and Beauty
Computers and the Imagination
Visions of the Future
Mazes for the Mind

Chaos in Wonderland

Visual Adventures in a Fractal World

Clifford A. Pickover

St. Martin's Press New York

 For information, write: St. Martin's Press, Inc., 175 Fifth Avenue, New York, New York 10010

First published in the United States of America in 1994

Printed in the United States of America

Library of Congress Cataloging-in-Publication Data

Pickover, Clifford A.
Chaos in Wonderland: visual adventures in a fractal world / Clifford A. Pickover. —1st ed.
p. cm.
Includes bibliographical references.
ISBN 0-312-10743-9
1. Chaotic behavior in systems. 2. Fractals. 3. Computer graphics. 4. Visualization. I. Title.
Q172.5.C45P53 1994 93-40399
003'.7 —dc20 CIP

First Edition: August 1994
10 9 8 7 6 5 4 3 2 1

To Elahe

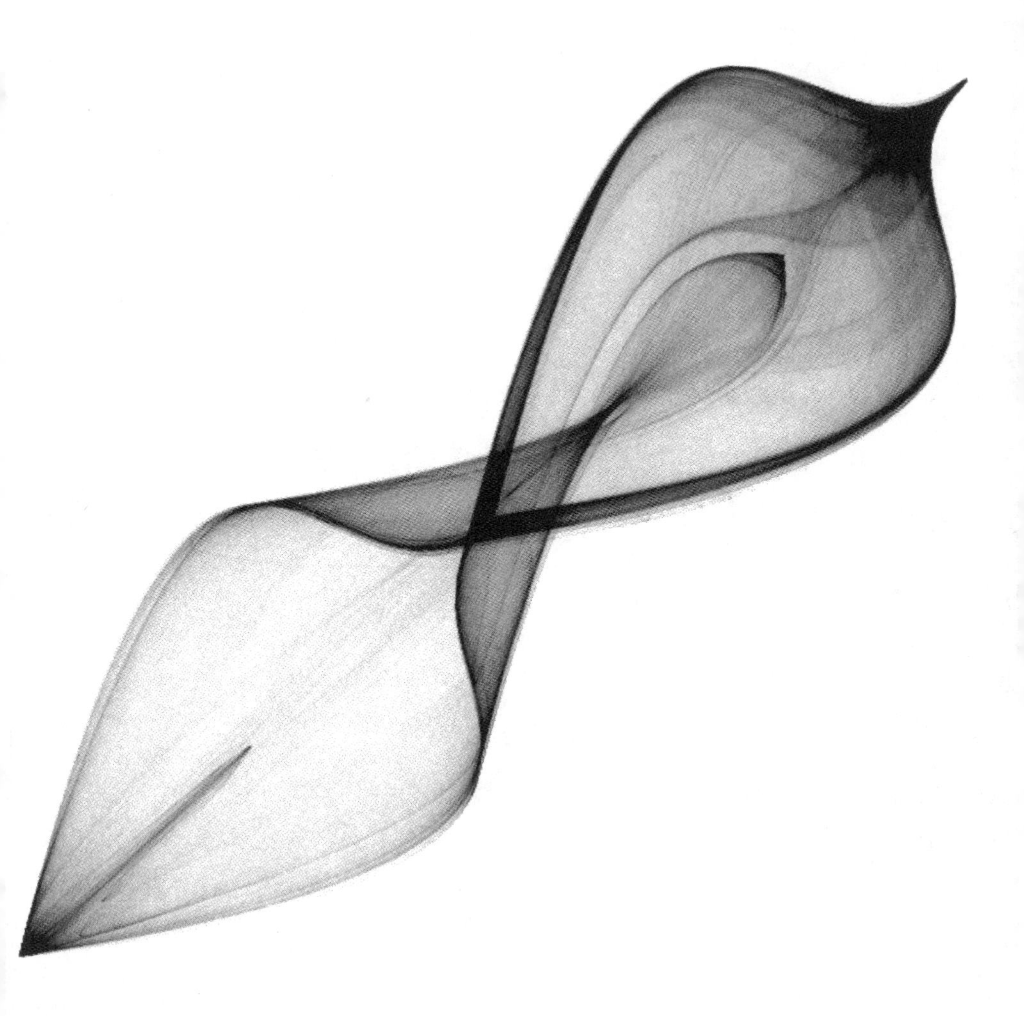

(Pictured here is a symmetrical chaotic attractor.)

Preface

"The future belongs to those who believe in the beauty of their dreams."
Eleanor Roosevelt

I sometimes dream that a hundred years from now a spaceship from earth discovers the remnants of advanced lifeforms on Ganymede, one of the water-containing moons of Jupiter. What happened to them? No one knows. They may have metamorphosed into structures not recognizable as life by human eyes, or may have become spores waiting to be awakened in some hidden crater. Perhaps they have been consumed by alien carnivores and then digested and expelled in another dimension.

Musings such as these led me to imagine the enigmatic Ganymede creatures you will meet on the following pages. They spend their days contemplating intricate mathematical patterns. Status in their society is determined by the beauty of their dream structures. On one level, this book catalogues their dreams and also describes a world filled with unusual biologies, technologies, and social structures. On a second level, this book, like my earlier books, is a computer-mathematical cookbook which allows readers to recreate the chaotic patterns produced by a single, simple set of equations. On a third level, this book introduces readers to *chaos science* – the science behind many intricate and unpredictable patterns in mathematics and nature. Our modern universe is no longer contemplated in terms of stable periodic planetary motions that were the heart of Newtonian classical mechanics. It is a chaotic universe of fluctuations and instabilities – where small changes may lead to amplified and unexpected effects. New tools and concepts have evolved in the past two decades to aid our understanding of chaotic forms. In particular, computer graphics has emerged as an important tool for analyzing chaotic patterns in fluid flows, the weather, the stock market, and the behavior of social insects and human crowds.

Research in chaotic patterns indicates that the line between art and science has become indistinct. Computer graphics makes this particularly apparent and adds a new element to the field of chaos and the related field of fractal geometry. The word "fractal" was invented by Benoit Mandelbrot to describe a set of curves rarely seen before the advent of computers with the ability to perform massive numbers of calculations quickly. Fractals are intricate geometrical objects which exhibit structure at any scale – regions of the pattern can be magnified again and again and not lose any detail or beauty. Chaotic behavior often leads to fractal patterns. The irregular patterns in this book are good examples of fractals.

The fascinating idea that there might be life on a Jovian moon has been discussed by several authors in the past. For example, Arthur C. Clarke,[1] Richard C. Hoagland,[2] and Dr. Roger Jastrow[3] have suggested that living forms could exist on Europa, beneath ice-covered oceans kept liquid by Jovian tidal forces. Personally, I find Ganymede (pronounced "Gah-nuh-meed") to be the most remarkable moon in our solar system. Ganymede's strange tangled bundles of *sulci* visually remind one of biological filaments or neuronal nets, and in fact the term *sulcus* is frequently used to label the shallow furrows on the surface of the brain. Ganymede is quite large, slightly larger than the planet Mercury, but it has a density almost three times less than Mercury. Most astronomers believe this indicates that Ganymede contains a huge amount of frozen water.

This book is divided into three main parts. *Part I* describes the biology, sociology, and mathematics of the Latöōcarfians so that the reader can appreciate the context in which *Part II* takes place. *Part II* is a sequence of science-fictional episodes describing the expedition of a human zoologist and his girlfriend as they explore a subterranean air chamber on Ganymede filled with bizarre lifeforms, civilizations, and technologies. Finally, *Part III* is a cornucopia of curiosities: games played on fractal boards, instructions on how to create globular star clusters using personal computers, a listing of "The 100 Strangest Mathematical Titles Ever Published," and additional puzzles to stimulate your imagination. A *Glossary* is provided to familiarize you with technical terms.

> *"Perhaps the most important fact about the times we live in is that they are going to be different soon. We live in a world of change – what Isaac Asimov has called 'a science-fiction world' – and anyone who wants to read a 'realistic' fiction turns naturally to science fiction, the literature of change."*
>
> James E. Gunn, *The New Encyclopedia of Science Fiction*

---

[1] Clarke, A. (1982) *2010: Odyssey Two*, Ballantine: NY.

[2] Hoagland, R. (1980) "The Europa Enigma," *Sky and Telescope*, January.

[3] Dr. Jastrow is from the NASA Institute of Space Studies.

In a Wonderland they lie,
Dreaming as the days go by,
Dreaming as the summers die:

Ever drifting down the stream –
Lingering in the golden gleam –
Life, what is it but a dream?

Lewis Carroll
Through the Looking-Glass

"Chaos, she decided, that's what love was,
or at least what attraction was.
Two people, a simple system,
but beyond the capacity
of the human mind to predict."

- *Dark Matter*, Garfield Reeves-Stevens

Contents

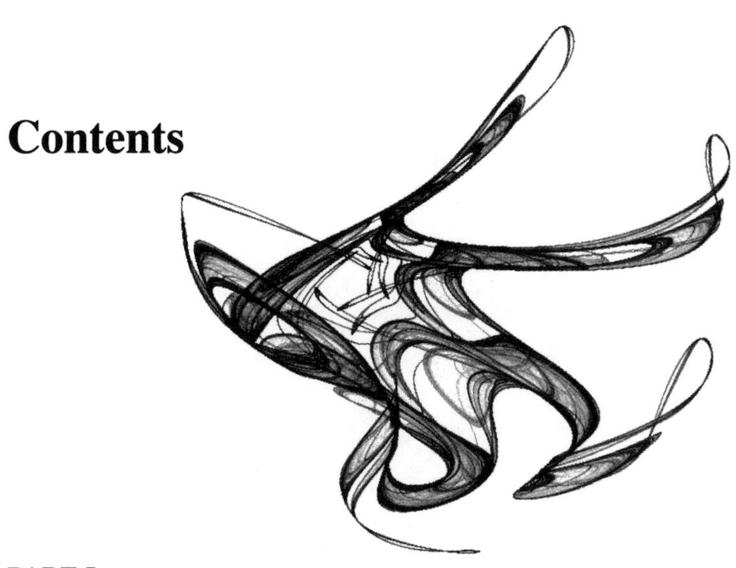

PART I
THE LATÖÖCARFIAN CIVILIZATION

Part I

THE LATÖÖCARFIAN CIVILIZATION

PLANETS

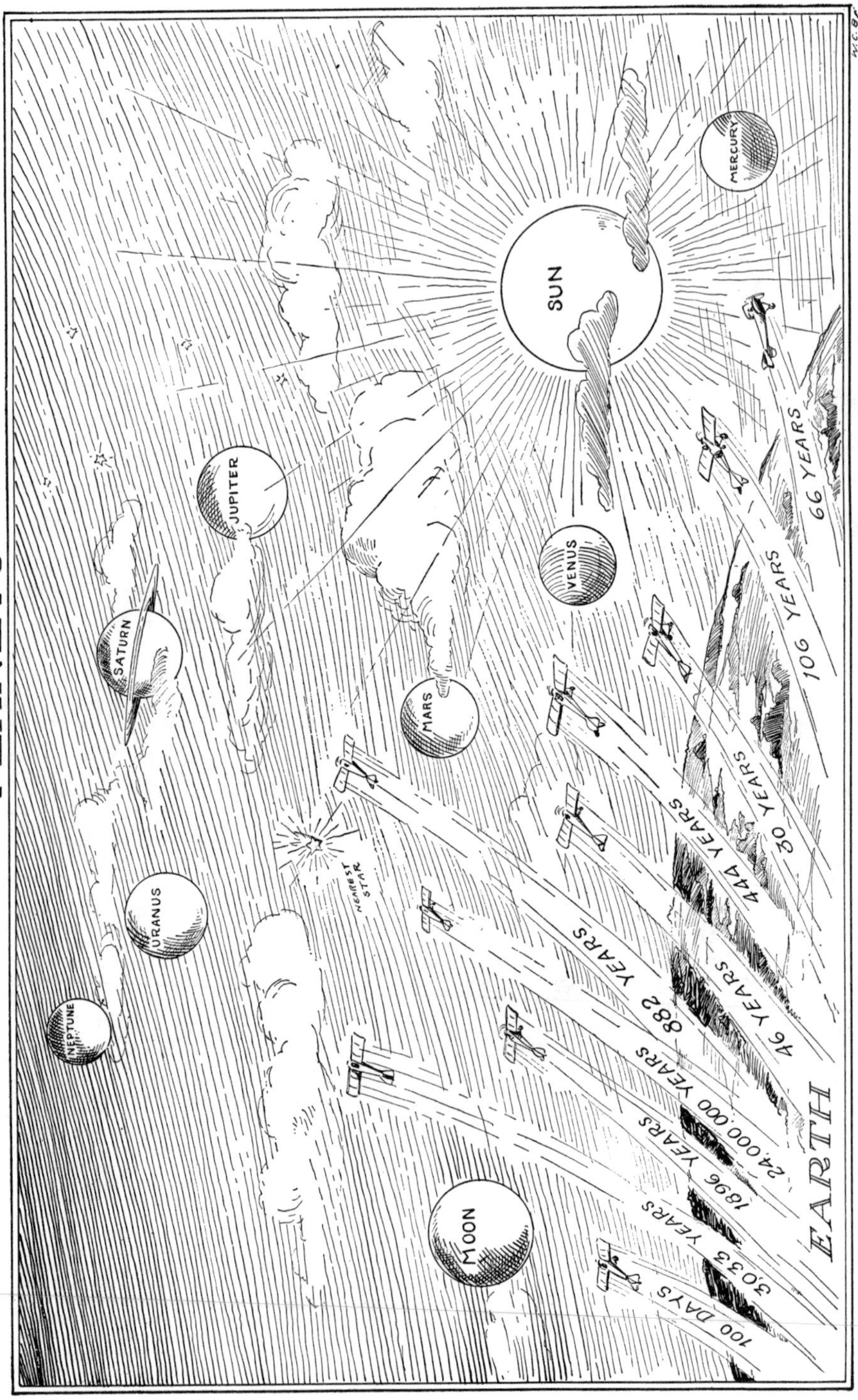

LENGTH OF TIME REQUIRED TO REACH THE VARIOUS PLANETS FROM THE EARTH VIA AEROPLANE TRAVELING 100 MILES PER HOUR

Chapter 1

Introduction

As a boy I often visited my father's study to examine his eclectic collection of old books. The book I recall most vividly was called *The Volume Library*, first published in 1911. Its Preface stated its goal:

> *The Volume Library* is a ready reference for the busy man where he may quickly and easily inform himself about matters of education, history, literature, science, biography, geography, trade, industry, art, and an indefinite number of other equally important subjects. It is as the same time a text where may be studied arithmetic, algebra, geometry, grammar, mythology, hygiene, among a long list of studies. *The Volume Library* is a library in one volume.

The edition of the book we had, published in 1928, contained all kinds of exotic facts and arcane minutiae.[1] For example, after its list of the "seven wonders of the world"[2] it provided a list of the "seven wonders of the *modern* world":

1. Wireless telegraphy, telephone and radio
2. Automobile
3. Airplane
4. Radium
5. Antitoxin
6. Spectrum analysis
7. X-ray

---

[1] The edition my father had was: Brubacher, A. (1928) *The Volume Library*. Educators Association, 303 Fifth Ave, New York, NY.

[2] The "seven wonders of the world" are: The Pyramids, the Colossus of Rhodes, Diana's Temple at Ephesus, the Lighthouse of Alexandria, the Hanging Gardens at Babylon, the Statue of the Olympain Jove, and the Mausoleum by Artemisia at Halicarnassus.

Which of these would remain on the list today? What would your list contain? Genetic engineering? Computers? Television?

One illustration in the book held a particular fascination for me, and kindled an early interest in astronomy, and later science fiction and science in general. The figure, shown facing this *Introduction*, clearly illustrated the length of time required to reach the various planets from the earth if one were to travel in a 100 mile-per-hour airplane.[3] Many questions came to mind as I studied the figure. Were there undiscovered planets? Was there life on other worlds? If so, what would their biology and technology be like? I imagined all kinds of aliens: heat-resistant spiders that lived in the craters of Mercury, gas-beings flying through the flaming prominences on the Sun, and more conventional lifeforms residing in Jupiter's water-containing moons. Later, as an adult, I continued to imagine and write about such possibilities. I think that this simple picture from *The Volume Library*, viewed as a boy, provided a seed from which my recent popular science books grew, and in particular provided an early stimulus for the Latööcarfian civilization in the current book.

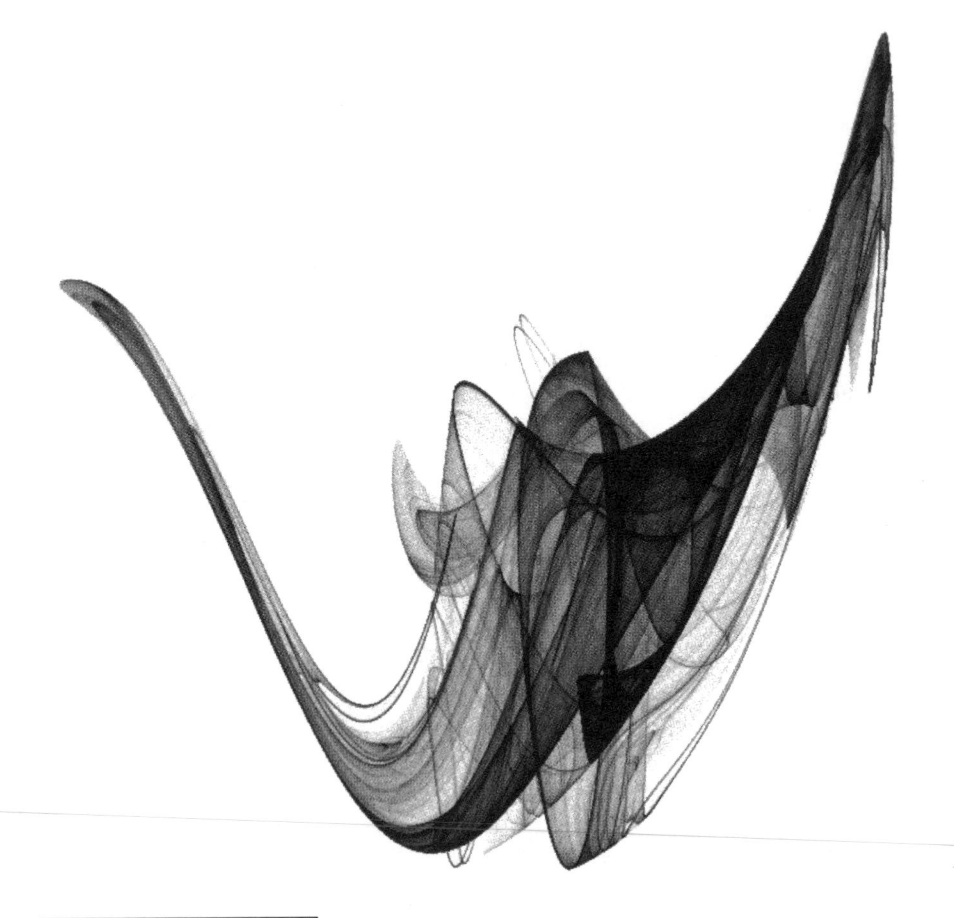

[3] Can you find any errors in this figure?

Chapter 2

The Ancient Latööcarfian Civilization

"Civilization begins where chaos and insecurity end."

Will Durant, *Our Oriental Heritage, 1954*

In the *Introduction* of this book, I told you about the times I visited my father's study to examine his collection of unusual books. *The Volume Library* was just one of many I discovered. Years after reading *The Volume Library*, I returned to my father's dusty library and found another odd book titled *The Latööcarfian Civilization*. It described a tribe of mathematicians living on a moon of Jupiter. I can't say where the book came from. No author or publisher was mentioned, and the book looked quite old. Much of the text is reprinted here. I've interspersed various scientific interludes to help you understand the mathematical and astronomical references.

Welcome to the world of the Latööcarfians.

2.1 The Air Pocket

Far from the bright twinkling city lights and the chaotic world of humans, lives a shy, sentient race of creatures known as the Latööcarfians. Their home is Ganymede, a moon of the planet Jupiter. Ganymede (radius 2,635 km / 1,636 mi) is the largest and brightest member of the Jovian family of moons. In fact, Ganymede is one of the largest satellites in the Solar System, rivalled only by Neptune's Triton, and Saturn's Titan. Ganymede has a rock and ice crust approximately 100 km thick, with a covering mantel of water or soft ice about 600 km thick. The icy surface has become dirty with age.

The Latööcarfian civilization developed long ago inside a huge air pocket within the ice of Ganymede. The ceiling of the subterranean air chamber is lined

Figure 2.1. *Jupiter and some of its moons.* Shown here are Io (top right), Europa (middle), Ganymede (top left), and Callisto (bottom left). These moons were first discovered in 1610 by Galileo.

with phosphorescent minerals and bioluminescent (glowing) bacteria which supplement the dim sunlight that penetrates the ice. Only in the last twenty years have the Latöökarfians travelled to the surface where they have built several beautiful cities and performed various mining activities.

The Latöökarfians' bodies are composed of aluminum gallium arsenide with traces of silicon from Ganymede's icy soil.[4] These materials cause their heads to be conductors of electrical signals, and their thoughts resemble the flow of electrons in computer chips. The Latöökarfians therefore think at speeds not achievable by Earthly, carbon-based lifeforms.

2.2 Ancient History

We know little about prehistoric Latöökarfian societies. We do know that the Neolithic Latöökarfian culture came to an end around 4000 B.C.[5] At this time this

4 Gallium is a soft, tough, bluish-white metal which is easily melted. On earth, it was discovered by M. Lecoq de Boisbaudran (1875) in a zinc-blend from the Pyrenees.

5 Dates are given in an Earth frame of reference. In actuality, a Jovian year is quite long – 11.8 earth years.

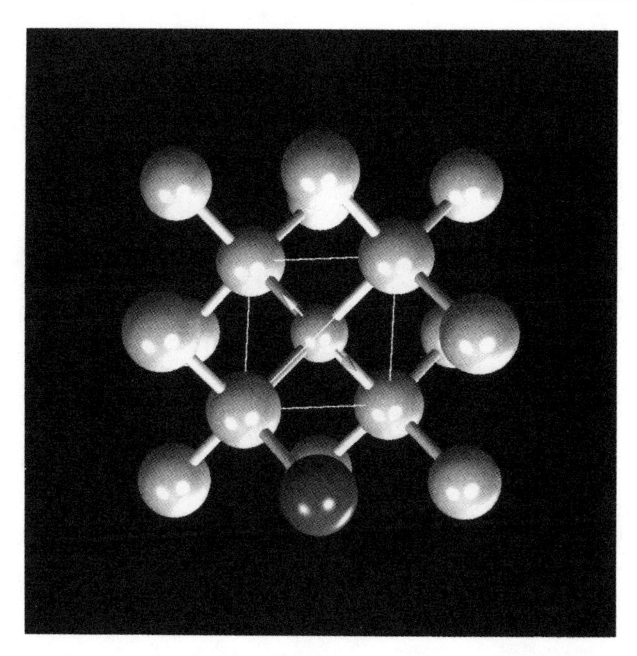

Figure 2.2. *Molecular model of aluminum gallium arsenide, a semiconductor.*

primitive society was gradually replaced by more complex patterns of culture based upon knowledge of writing and the replacement of stone tools by metallic implements. Calendars were devised. Religion, the state, and other institutions became more highly developed. Science, commerce, and industry flourished.

Since there was no enduring and unified culture on Ganymede until about 3100 B.C., the centuries from 4000 to 3100 are referred to as the pre-dynastic period. During this time, Ganymede was divided into a number of city-states cooperating with each other for economic purposes. Eventually there was a fusion of states to form two large kingdoms, one in the north and the other in the south. Ganymedean historians believe that consolidation took place though voluntary agreement since there is little evidence of military conquest.

Significant progress was made during the pre-dynastic Ganymede culture. For example, ornaments and tools were expertly fashioned from aluminum and niobium. Various new niobium compounds were discovered, allowing the Latöocarfians to create superior sculpture, pottery, and engraving. Ceramic and gem-carving arts also reached a high level of elegance.

In about 3200 B.C., the northern and southern kingdoms were combined into a single political unit under the rule of Senem. Senem was considered divine and held in such high respect that he could not be mentioned by name. He was forbidden to marry outside of his immediate family, lest his gallium arsenide compounds be contaminated by inferior mixtures – and therefore slow his thoughts. The King's chief subordinates were called viziers, and the government was founded upon a policy of peace and non-aggression, although wars sometimes arose. No

Figure 2.3. *Galileo Galilei, the Italian scientist who discovered Ganymede in 1610.*

separation of church and state existed. Later a Saracenic philosophy developed which taught that reason was superior to faith as a source of knowledge and that religion should be interpreted in a figurative or allegorical sense. The leading types of Latöócarfian poetry during this time were the drama, the pastoral, and the mime. Drama was almost exclusively comedy. These ancient Lactöócarfians were also capable mathematicians and developed algebra and trigonometry.

The Latöócarfians made use of a large, intelligent species of animal called a "Prohaptor" to perform most of the physical labor on Ganymede. See section 9.6, "Prohaptors," for more information.

> *"... An element of chance, perhaps of freedom, seems to enter into the conduct of metals and men. We are no longer confident that atoms, much less organisms, will respond in the future as we think they have responded in the past. The electrons, like Cowper's God, move in mysterious ways their wonders to perform, and some quirk of character or circumstance may upset national equations...."* Will and Ariel Durant, *The Lessons of History*

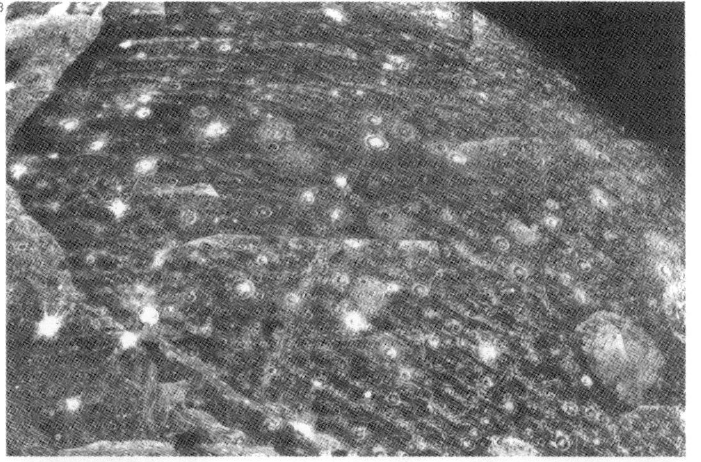

3. Galileo Regio
The most important of the large, dark types of terrain, Galileo Regio, covers about one-third of the hemisphere turned permanently away from Jupiter. It is heavily cratered, but its northern part is less dark than the rest of the interior, and may indicate some kind of condensate. Crossing Galileo Regio may be seen a series of parallel, gently curved bright streaks. These could be the result of a vast impact some distance away, but no trace of any impact center has been found, and the cause of the features may, in fact, be internal. Galileo Regio shows comparatively little vertical relief, because of the glacier-like "creep" in a crust composed largely of ice. The region is about 3,200 km in diameter, and probably represents the oldest surface on view on the satellite.

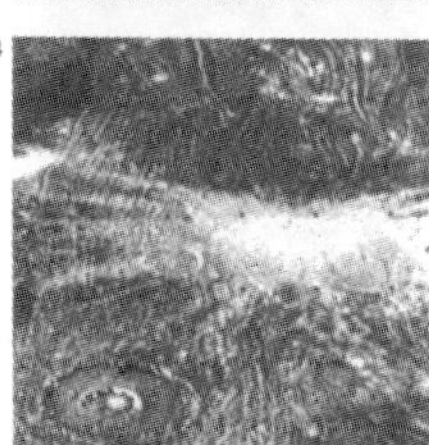

4. Ray crater (18 S, 192 W)
The ice surrounding ray craters is almost white; fresh material has been thrown up as a result of a meteoritic impact. This image shows one of the smaller unnamed ray craters; the crater at the lower left is called Eshmun.

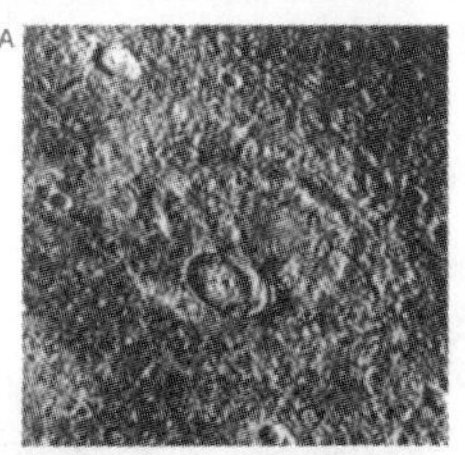

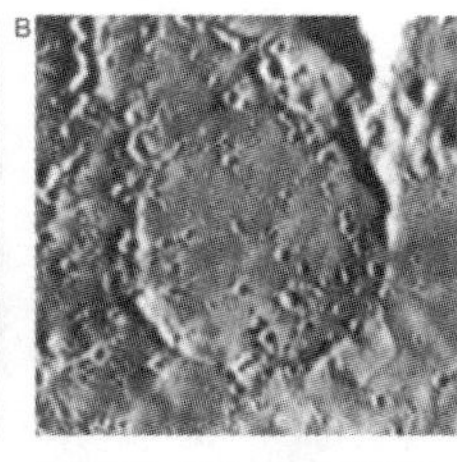

5. Ghostly crater (20 N, 120 W)
A large crater (**A**) has been almost obscured by fresher material, and a smaller, overlapping crater has been formed more recently. A photograph of the ghostly lunar crater Stadius is reproduced for comparison (**B**).

6. Tiamat Sulcus (3°S, 210°W)
A bright band of grooved terrain is shown here dividing areas of darker terrain in the Marius Regio. The bright band, Tiamat Sulcus, appears to be fractured by a fault extending from Kishar Sulcus. There is a discrepancy in the number of grooves on either side of the fault: fourteen on the northern side, as compared with twenty on the southern. The width of the grooves also differs. One explanation is that the grooves resulted from fractures that took place at different times on either side of the fault. The large number of craters suggests that the dark areas in the photograph are extremely ancient, while the brighter grooves, which exhibit fewer craters, are likely to have been formed more recently.

Map of Ganymede

Craters

Achelous	66°N, 4°W
Adad	62°N, 352°W
Adapa	83°N, 22°W
Ammura	36°N, 337°W
Anu	68°N, 332°W
Asshur	56°N, 325°W
Aya	67°N, 303°W
Ba'al	29°N, 326°W
Danel	4°N, 21°W
Diment	29°N, 346°W
Enlil	52°N, 301°W
Eshmun	22°S, 187°W
Etana	78°N, 310°W
Gilgamesh	58°S, 124°W
Gula	68°N, 1°W
Hathor	70°S, 265°W
Isis	64°S, 197°W
Keret	22°N, 34°W
Khumbam	15°S, 332°W
Kishar	78°N, 330°W
Melkart	13°S, 182°W
Mor	35°N, 323°W
Nabu	36°S, 2°W
Namtar	49°S, 343°W
Nigitsu	48°S, 308°W
Nut	61°S, 268°W
Osiris	39°S, 161°W
Ruti	15°N, 304°W
Sapas	59°N, 31°W
Sebek	65°N, 348°W
Sin	56°N, 349°W
Tanit	59°N, 32°W
Teshub	2°N, 16°W
Tros	20°N, 28°W
Zaqar	60°N, 31°W

Regiones

Barnard Regio	22°N, 10°W
Galileo Regio	35°N, 145°W
Marius Regio	10°S, 200°W
Nicholson Regio	20°S, 0°W
Perrine Regio	40°N, 30°W

Sulci

Anshar Sulcus	15°N, 200°W
Apsu Sulci	40°S, 230°W
Aquarius Sulcus	50°N, 10°W
Dardanus Sulcus	20°S, 13°W
Harpagia Sulci	0°, 317°W
Kishar Sulcus	15°S, 220°W
Mashu Sulcus	22°N, 200°W
Mysia Sulci	10°N, 340°W
Nun Sulci	50°N, 320°W
Philus Sulci	37°N, 215°W
Phrygia Sulcus	20°N, 5°W
Sicyon Sulcus	44°N, 3°W
Tiamat Sulcus	3°S, 210°W
Uruk Sulcus	0°, 157°W

Chapter 3

Modern History

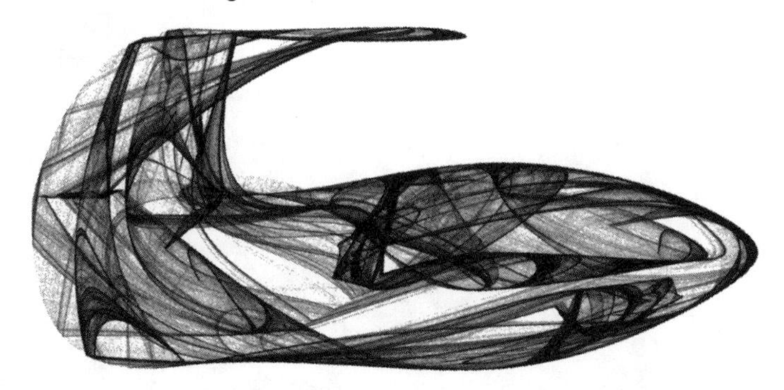

"Seconds later, he was circling Ganymede. Now there was a far more complex and interesting place. The most extraordinary feature of the Ganymedean landscape was the presence of meandering stripes. This grooved terrain looked as if it had been produced by armies of intoxicated ploughmen, weaving back and forth across the face of the satellite." Arthur C. Clarke, *2010*

3.1 Mathematical Art

Today the Ganymedean landscape consists of ancient, dark, heavily-cratered terrain, intermingled with younger regions with bundles of long parallel grooves. The term "sulcus," meaning furrow, is often used to describe these patterns created by mining operations of the Latöócarfians. Ganymede's surface was originally dark and composed partly of soft ice. Most of the ice cooled and froze during the first few hundred million years of Ganymede's evolution.

Due to various atmospheric changes in the underground ice pockets inhabited by the Latöócarfians, the Latöócarfians began to evolve and mutate. By the 1500's[6] those Latöócarfians residing beneath the Barnard Regio basin (22° N, 10° W – see frontispiece) and the Tiamat Sulcus (3° S, 210° W) had no appendages with which to manipulate their environment.[7] Through eons of evolution the Latöócarfians learned to spend their days in peaceful contemplation of mathematics. In the last century, Latöócarfian mathematical art was not mere decoration but was

[6] Dates are given in Earth notation.

[7] There are many examples of evolutionary appendage loss on Earth. Consider, for example, the snake and the whale which evolved from long-limbed ancestors, and humans who have lost their ancestral primate tails.

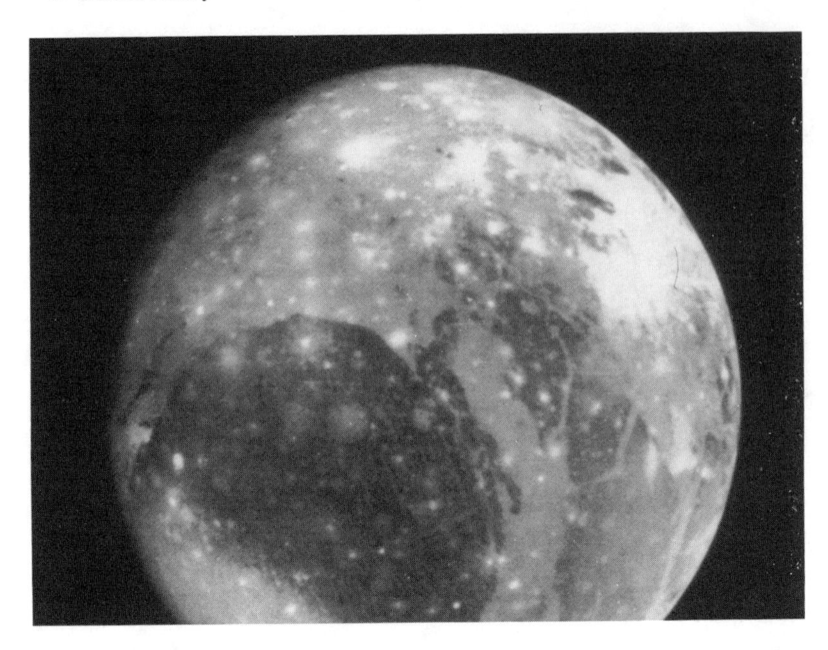

Figure 3.1. *Ganymede from 1.2 million km (746,000 miles).*

generated for the spiritual ennoblement of Latööcarfians. They drew no distinction between esthetic and ethical realms: the beautiful and the good were identical.

Today, their mathematical dreams are not of the simple parabolas and sine waves well-known to mathematics students on Earth. Rather, all through the lunar nights and days they dream of chaotic mathematical patterns produced by the formulas:

$$x_{t+1} = \sin(y_t b) + c \sin(x_t b) \tag{3.1}$$

$$y_{t+1} = \sin(x_t a) + d \sin(y_t a) \tag{3.2}$$

They study these equations but no others,[8] and never grow weary of their search for intricacy and attractiveness. These equations determine the position of points in intricate patterns as a function of time, represented by the subscript symbol t. For example, repeating the formulas a million times can lead to extraordinary designs composed of a million dots. The Latööcarfians communicate their visual dreams of these formulas to one another using infrared and/or electromagnetic signals in order to share their wonder at the unpredictable and beautiful behavior.

[8] See Appendix A, "Mutations of Equations," for minor, rare mutations of these equations.

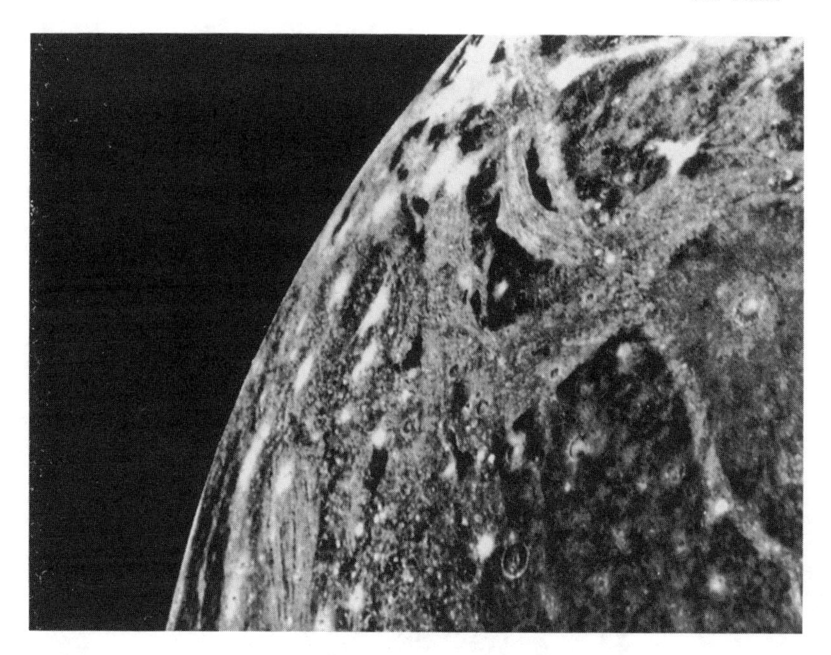

Figure 3.2. *Ganymede from 244,000 km (151,800 miles).*

3.2 Status

Status in the Latööcarfians society is not achieved through political prowess or financial fritinancy, but rather through the mental simulation of these geometrical patterns. Their purpose is partly to symbolize the chaos of modern life and to express defiance of traditional notions of forms – to repudiate the conception of art as mere ornament.

Latööcarfians use the designs to identify one another, in the same way names are used on Earth. For example, symmetrical patterns are signatures of the royalty, and the lower classes do not have the capacity to generate these patterns. In effect, this means that the lower classes do not have the capacity to "pronounce" the King's name.

Today, the Latööcarfian empire is vast and populous, encompassing some ten million individuals. About a tenth of that population lives in the subterranean capital Tenochtitlãn. This city is unusually fertile and productive, renowned for its excellent crystal gardens and abundant gallium arsenide.

> *"No live organism can continue for long to exist sanely under conditions of absolute reality. Even larks and katydids are supposed, by some to dream."*
>
> Shirley Jackson

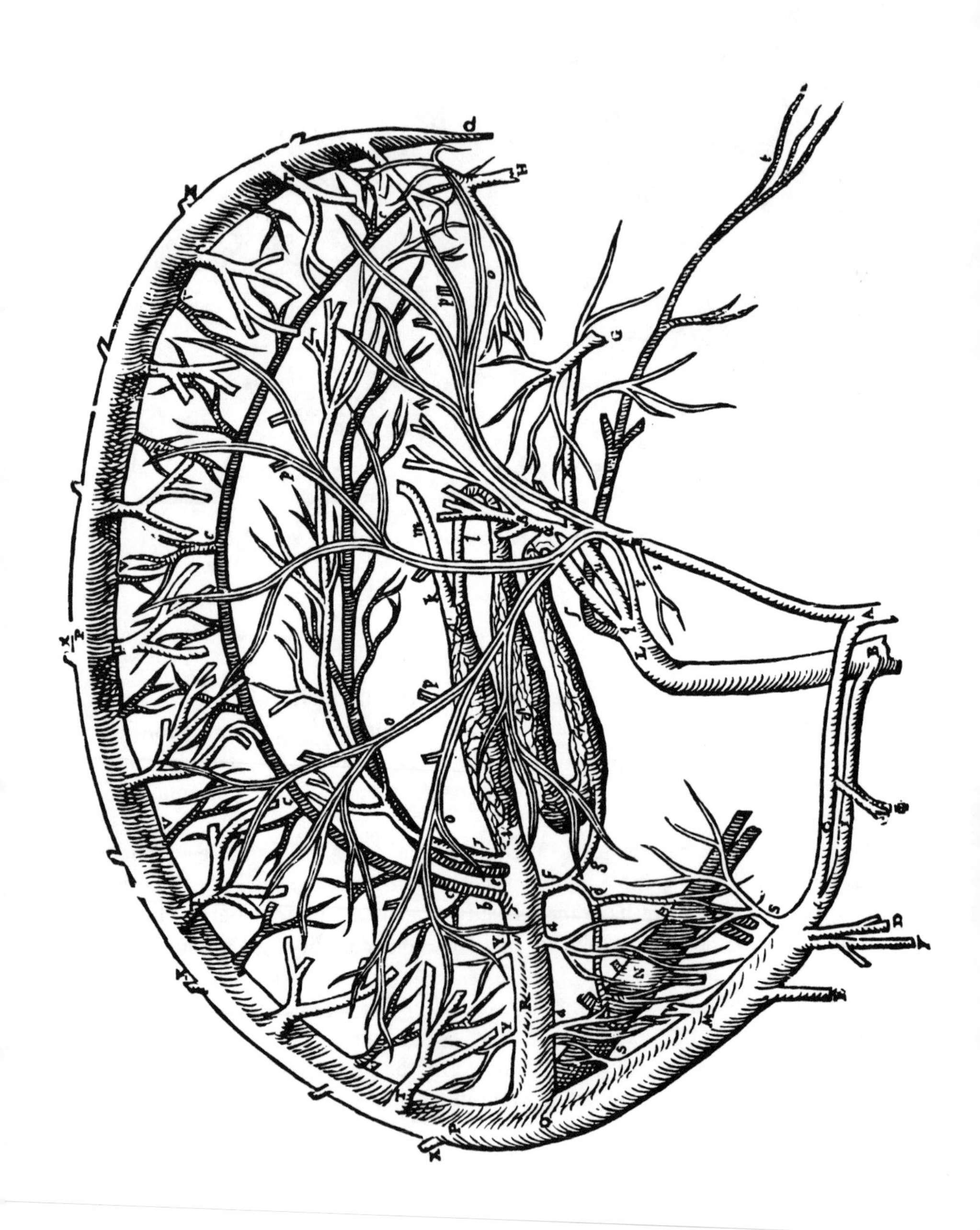

Chapter 4

Biology of the Latööcarfians

"Omnia quia sunt, lumina sunt. All things that are, are lights."

Scotus Erigna, *11th C*

4.1 Light-Emitting Heads

"All we are is light made solid." Anonymous

Since gallium arsenide emits light when subjected to an electric current, the Latööcarfians display intricate patterns (discussed in the previous chapter) on their heads. Their heads resemble myriads of tiny glittering lights. Like a million fireflies dancing to some unheard rhythm, the beautiful head-displays light up the dark Ganymedean evenings. Technical readers may be interested in the details of light production. The aluminum in the Latööcarfians' aluminum gallium arsenide heads produces an *oxide* used to insulate billions of light-emitting gallium arsenide tubes. Like a fire hose carrying water, the oxidized material channels light through an aluminum gallium arsenide core. Since the oxidized regions have a lower index of refraction, they deflect light, confining it to tiny fibers which channel electrons as well as photons in the Latööcarfians' heads.

Within the capital city, the Latööcarfians employ a pictographic writing system particularly suited to expressing mathematical ideas, and they frequently use simple symbols to represent some of the common intricate fractal curves. The pictographs are displayed on their heads.

4.2 Circulatory System

The Latöökarfian torso and head contain many small "blood vessels" through which *electrorheological fluids* flow. Like a kind of schizophrenic jello, these liquids rapidly transform from a liquid to a solid and back again by varying an electric field. On a microscopic level, the electrorheological fluids are suspensions of graphite particles in an oil. When the suspended particles are aligned, the fluid solidifies into a jellylike substance, and the effect can be turned on and off more than a thousand times a second. The purposes of the electrorheological fluids are many: shock absorption, lubrication, cooling, and an aide to the *piezoelectric* musclelike organs described in 4.3.4, "Body and Mouth Motions."

4.3 Communication

4.3.1 Saying Hello

How does one Latöökarfian citizen "know" which other Latöökarfian citizen to obey when there are so many infrared and electromagnetic communication signals in the atmosphere? When facing one another they can sometimes use head-display patterns to convey information, although the accuracy and usefulness of this approach is limited. For example, a common problem arises from the fact that lower classes cannot produce the signature patterns of the upper classes. More precise communication and identification is accomplished using a 48-bit identification code embedded in each Latöökarfians' head. This large number of code bits ensures there are about 300 trillion possible numbers, giving each Latöökarfian a unique code or address. When a signal is sent, its digital message can be addressed only to the serial number of the proper receiving Latöökarfian.

Latöökarfian brains contain, among other things, two masses functionally resembling 8-bit microprocessors.

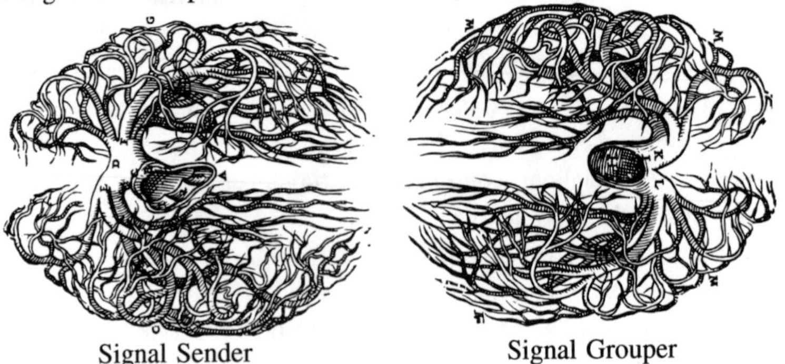

Signal Sender Signal Grouper

One processor manages the sending of infrared and electromagnetic signals. The other processor handles a special neurological protocol permitting the brain to group signals into subsets so the sender can address several Latöökarfian serial numbers simultaneously.

4.3.2 Secrecy

Although their society is largely peaceful, wars and skirmishes sometimes arise. Occasionally supersecret communication is a need not only for military commanders and intelligence agents, but also for the citizens who wish to relay confidential patterns. In a society where status depends on the beauty of one's patterns, and where communications are broadcast through the air, individuals often don't want their neighbors to intercept their thoughts or see their pattern experiments. The tremendous quantities of information exchanged in the Latöócarfian society make it imperative that confidentiality be guaranteed. This secrecy occurs through a system of encryption which is computational easy and yet difficult to decrypt should the messages or patterns be intercepted. The Latöócarfians use an encryption process involving two elaborate transformation algorithms, T and T^{-1}, operating on their communication signals. (Special organs in the head have evolved for carrying out the enciphering (T) and deciphering (T^{-1}) in an efficient manner.)

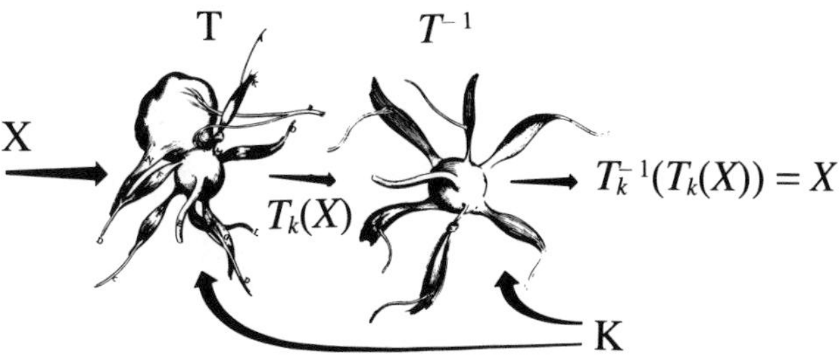

If the receiver and the sender of the message know both the algorithm and a mathematical key k, then enciphering and deciphering a message are both easy operations. Even if the transformation algorithm becomes known, the key must also be known to decode the message. A secret numerical key is not broadcast through the air but rather is exchanged verbally in face-to-face meetings so that there is less chance that the key be intercepted. The transformation algorithms themselves rely on chaotic formulas such as those used to create the Latöócarfian patterns.

4.3.3 Flipping Coins

Decisions involving chance are easy to carry out on Earth. For example, to randomly choose between two options, one might flip a coin and see whether it lands "heads" or "tails." One might also roll a dice for more choices. This presents a problem for the limbless Latöócarfians. The following true scenario is given as an example of how Latöócarfians use randomizing protocols in game playing and in daily life.

One cold day, a Latöócarfian boy was competing with his sister for the only remaining empty seat at a mathematics convention. Because both of them wanted to attend and only one could, they decided the only fair thing was to leave it to chance.

"OK," the boy said. "I just flipped a bit in my head. Is it 1 or 0? If it is 1, I go to the convention, otherwise you can go."

"Wait," his sister said with a creeping uneasiness at the bottom of her heart. "That's not fair. You could be lying."

"Listen sis, don't you trust your own brother? OK, I have an idea. I give you the number

```
13,407,807,929,942,597,099,574,024,998,204,846,127,
479,365,820,592,393,377,723,561,443,721,764,030,073,546,
976,801,874,298,166,903,427,690,031,858,186,486,050,
853,753,882,811,946,569,946,433,649,006,084,097
```

and if the largest prime factors contain an even number of 7's, you get to go to the conference. Otherwise I go."

"OK," the sister agreed. After a few minutes of computation, they determined that this large number could be represented as the product of three smaller ones: 2,424,833 (a prime) and a 99- and a 49-digit prime number:

$$2{,}424{,}833 \times$$

$$7{,}455{,}602{,}825{,}647{,}884{,}208{,}337{,}395{,}736{,}200{,}454{,}918{,}783{,}366{,}342{,}657 \times$$

$$741{,}640{,}062{,}627{,}530{,}801{,}524{,}787{,}141{,}901{,}937{,}474{,}059,$$
$$940{,}781{,}097{,}519{,}023{,}905{,}821{,}316{,}144{,}415{,}759,$$
$$504{,}705{,}008{,}092{,}818{,}711{,}693{,}940{,}737.$$

A blush of pleasure rose to her electrorheological cheeks. She smiled at her brother as a lachrymose little sigh escaped from her brother's lips.[9]

It should be pointed out that similar methods are used for determining chance outcomes in poker-like gambling games on Ganymede.

[9] On Earth, in 1990, Arjen Lenstra of Bellcore and Mark Manasse of the Digital Equipment Corporation were able to compute these same three prime factors of the 155-digit number. The task required 200 volunteers and nearly 1000 computers!

4.3.4 Body and Mouth Motions

Although they are limbless, Latöocarfians have the ability to wriggle their bodies and open and close their mouths. *Piezoelectric materials*, such as quartz and zinc oxide, line their mouths, torso, and alimentary canals. The oxides expand or contract when they are exposed to a voltage, as their molecules twist to align their internal charges with the electrical field. As a result, these substances act as mechanical devices that curl or extend in response to electrical signals from the Latöocarfians' heads. The piezoelectric materials also initiate peristaltic motion in the Latöocarfians' esophagi, stomachs, and intestines. Other wildlife on Ganymede use piezoelectric materials in a similar manner for digestion, locomotion, and for creating facial expressions. (For example, see the description of zinc oxide ants in section Chapter 36, "Death-Fungi and Zinc Ants.")

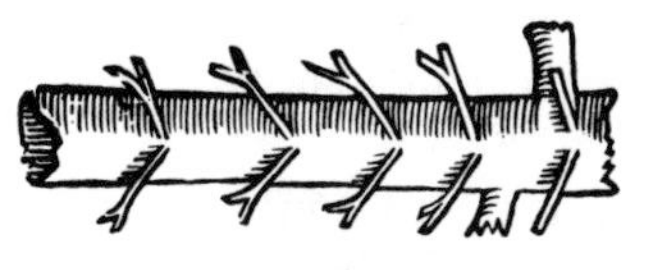

Piezoelectric Esophagus

Like humans and Earthly heterotrophic organisms, Latöocarfians derive their power and energy by metabolizing substances in the plants and animals they consume. To a lesser extent, during daylight hours they can also utilize the radiant energy of light in a chemical process similar to that used in solar cells.[10] To accomplish this, they rely on the photovoltaic effect to produce electricity: Light dislodges electrons from atoms in the linings of their heads. As the electrons flow from one layer to another, a charge builds, and electrons then flow through their brain circuitry. Their two-layered gallium arsenide and gallium antinomide skin converts 40% of the light to electricity. This double layer is particularly useful in utilizing the unusual spectrum of light produced by the bioluminescent bacteria and phosphorescent chemicals in the ice ceiling.

4.4 Alzheimer's on Ganymede

Possessing semiconductor brains not unlike modern computers, the Latöocarfians sometimes exhibit failure of their gallium arsenide materials. This is similar to human brain disease and occurs in the later years of their lives (approximately 200 years of age). Before modern medicine was developed on Ganymede, their lapses in memory were a mystery to Latöocarfian physicians. Recently, however, they determined that the recording section of their heads sometimes "forgot" random bits of information in the 32-kilobit-per-second data stream they were storing. Portions of the Latöocarfian heads act like a 400-megabyte digital tape recorder, and, sadly, 30% of the elders' heads acquired the debilitating habit of dropping random bits. Luckily, many Latöocarfians can interpolate the missing data. Others spend time oversampling data and less time processing data.

---

[10] Note: a day on Jupiter is about 9 hours.

A few years ago when the King himself displayed symptoms, his error rate became so high that medical practitioners were forced to find a way to bypass the recording section of his brain. They did so by gluing a huge hunk of aluminum gallium arsenide material to the side of his head. It jutted grotesquely from his body, like a deformed, rotting pear that had somehow taken root in his head. Today the King wears the protruding globule while in private, and removes the mucilagenous mass for short times when making public appearances, where cosmetic appearance is of importance.

As the Latöőcarfians age, their gallium arsenide lattices begin to accumulate damage from cosmic rays that penetrate the ice of Ganymede. The head displays of older individuals are dim as a result of this deterioration.

Alzheimer's disease on Ganymede is characterized by various pathological markers in the Latöőcarfian's brain – large numbers of aluminum gallium arsenide plaques surrounded by neurofibrillary tangles, and loss of semiconductor material from various regions. The normal Latöőcarfian head contains an extensive network of tubes and pores to prevent overheating of the head during computations. In Latöőcarfian Alzheimer's disease, this vascular system also becomes damaged from extensive plaque deposition. Note that if the Latöőcarfians spend too much time in the daylight, their heads can overheat causing damage to their brains. Therefore, ice is a favorite "hang out" for relaxing and recuperating. Some of their best thoughts occur during their ice baths. The activities of the lower classes of Latöőcarfians cause them to spend more time in the light, and thus they are more susceptible to damage by heat.

4.5 The Ice Plankton of Ganymede

Tiny protozoa live within the ice of Ganymede and form an integral part of the Latöőcarfian ecosystem. These planktonic creatures, called öös, migrate between the surface of Ganymede and the ceiling of the subterranean Latöőcarfian air chamber in a journey that lasts 17 years. The öös derive some of their initial energy from direct sunlight shining upon Ganymede's surface.

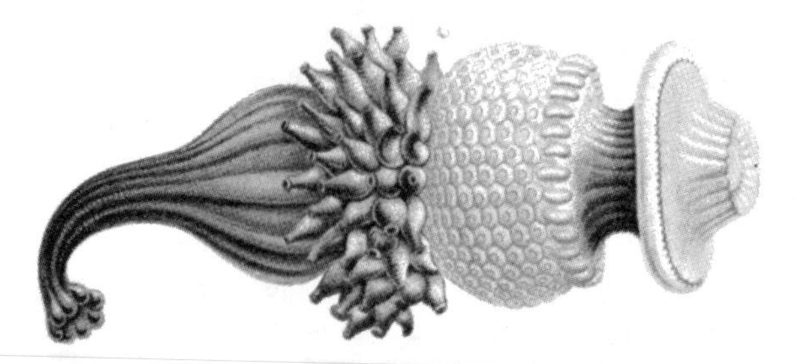

While on the surface, öös concentrate palladium in small membrane-enclosed organs located in the rear of their bodies. These mitochondria-like globules bring together palladium alloys and aqueous hydrogen ions, forming miniature "cold fusion" devices. The palladium-aided fusing of hydrogen heats the öös' rear end.

The hot rears enable the plankton to tunnel down through the ice by melting tiny channels as they travel with their rear ends first. The tunnels quickly freeze up behind the planktonic wanderers as they descend. When the öös finally arrive at the ceiling of the air pocket, they find that the rich atmosphere is conducive to their breeding. Mating lasts an hour. Baby öös are born within days, at which point they begin their 17 year journey to the surface. Several weeks after mating, the parent öös die and become food for the bioluminescent bacteria which line the air pocket's ceiling.

In addition to palladium, the öös concentrate traces of gallium arsenide on the surface of Ganymede, allowing their bodies to form gallium arsenide spicules secreted as an internal skeleton. The skeleton is built into beautiful, intricate patterns. Some öös contain shells which are perforated by miniature pores. The pores permit the extrusion of piezoelectric pseudopods to aid in tunneling through the ice. One species of öös has a spiraled series of diminishing chambers which are covered with tiny pores. Piezoelectric contractile vacuoles remove excess water from the protoplasm by pushing it out of the plankton's body at regular intervals.

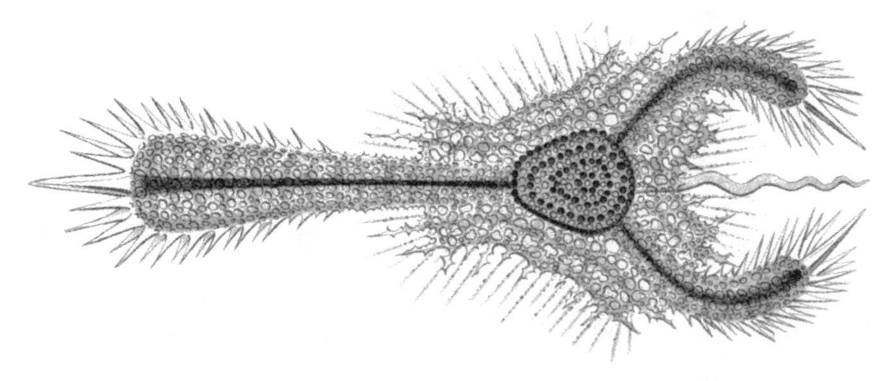

4.6 Pattern Smorgasbord

This book is a catalog displaying small morsels from the infinite smorgasbord of patterns which spring to life through the Latöocarfian system of equations. Although we humans cannot produce the designs with the same lightning speed the gallium arsenide Latöocarfians can dream the formulas in their minds' eye, the patterns are still very easy to compute using personal computers. To encourage your involvement, computational hints and parameters are given throughout the book.

4.7 Cross References

Several unusual Ganymedean technologies and creatures are discussed in Part II, "The Dream-Weavers of Ganymede." These include:

1. Fractal spiders – see section Chapter 23, "Fractal Spiders."
2. Neptune brooms – see section Chapter 24, "Kalinda."
3. Fractal swords – see section Chapter 26, "Battle."

4. A hyperbolic pipe world – see section Chapter 28, "Pipe World."
5. The Navanax slugs – see section Chapter 29, "The Navanax People."
6. Mole people – see section Chapter 30, "The Underground Association."
7. Liquid computers – see section Chapter 31, "The Water Beings."
8. Fractal palaces – see section Chapter 34, "The Fractal Palace of Ice."
9. Piezoelectric ants – see section Chapter 36, "Death-Fungi and Zinc Ants."
10. The Zooz – see section Chapter 36, "Death-Fungi and Zinc Ants."
11. The Kinorhyncha piezoelectirc worms – Chapter 37, "Starfish Soup."

> *"Man is a moment in astronomic time, a transient guest of the earth, a spore of his species, a scion of his race... Only a fool would try to compress a hundred centuries into a hundred pages of hazardous conclusions. We proceed."*
>
> Will and Ariel Durant, *The Lessons of History*

Pictured below is Ganymede, from 145,000 km (87,000 mi). The picture shows complex patterns of ridges and grooves which are the result of deformation of the thick icy crust.

Chapter 5

Interlude: Graphics and Mathematics

> *"The computer provides a crucial jump in the way we do science because it allows us to move in a realm where not all equations have solutions. Nonlinear systems can display their feathers, and all of a sudden we see they have a whole bunch of plumage that linear systems didn't have."* Norman Packard

As background to the Latöőcarfian formulas, let's first consider how scientists on Earth explore and study complex systems.

Traditionally when physicists or mathematicians saw complicated results, they often looked for complicated causes. In contrast, many of the shapes in this book describe the fantastically complicated behavior of the simplest of formulas. The results should be of interest to artists and non-mathematicians, and anyone with imagination and a little computer programming skill. Some readers may wonder why *scientists* and *mathematicians* use computer graphics to display mathematical results. Science writer James Gleick said it best in his 1987 book:

> *"Graphic images are the key. It's masochism for a mathematician to do without pictures... [Otherwise] how can they see the relationship between that motion and this. How can they develop intuition?"*

Therefore one of the purposes of this book is to illustrate simple graphics techniques for visualizing graphically interesting manifestations of chaotic, or irregular, behavior arising from the Latöőcarfian formulas. Readers new to the field of chaos may be amazed to find that even simple systems that have precise rules for going from one position to another, if projected far enough into the future, develop spirals, loops, and folds – all telltale signs of order giving way to chaos. This same kind of unpredictable behavior, the hallmark of chaos, can be found in weather patterns, dripping faucets, epileptic seizures, and heart attacks. Without the computer, who among us would have the patience to project the results of formulas far enough into the future where chaos begins to wobble its ugly head?

Get set for the erratic side of nature and mathematics – the discontinuous, chaotic monstrosities. Other interesting formulas for chaotic pattern generation are referenced in section Appendix K, "Other Chaotic Attractor Equations."

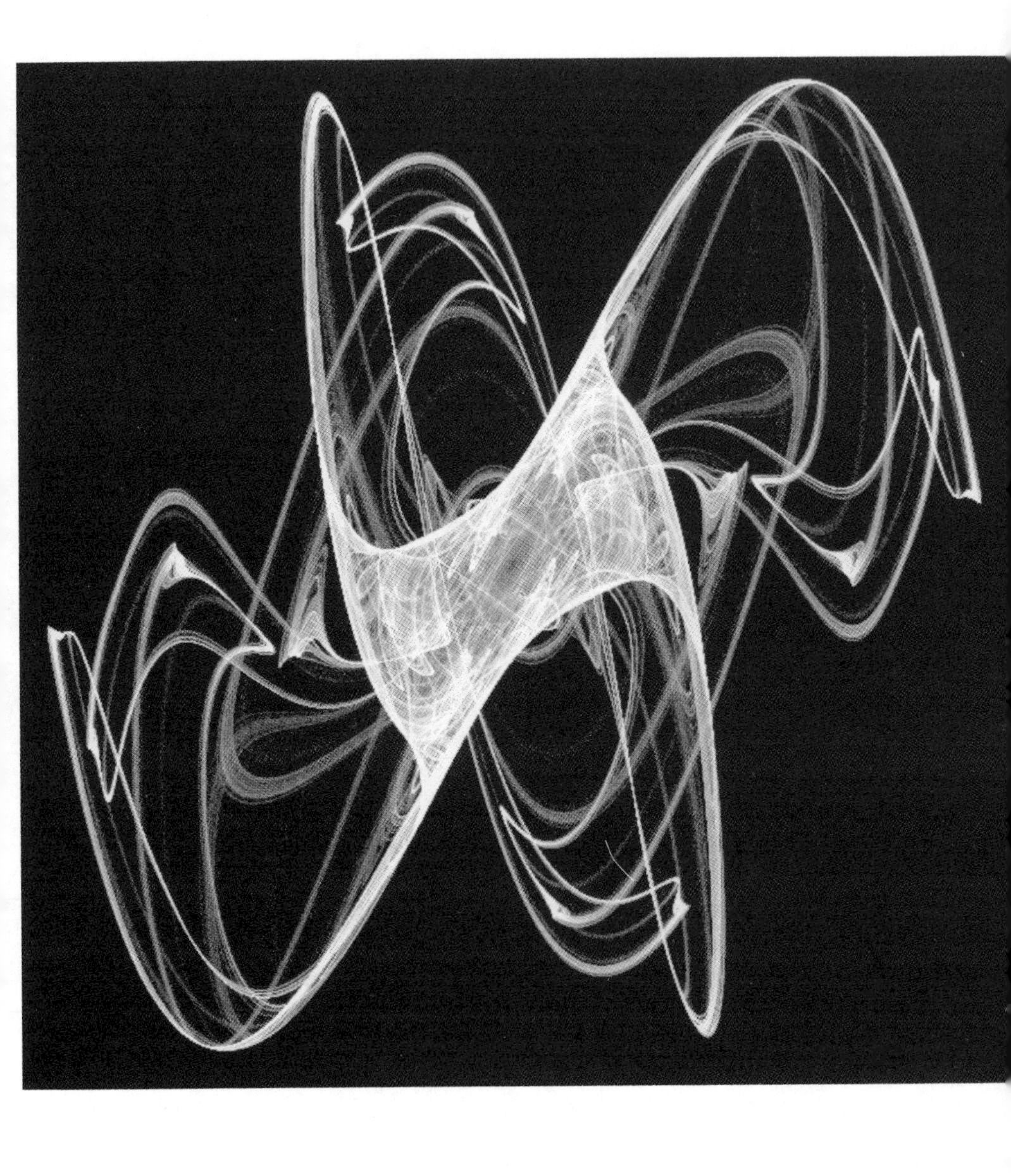

Chapter 6

The King

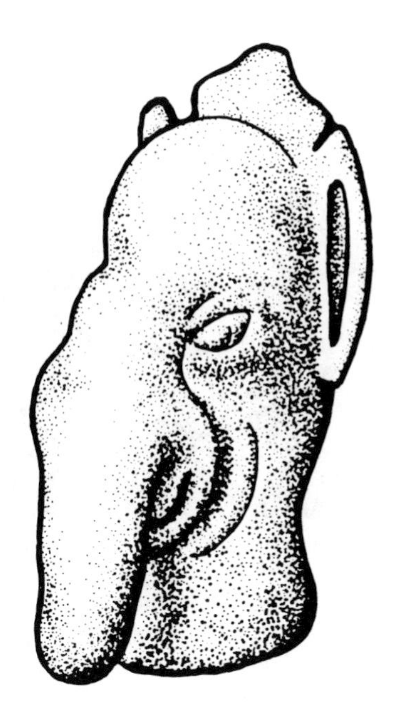

"Mathematics has always shown a curious ability to be applicable to nature, and this may express a deep link between our minds and nature. We are the universe speaking out, a part of nature. So it is not surprising that our systems of logic and mathematics sing in tune with nature." George Zebrowski

"The deepest need of man is to overcome his separateness, to leave the prison of his aloneness." Eric Fromm, *The Art of Loving*

6.1 Riding the Waves

The humble sine waves that lie at the very foundation of trigonometry have a special beauty all their own. It takes just a little coddling to bring the beauty out. But who would guess, for example, that intricate fractal patterns lurk within the sine operation applied to real numbers? The Latööcarfians use sine waves for producing designs of infinite detail and beauty. Below is a graph of $y = \sin x$. It repeats itself every 2π radians (360°).

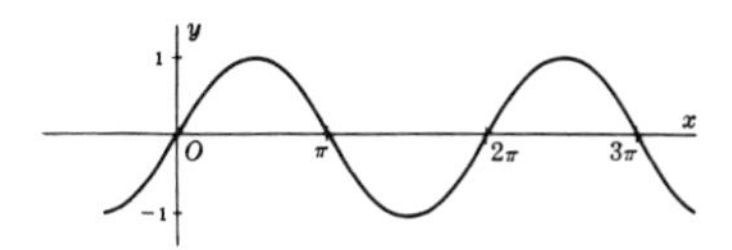

As mentioned in Chapter 3, "Modern History," social status in Latööcarfian society is founded upon thoughts of mathematical beauty. The more beautiful the pattern,

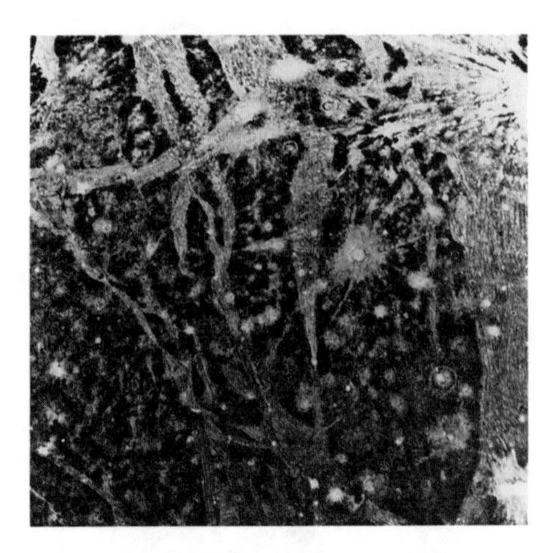

Figure 6.1. *Ganymede.* On the afternoon of March 5, 1979, Voyager 1 took this picture from a range of 158,400 miles (246,000 km). The height of the frame represents a distance of 600 miles on the surface. The surface displays numerous impact caters, many of which have extensive bright ray stems. Bright bands traverse the surface in various directions, and these bands contain an intricate system of alternating linear bright and dark lines which may represent a deformation of the crusted ice layer.

the greater the individual's prestige and position. The current King came to power as a result of his discovery of the pattern facing this chapter. His name is Yars Kotheck, and he teaches that all the miseries in life are due to selfishness and greedy desire. There is no social order, no security, no happiness, no peace, no kingship or righteous leadership unless Latöócarfians lose themselves to the study of mathematically chaotic patterns. Yars Kotheck has often said that to forget oneself in mathematics is to escape from the prison of aloneness and to free oneself from near immobility due to the Latöócarfian lack of appendages.

6.2 Personal Computers

For human readers interested in reproducing this pattern on a personal computer, the following steps are required:

```
x = 0.1; y = 0.1; /* starting point */
DO 10 Million Times
   xnew = sin(y*b) + c*sin(x*b)
   ynew = sin(x*a) + d*sin(y*a)
   x = xnew; y = ynew; PlotDotAt (x, y)
END
```

(See M.2, "Latöócarfian Explorer Program" for a listing in BASIC, and see B.2, "Some Figure Parameters and Descriptions" for information on related patterns.) The values of the real number constants *a*, *b*, *c*, and *d* may be chosen at random in

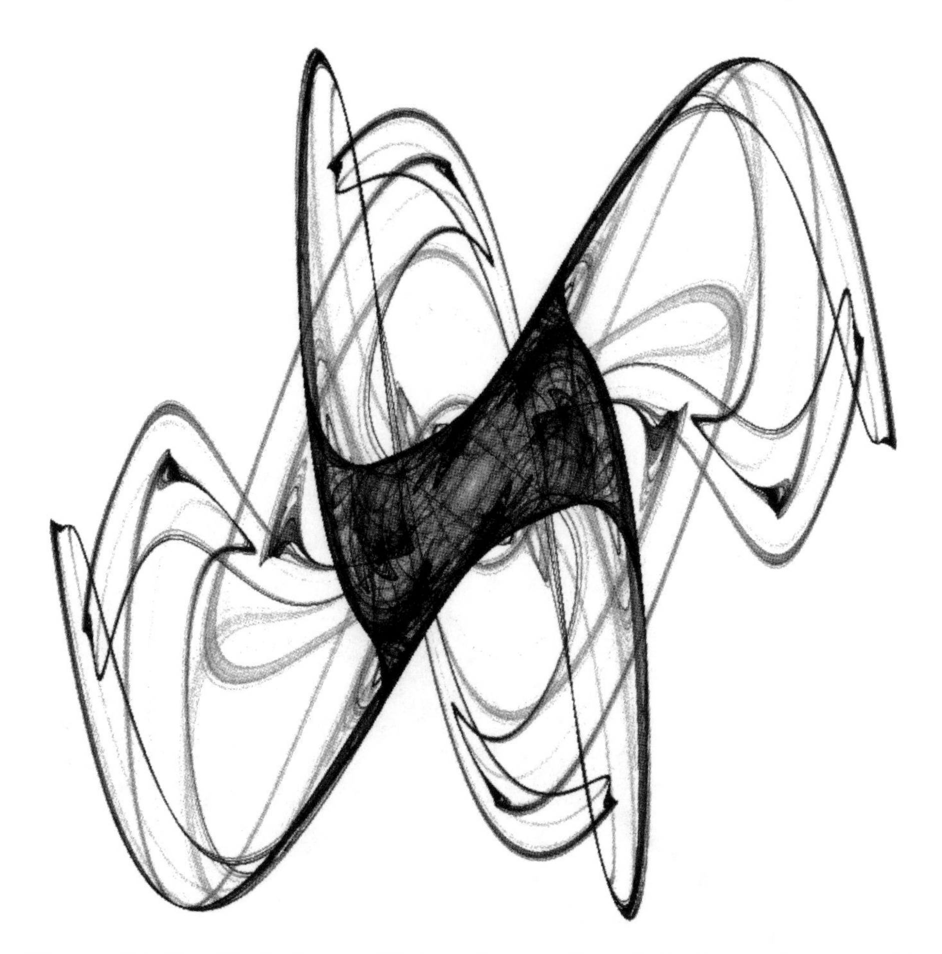

Figure 6.2. *The King's Dream.* (Photographic negative of the image shown on this chapter's frontispiece.)

a range $(-3 < a, b < 3)$ and $(0.5 < c, d < 1.5)$. These simple systems generate information as the system evolves. To see the patterns unfold, use the rules and starting conditions, repeat the equations over and over again, stand back, and watch the visually exciting behavior evolve on the computer screen. Each new value of x and y determines the position of points on a plane.

To produce the King's dream, use the following constants: $(a = -0.966918$, $b = 2.879879$, $c = 0.765145$, and $d = 0.744728)$. The picture boundaries are $(-1.86 < x < 1.86)$ and $(-1.51 < y < 1.51)$. The Lyapunov exponent, which is explained in detail in Chapter 14, "Interlude: Lyapunov Logomania," characterizes the degree of chaos in the pattern. For the King's dream, the value of the Lyapunov exponent is 0.48. If you magnify the center of the pattern, you will find additional intricate plumage.

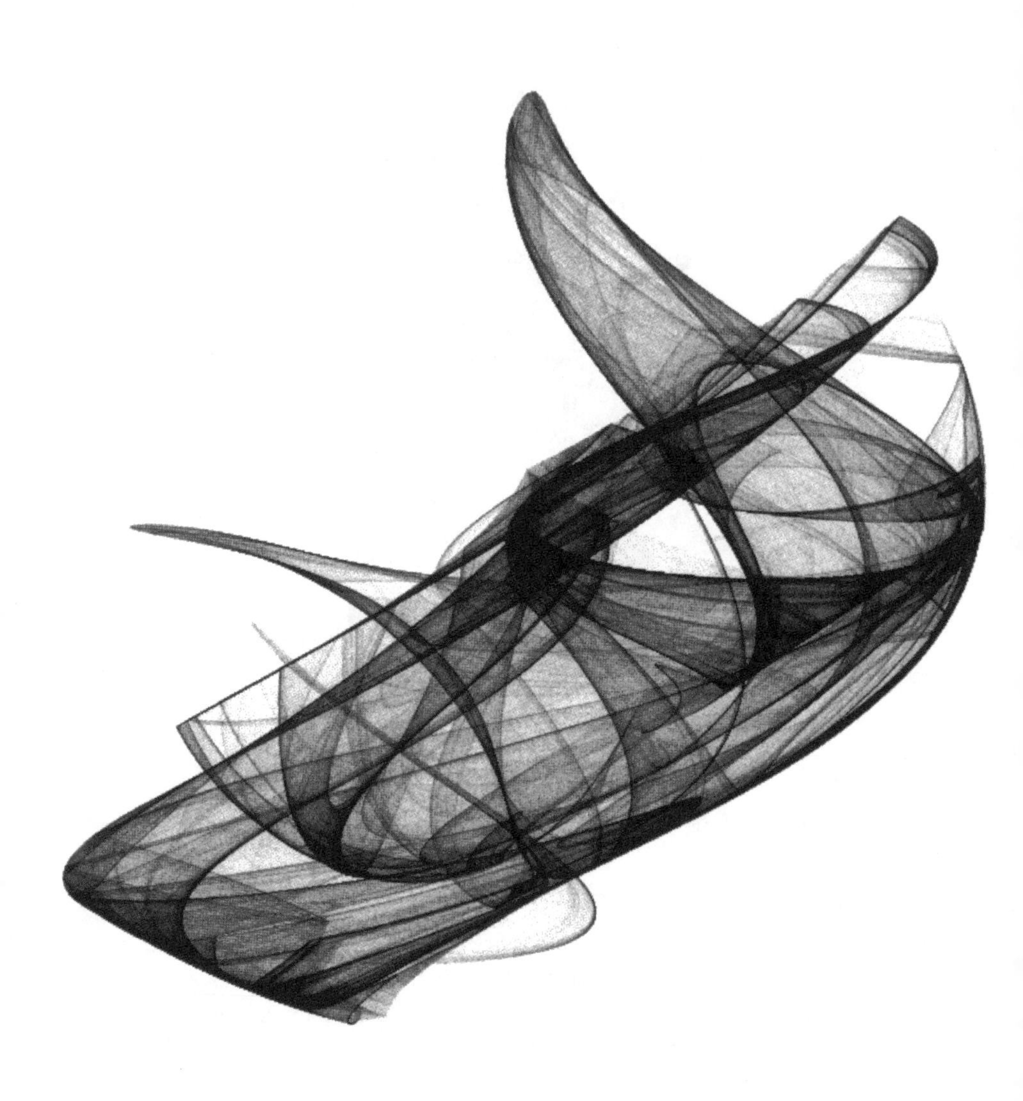

Chapter 7

Interlude: Chaos and Dynamical Systems

"If we wish to make a new world we have the material ready. The first one, too, was made out of chaos." Robert Quillen

"I believe that fractals respond to a profound uneasiness in man."
Benoit Mandelbrot, 1990 *Newsweek*

"Why all the fuss about fractals?" L. Kadanoff, 1986 *Physics Today*

7.1 Attractors

The bizarre and beautiful Latööcarfian patterns represent the behavior of, what chaos experts call, *dynamical systems* and *strange attractors*. This short chapter should familiarize you with these and related terms.

To ancient humans, Chaos represented the unknown, the spirit world – menacing, nightmarish visions that reflected man's fear of the irrational and the need to give shape and form to his apprehensions. Today chaos usually involves the study of a range of phenomena exhibiting a sensitive dependence on initial conditions. This means that if you very slightly change a parameter in an equation or system, very different behavior can result. As mentioned in previous chapters, there are many examples of chaotic behavior. From chaotic toys with randomly blinking lights to wisps and eddies of cigarette smoke, chaotic behavior is generally irregular and disorderly. Other examples include weather patterns, some neurological and cardiac activity, the stock market, and certain electrical networks of computers. Although chaos often seems totally "random" and unpredictable, it actually obeys strict mathematical rules that derive from equations that can be formulated and studied. Today, there are several scientific fields devoted to the study of how complicated behavior can arise in systems from simple rules and how minute changes in the input of a nonlinear system can lead to large differences in the output; such

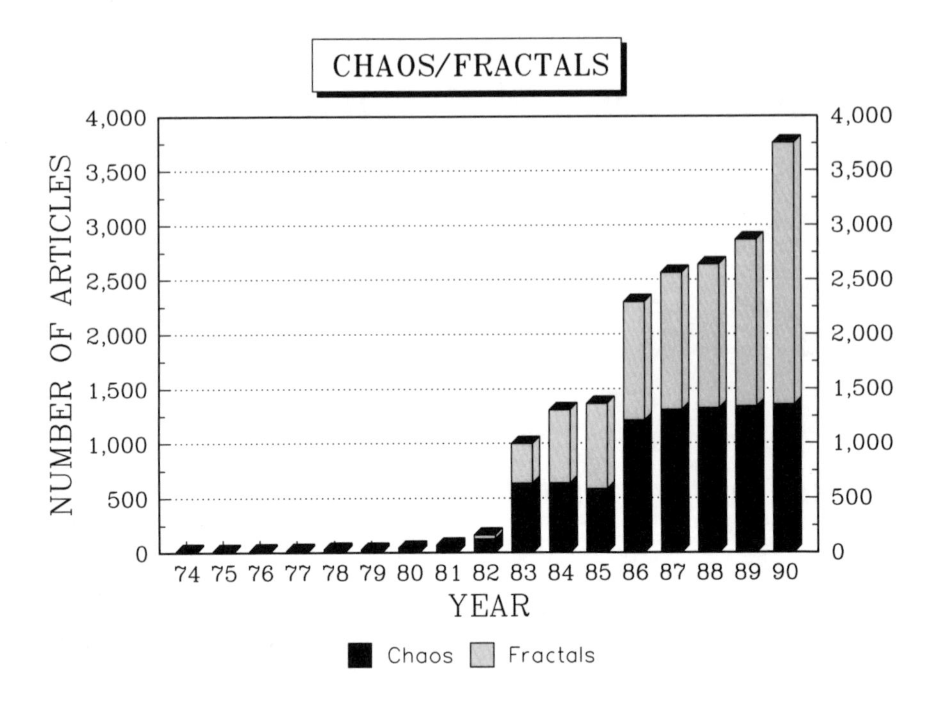

Figure 7.1. *Chaos and fractal articles.* A review of the world scientific literature between 1973 and 1990 shows the number of chaos and fractal articles rising dramatically between the years 1982 and 1990. (Reprinted from Pickover, C. (1991) *Computers and The Imagination*, St. Martin's Press: New York.)

fields include chaos, fractal, complexity, and cellular automata theory. Many chaotic systems that people investigate are expensive to set up and rather complicated to study. Of course, there is special interest in systems, such as those presented in this book, from which researchers can easily collect experimental data.

The inexhaustible reservoir for the images in this book is the *dynamical system.* Dynamical systems are models comprised of the rules describing the way some quantity undergoes a change through time. For example, the motion of planets about the sun can be modeled as a dynamical system in which the planets move according to Newton's laws. Many interesting mathematical images also represent the behavior of mathematical expressions called *differential equations.* Think of a differential equation as a machine that takes in values for all the variables at an initial time and then generates the new values at some later time. Just as one can track the path of a jet by the smoke path it leaves behind, computer graphics provides a way to follow paths of particles whose motion is determined by simple differential equations. The practical side of dynamical systems is that they can sometimes be used to describe the behavior of real-world things such as planetary motion, fluid flow, the diffusion of drugs, the behavior of inter-industry relationships, and the vibration of airplane wings. Often the resulting graphic patterns resemble smoke, swirls, candle flames, and windy mists.

The patterns in this book are also good examples of *strange attractors.* As background, *predictable attractors* represent the behavior to which a system settles down or is "attracted" (for example, a point or a looping closed cycle). An example of a *fixed point attractor* is a mass at the end of a spring, with friction. It eventually arrives at an equilibrium point and stops moving. A *limit cycle* is exemplified by a metronome. The metronome will tick-tock back and forth – its motion always periodic and regular. A *strange attractor* has an irregular, unpredictable behavior. Its behavior can still be graphed, but the graph is much more complicated. With "tame" attractors, initially close points stay together as they approach the attractor. With strange attractors, initially adjacent points eventually follow widely divergent trajectories. Like leaves in a turbulent stream, it is impossible to predict where the leaves will end up given their initial positions.[11]

So extensive is the interest in fractals and chaos that keeping up with the literature on the subject is rapidly becoming a full-time task. In 1989 the world's scientific journals published about 1,200 articles with the words "chaos" or "fractal(s)" in the title. Figure 7.1 shows the number of papers with titles containing the words "chaos" or "fractal(s)" for the years 1975-1990, the 1990 values estimated from data for January-June 1990.

In physics, there are certain famous and clear examples of chaotic physical systems. Here are a few examples: thermal convection in fluids, supersonic panel flutter in supersonic aircraft, particles impacting on a periodically vibrating wall, various pendula and rotor motions, nonlinear electrical circuits, and buckled beams. Moon's book in Appendix J, "For Further Reading," gives many more examples.

These days computer-generated fractal patterns are everywhere. From squiggly designs on computer art posters and tee-shirts, to illustrations in the most serious of physics journals, interest continues to grow among scientists, artists, and designers. Fractals are bumpy objects which usually show a wealth of detail as they are continually magnified. Some of these shapes exist only in abstract geometric space, but others can be used to model complex natural objects such as coastlines and mountains.

Chaos and fractal geometry go hand-in-hand. Both fields deal with intricately shaped objects, and chaotic processes often produce fractal patterns. A good example is turbulent (chaotic) fluid flow which produces similar-looking patterns at different size scales.

---

[11] Note that some scientists look for strange attractors *wherever* nature is irregular. Some argue that the earth's weather might lie on a strange attractor. Even though it may never be possible to precisely predict phenomena like the weather or the stock market, one might foresee the global patterns of their behavior – the "order within the chaos."

7.2 The Lorenz Attractor

> *"It is interesting that people try to find meaningful patterns in things that are essentially random."* Lt. Commander Data, *Star Trek: The Next Generation*

Here I want to encourage your involvement by providing a simple computer recipe. Consider, for example, the famed Lorenz Attractor. In 1962, MIT meteorologist E.N. Lorenz was attempting to develop a model of the weather. Lorenz simplified a weather model until it consisted of only three differential equations.

$$dx/dt = 10(y - x) \tag{7.1}$$

$$dy/dt = -xz + 28x - y \tag{7.2}$$

$$dz/dt = xy - (8/3)z \tag{7.3}$$

t is time, and d/dt is the rate of change with respect to time. If we plot the path that these equations describe using a computer, the trajectories seem to trace out a squashed pretzel. The surprising thing is that if you start with two slightly different initial points, e.g. (0.6, 0.6, 0.6) and (0.6, 0.6, 0.6001), the resulting curves first appear to coincide, but soon chaotic dynamics leads to independent, widely divergent trajectories. This is not to say that there is no pattern, although the trajectories do cycle, apparently at random, around the two lobes. In fact, the squashed pretzel shape always results no matter what starting point is used. This is the behavior to which the system is attracted.

To create the Lorenz attractor, one needs to solve the system of differential equations given above (see Code 7.1). Several numerical techniques can be used that yield accurate values for x, y, and z as a function of time. The most straightforward approach, which I have used to get a rough idea about the Lorenz attractor, simply replaces dx with *(xnew - x)*, and replaces dt by a time step, called h. Other higher accuracy approaches, such as Runge Kutta methods, can be used but only with consequent increase in computer time. To create a projection of this 3-D figure in the x-y plane, simply plot (x,y) points and omit the z-value (Figure 7.2).

7.3 Build Your Own Chaos Machine

One of the best non-computer projects I know for observing chaos is to build a double pendulum – a pendulum suspended from another pendulum. The motion of the double pendulum is quite complicated. The second arm of the pendulum sometimes seems to dance about under its own will, occasionally executing graceful pirouettes while at other times doing a wild tarantella. You can make the double pendulum from wood. At the pivot points, you might try to use ball bearings to insure low friction. (Ball bearings can be obtained from hobby shops or from discarded motors and toys.) Place a lead weight at the bottom of the first pendulum so that the pendulum will swing for a longer time. (The weight stores potential energy when the pendulum is lifted.) The second pendulum arm can be about half the length of the first. You can

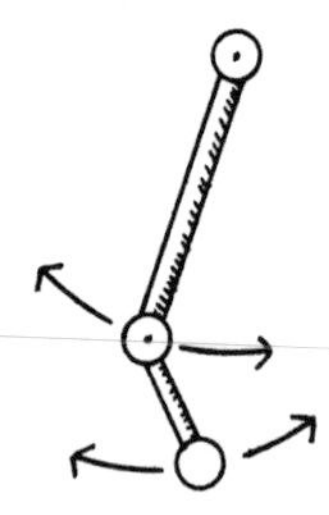

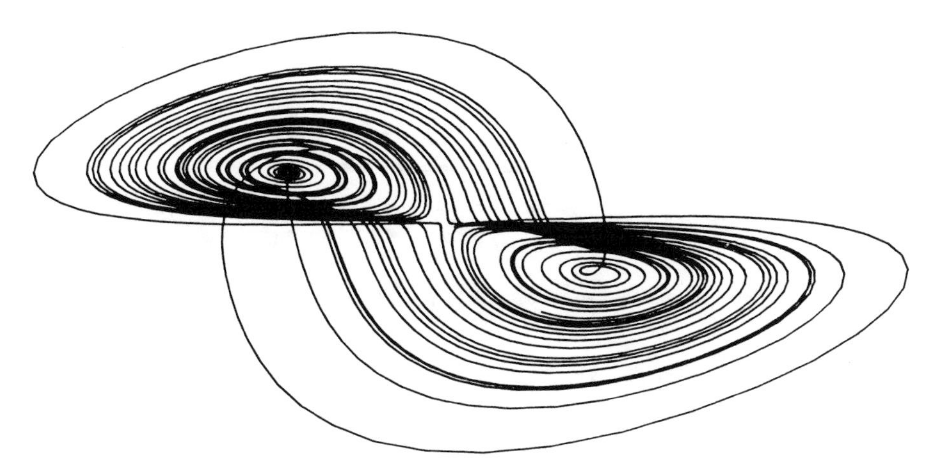

Figure 7.2. *The Lorenz Attractor.*

place a bright red dot, or even a light, on one end of the second pendulum so that your eye can better track its motion. Note that your pendulum will never trace the same path twice. This is because you can never precisely reposition it at the same starting location due to slight inaccuracies in knowing where the starting point is. These small initial differences in position are magnified through time until the pendulum's motion and position becomes unpredictable. Can you predict where the lower pendulum will be after two or three swings? Could the most powerful supercomputer in the world predict the position of the pendulum after 30 seconds, even if the computer were given the pendulum's precise equations of motion? Unlike the strange attractor patterns in this book, your pendulum's pattern will eventually come to rest at a point due to friction.

```
ALGORITHM: Code to Generate the Lorenz Attractor

Typical values: h=0.01, npts=4000;
Typical starting values: x,y,z = 0.6;

 frac=8/3;
 do i = 1 to npts;
  xnew = x + h*10*(y-x);
  ynew = y + h*((-x*z) + 28*x-y);
  znew = z + h*(x*y - frac*z);
  x=xnew; y=ynew; z=znew;
  MovePenTo(x,y);
 end;
```

Pseudocode 7.1. *How to generate the Lorenz attractor*

The number π

$\frac{\pi}{2}$ = limit of product: $\frac{2\cdot 2}{1\cdot 3}\cdot\frac{4\cdot 4}{3\cdot 5}\cdot\frac{6\cdot 6}{5\cdot 7}\cdot\frac{8\cdot 8}{7\cdot 9}\cdot\frac{10\cdot 10}{9\cdot 11}\cdots\cdots$

$\frac{\pi}{4}$ = limit of series: $1-\frac{1}{3}+\frac{1}{5}-\frac{1}{7}+\frac{1}{9}-\frac{1}{11}+\frac{1}{13}-\frac{1}{15}+\frac{1}{17}-\frac{1}{19}+\frac{1}{21}-\cdots\cdots$

$\frac{\pi}{2}$ = limit of series: $1+\frac{1}{3}\left(\frac{1}{2}\right)+\frac{1}{5}\left(\frac{1\cdot 3}{2\cdot 4}\right)+\frac{1}{7}\left(\frac{1\cdot 3\cdot 5}{2\cdot 4\cdot 6}\right)+\frac{1}{9}\left(\frac{1\cdot 3\cdot 5\cdot 7}{2\cdot 4\cdot 6\cdot 8}\right)+\cdots$

$\frac{4}{\pi}$ = limit of fraction: $1+\cfrac{1^2}{2+\cfrac{3^2}{2+\cfrac{5^2}{2+\cfrac{7^2}{2+\cfrac{9^2}{2+\cfrac{11^2}{2+\cdots}}}}}}$

The number e

e = limit of sequence: $(1\tfrac{1}{2})^2, (1\tfrac{1}{3})^3, (1\tfrac{1}{4})^4, (1\tfrac{1}{5})^5, \cdots\cdots$

e = limit of series: $1+\frac{1}{1!}+\frac{1}{2!}+\frac{1}{3!}+\frac{1}{4!}+\frac{1}{5!}+\frac{1}{6!}+\cdots\cdots$

e = limit of fraction:

$$e=2+\cfrac{1}{1+\cfrac{1}{2+\cfrac{1}{1+\cfrac{1}{1+\cfrac{1}{4+\cfrac{1}{1+\cfrac{1}{1+\cfrac{1}{6+\cfrac{1}{1+\cfrac{1}{1+\cfrac{1}{8+\cfrac{1}{1+\cdots}}}}}}}}}}}}$$

Chapter 8

Research at the King's Fractal Palace

"Geometry aims at knowledge of the eternal." Plato, 347 BC

8.1 The Pi Slaves

"They wonder whether the digits of pi contain a hidden rule, an as yet unseen architecture, close to the mind of God." Richard Preston, 1992, *The New Yorker*

Yars Kotheck, the King of Ganymede, lives in a beautiful fractal palace attended to by obsequious servants and mathematicians of lesser standing. (See Chapter 34, "The Fractal Palace of Ice," for a description of his castle.) The castle often bustles with activity, its various meeting rooms filled with the noisy sounds of heated mathematical debate as scientists discuss mathematical minutia and computational techniques. For his private projects, the King assigns the lower classes to carry out the more mundane calculations, such as the computation of $\pi = 3.1415 \ldots$ to a googol (10^{100}) decimal places.[12] Several years ago, a group of servile Latööcarfians was given this task and spent years on the π computation while confined to a

---

[12] The mathematical constant pi, denoted by the Greek letter π, represents the ratio of the circumference of a circle to its diameter. It is the most famous ratio in mathematics both on Earth and Ganymede. π, like other fundamental constants of mathematics such as $e = 2.718 \ldots$, is a transcendental number. The digits of π and e never end, nor has anyone detected an orderly pattern in their arrangement. Humans know the value of π to over a billion digits. Transcendental numbers cannot be expressed as the root of any algebraic equation with rational coefficients. This means that π could not exactly satisfy equations of the type: $\pi^2 = 10$ or $9\pi^4 - 240\pi^2 + 1492 = 0$. These are equations involving simple integers with powers of π. The numbers π and e can be expressed as an endless continued fraction or as the limit of an infinite series. The remarkable fraction 355/113 expresses π accurately to six decimal palaces. See Appendix H, "Meditations on Transcendentals" for more information on transcendental numbers.

Ueber die Zahl π.*)

Von

F. LINDEMANN in Freiburg i. Br.

Bei der Vergeblichkeit der so ausserordentlich zahlreichen Versuche**), die Quadratur des Kreises mit Cirkel und Lineal auszuführen, hält man allgemein die Lösung der bezeichneten Aufgabe für unmöglich; es fehlte aber bisher ein Beweis dieser Unmöglichkeit; nur die Irrationalität von π und von π^2 ist festgestellt. Jede mit Cirkel und Lineal ausführbare Construction lässt sich mittelst algebraischer Einkleidung zurückführen auf die Lösung von linearen und quadratischen Gleichungen, also auch auf die Lösung einer Reihe von quadratischen Gleichungen, deren erste rationale Zahlen zu Coefficienten hat, während die Coefficienten jeder folgenden nur solche irrationale Zahlen enthalten, die durch Auflösung der vorhergehenden Gleichungen eingeführt sind. Die Schlussgleichung wird also durch wiederholtes Quadriren übergeführt werden können in eine Gleichung geraden Grades, deren Coefficienten rationale Zahlen sind. Man wird sonach die Unmöglichkeit der Quadratur des Kreises darthun, wenn man nachweist, dass *die Zahl π überhaupt nicht Wurzel einer algebraischen Gleichung irgend welchen Grades mit rationalen Coefficienten sein kann.* Den dafür nöthigen Beweis zu erbringen, ist im Folgenden versucht worden.

Figure 8.1. *First page from a mathematical classic.* In 1882, German mathematician F. Lindemann proved that π is transcendental, finally putting an end to 2,500 years of speculation. In effect, he proved that π transcends the power of algebra to display it in its totality. It can't be expressed in any finite series of arithmetical or algebraic operations. It can't be written on a piece of paper as big as the universe.

dusty Koch-curve[13] shaped room in the King's palace. After a year of computation, the π slaves, as they were called, came up with a startling discovery. Gradually the frequence of occurrence of the digit 0 began to increase once they had computed beyond the trillionth decimal place. For example, at the five trillionth place, the zeros obviously dominate:

000000100006000000033000000000000000

The other digits just started to peter out, like exhausted marathon runners dropping out of an infinitely long mathematical race. After another few months of computation, the gush of non-zero digits began to drip like water from a turned-off fire hose. The π slaves were sobered instantly by the frightening possibility that π would eventually turn itself off. There fears were justified. π suddenly stopped at the quadrillionth decimal place:

3.1415 ... 0000000000000000000000

at which pointed it continually repeated the digit zero. One of the π slaves asked a Prohaptor to carry him to Yars Kotheck so that he could report the sad news on the demise of π.

---

[13] See Glossary.

* * *

The King sat stiffly in a straight-backed throne as the π slave prepared himself to bear the bad news. The slave coughed, and he noticed that the King's magenta eyes possessed a peculiar disturbing power; they seemed to be looking into him rather than at him, as if the King were carefully examining his innermost secrets.

After the slave described the discovery, Yars Kotheck looked down, his posture militant. A minute passed in silence. The slave's tension rose a few more percentage points. Two minutes passed. The King stared at the slave, leaving the little Latööcarfian in a paroxysm of fear. The slave's piezoelectric mouth began to open and close spasmodically. The King's face also betrayed a certain tension, a secret passion held rigidly under control. The electrorheological fluid in the blood vessels on the King's face throbbed convulsively. The King said only four words:

"Get out of here."

A national day of mourning was declared, and the capital city of Tenochtitlãn began to wear a perpetual sulk. The Latööcarfian flag was flown at half-mast over public buildings.

The π slaves began new computations on the decimal expansion of the transcendental number e.

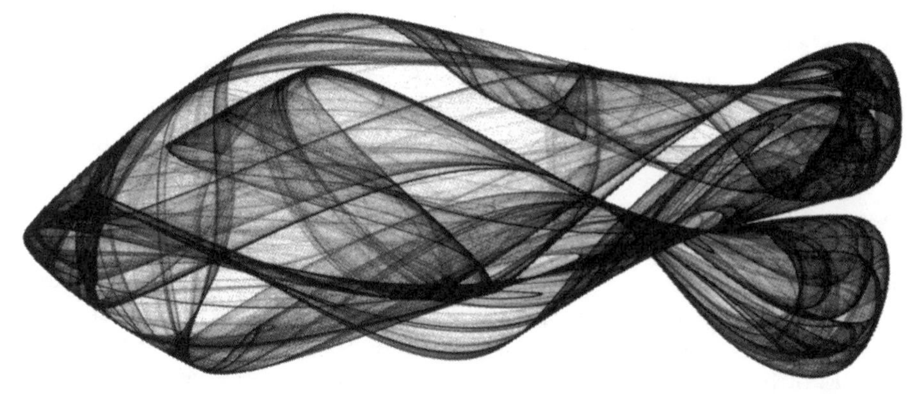

8.2 Digressions

1. Here is an amusing problem concerning π and e which dates back a few years. Determine, without using tables or making actual computations, which is larger: e^{π} or π^{e}? Does a transcendental number raised to a transcendental power produce a transcendental result?

2. The best known formula relating e and π is Euler's formula: $e^{i\pi} + 1 = 0$, which was considered by some of Euler's metaphysically inclined contemporaries to have mystical significance. Edward Kasner and James Newman in *Mathematics and the Imagination* note, "We can only reproduce [the equation] and not stop to inquire into its implications. It appeals equally to the mystic, the scientists, the mathematician." Martin Gardner in *The Unexpected Hanging* observes that the formula unites the five most important symbols of mathematics: 1, 0, π and e and i (the square root of minus one). This union was regarded as a *mystic union* containing representatives from each branch

of the mathematical tree: Arithmetic is represented by 0 and 1, algebra by the symbol i, geometry by π, and analysis by the transcendental e. Harvard mathematician Benjamin Pierce said about the formula, "That is surely true, it is absolutely paradoxical; we cannot understand it, and we don't know what it means, but we have proved it, and therefore we know it must be the truth." Is there a more compact formula relating π and e? Is there a compact formula for relating π, e, i, and ϕ, the golden ratio? (See Appendix H, "Meditations on Transcendentals" for answer.) The number 1.61803 ..., called the golden ratio,[14] appears in the most surprising places, and because it has unique properties, mathematicians have given it a special symbol, ϕ. Are there other simple formulas which relate e and π?

3. Various books have described the possibility of alien messages coded within the digits of π, including Carl Sagan's *Contact* and my *Mazes for the Mind*. What other books have described the coding of messages in transcendental numbers? How would you go about coding a message in π?
4. The numbers π and e are two famous transcendental numbers. Are there others of nearly equal fame? (See Appendix H, "Meditations on Transcendentals" for answers.)
5. Is i^i transcendental? (As mentioned before, the symbol i stands for the imaginary number $\sqrt{-1}$.) Is i^i a real or an imaginary number? (See Appendix H, "Meditations on Transcendentals" for answers.)

8.3 Parallel Processing

"As regards my means of expression, I try my hardest to achieve the maximum of clarity, power, and plastic aggressiveness; a physical sensation to begin with, followed up by an impact on the psyche." Joan Miro, 1920

Occasionally the King organizes huge assemblages of Latööcarfians to compute in parallel at some assigned task usually concerning pattern generation. This is similar to the simultaneous execution of many processes in multiple computers.[15] The final patterns are channeled to the King, who stands in the center of the regular array of individuals, at which point he displays the designs on his head as patterns of light. His most ambitious project required the Prohaptors to arrange a 1000x1000 array of Latööcarfians in a huge outdoor stadium. Each individual in the array communicated with several neighbors using an infrared link. The spectacular crystalline arrangement of mathematician creatures was a sight talked about for years. The aim of the project was to produce a neural net in order to determine what a universal mind, a supercomputer, would consider as beautiful. Yars Kotheck asked the question: What would a computer find as beautiful?

---

[14] $((1+\sqrt{5})/2 = 1.61803 \ldots$.

[15] Note that on Earth, in 1992, the CM-2 Connection machine is an example of a massively parallel computer with 65,536 processors. In a parallel machine, many processors work simultaneously on a problem, whereas in a serial machine – a normal computer – the problem is solved one step at a time.

After one month of computation, the King determined that the 1000x1000 array of mathematicians performing parallel computations was not sufficient to accomplish his goal. He therefore commanded his subjects to form a 3-D array of mathematicians with dimensions 1000 x 1000 x 1000. Prohaptors worked night and day positioning the millions of immobile Latöócarfian bodies. In order to do this the mathematicians had to stand on the heads of other mathematicians. Once the cubical array was assembled, Prohaptor children worked continuously, weaving in and out of the rows and columns of mathematicians, providing them food and psychological counseling as required. After a month of computation, the 3-D neural net of Latöócarfians produced a single picture. It looked like a combination of graffiti-like elements with figures echoing the art of children and primitive societies. In short, it looked like a black-and-white lithograph done by Joan Miro. The King upon seeing the result opened his piezoelectric mouth wide and screamed for a whole minute, as his subjects sat silently in rapt attention. They could not ascertain if the King's scream was one of happiness or fury.

The King then looked at the cubical array of mathematicians. In the fading light, his magenta eyes, usually the color of dirty purple ice, looked almost orange now, bright and utterly alien. Some of the mathematicians at the bottoms of the columns were clearly exhibiting signs of physical stress as they bore the weight of their comrades. The King asked his personal Prohaptor to give a shove to one of the columns of mathematicians. A few of the nearby Latöócarfians looked a little wary and haunted after hearing this. Some of their faces had the withered look of empty balloons. With sinuous brilliance, honed by years of training, the Prohaptor leaped into the air and kicked the nearest column of creatures. As the column fell it started a chain reaction, taking other columns with it. The entire 3-D Latöócarfian crystal disintegrated as the living columns toppled to the ground with a crunch.

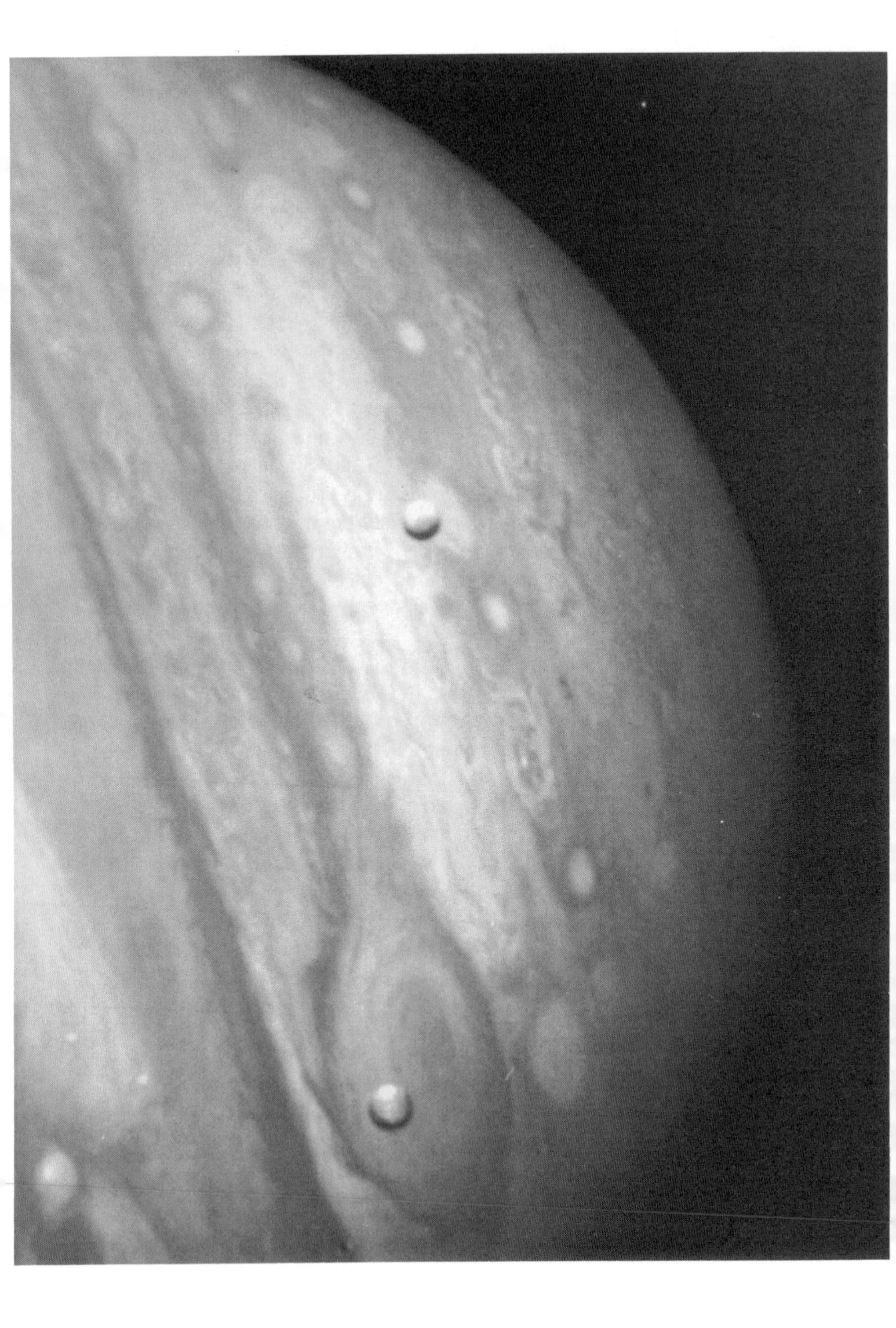

Chapter 9

Latööcarfian Classes

"Simple shapes are inhuman. They fail to resonate with the way nature organizes itself or with the way human perception sees the world." George Zebrowsik

"If you go too far in fantasy and break the string of logic, and become nonsensical, someone will surely remind you of your dereliction ... Pound for pound, fantasy makes a tougher opponent for the creative person."
Richard Matheson, 1984 *Masques*

The main political and economic unit of the Latööcarfian regime is the manorial estate. In most of the sulci, and in the Barnard and Galileo Regios (Figure 3.2), Lords of higher rank hold many manors, the number usually between 100 and 1000. The land of the Ganymedean manor is partly arable, partly crystalline meadow (very highly valued for aesthetic reasons), and partly pasture. The arable land is divided into four large fields, arranged for a four year rotation of crystal crops and fallow. Each manorial estate comprises one or more villages, the lands occupied by the Latööcarfian peasants. Except for the nobles and their families, the manor consists of Latööcarfians of servile status. The four classes having this status are the cotters, crofters, villeins and serfs. The Lords, villeins, and serfs all hold defined rights to share in the meadow.

9.1 Lords

Lords are those Latööcarfians who dreamed of chaotic attractors almost as beautiful as the King's, as judged by a special review committee. Their patterns, like the King's, are always symmetrical, and have a center of symmetry at the origin of the coordinate system. A point at position (x, y) always has a counterpart at $(-x, -y)$.

9.2 Villeins and Serfs

Villeins and serfs also dream of symmetrical attractors with a center of symmetry, but their patterns are disconnected, often separated in space by a large gap.

9.3 Crofters and Cotters

The crofters and cotters are wretchedly poor Latööcarfians who have no definite status under the Latööcarfian feudal regime. Their patterns, while often beautiful to an Earthly eye, have no symmetry. These Latööcarfians occupy small shanties and hire themselves out to the richer serfs – lending parts of their aluminum gallium arsenide heads to aid the serfs, villeins, and Lords in quicker computations.

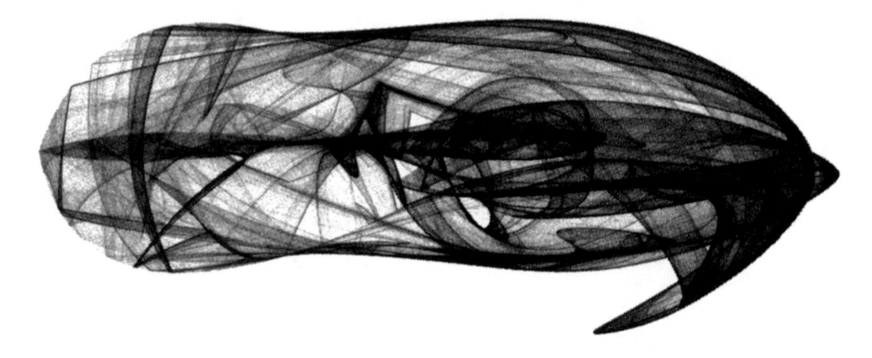

9.4 Women

"And what is love without the eternal enmity between the sexes." H. Hesse

For a time, female Latööcarfians residing in the Osiris crater were treated with indifference and sometimes with brutality and contempt. King Yars Kotheck put an end to such chauvinism, and today gender is not a factor in status or achievement in society. Gender is determined primarily by size and personality, and from a biological standpoint there is little difference between the sexes. Reproduction is accomplished by exchanging gallium arsenide regions of the heads, allowing intimate sharing of thoughts. "Baby" Latööcarfians are created by assembling pieces, or "modules," contributed by the two parents.

9.5 Slaves

"As his studies come to a close the historian faces the challenge: Of what use have your studies been? Have you found in your work only the amusement of recounting the rise and fall of nations and ideas, and retelling 'sad stories of the death of kings'? Is it possible that, after all, history has no sense, that it

teaches us nothing, and that the immense past was only the weary rehearsal of mistakes that the future is destined to make on a larger stage and scale?"

Will and Ariel Durant, *The Lessons of History*

A few slaves continue to be held during the reign of Yars Kotheck. These are individuals who dream of non-chaotic patterns with negative Lyapunov exponents (see Chapter 14, "Interlude: Lyapunov Logomania"). Slaves are household servants and scarcely considered as real Latöōcarfian beings. They are instruments of burden, like cattle or horses on Earth, to be worked for the profit of their masters. It is a sad commentary on Latöōcarfian civilization that nearly all of the productive labor on Ganymede, such as ditch digging, is done by the slaves whose limbless bodies wriggle on the soil in order to excavate and build residencies. Many travel to the surface of Ganymede through sophisticated tunnels and air locks built of ice. Here they help perform difficult mining operations and excavations. The resulting disturbances to the surface ice and soil form the Ganymedean sulci.

The heads of many of the slaves are composed of the semiconductor silicon, rather than of gallium arsenide which permits faster computations. Therefore, they are not as well suited to the mathematical pursuits of the higher classes. On the other hand, their silicon heads require less power, and do not have the overheating problems associated with the upper classes.

9.6 Prohaptors

The sedentary Latöōcarfians make use of a race of large, intelligent animals to protect them from some of the fierce Ganymedean animals, from alien invaders, and from each other. These six-foot-tall creatures are called Prohaptors.[16] They carry long swords and possess several feet resembling horse-shoe crabs. As a reward for service, Prohaptors are stimulated in the pleasure center of their brains. All physically fit Prohaptors are considered warriors and must wear martial trappings, no matter what their disposition and occupation.

According to their early migration myths, the Prohaptors originally lived as nomadic hunter-gatherers in the arid regions of the Ganymedean air pocket. Then, early in the 11th century, they began migrating northward to the valley of the Latöōcarfians.

---

[16] On Earth the term "prohaptor" refers to a sucker on the head of certain trematode worms. The Prohaptors of Ganymede get their name from their two large sucker-like snouts.

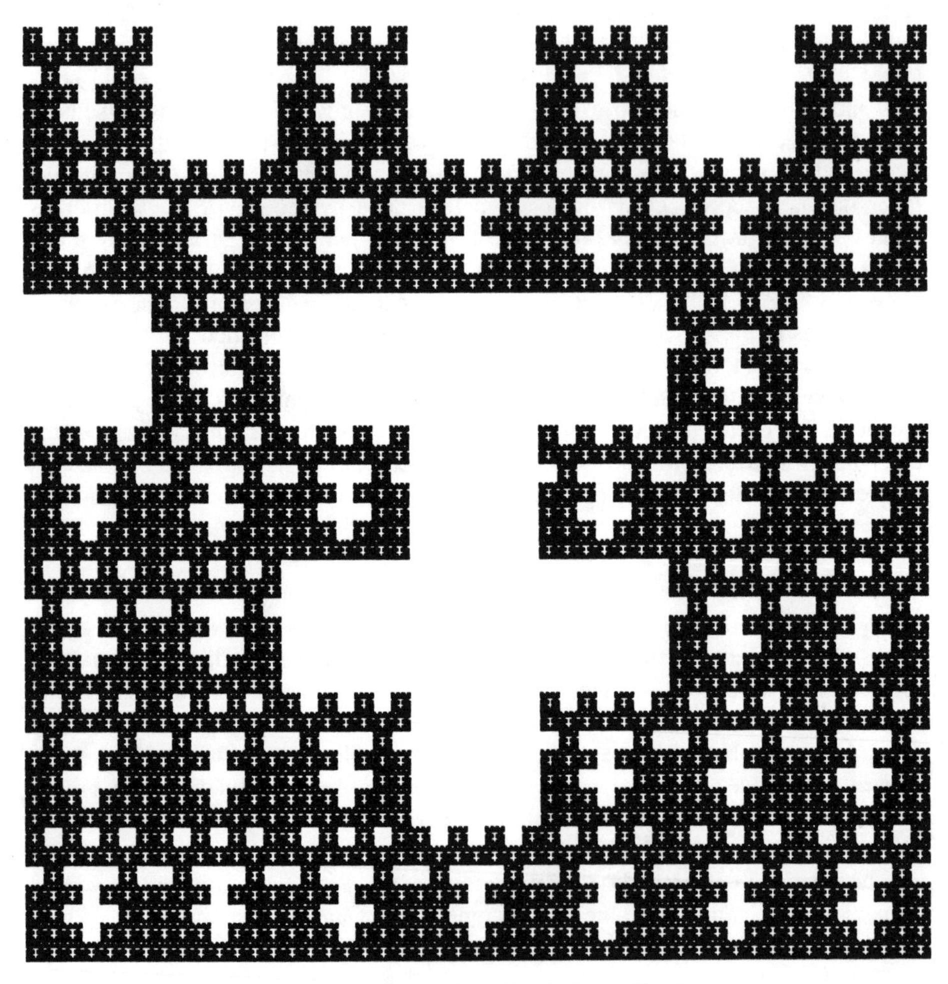

A Lord's Recursive Castle (Dwelling)

(For a computational recipe, see Pickover, C. (1993) Recursive worlds. *Dr. Dobb's Software Journal*, Sept., 18(9): 18-26.)

Chapter 10

Interlude: What is Iteration?

"We fear death, yet we long for slumber and beautiful dreams." Kahil Gibran

The patterns in this book were created using mathematical feedback loops. Feedback is a term we often hear today in a variety of settings – for example, amplifier feedback during rock concerts, biofeedback in medicine and psychology, and chemical feedback in the field of biochemistry. Generally, feedback means that a portion of the output of a system or machine returns to the input. In electronic amplifiers feedback can occur if the microphone is placed too close to the speaker. In the world of mathematics, feedback is often the result of an "iteration" or "recursion." By iteration, we mean the repetition of an operation or set of operations. In mathematics, composing a function with itself, such as in $f(f(x))$, is an iteration. The computational process of determining the value of x at some future increment in time (x_{t+1}) given the value of x at the present time (x_t) is called an iteration.

When mathematical feedback loops are implemented using computer programs, all kinds of fascinating numerical challenges arise. Because the internal accuracy of computers is limited, numerical values are represented with a finite precision. Therefore, during each multiplication, digits are lost. This implies that the results obtained will depend on the order of operations, and the usual associative property of multiplication will no longer exist. Jean-Francois Colonna has demonstrated (in "The Subjectivity of Computers," *Communications of the ACM* (1993) August, 36(8): 15-18) that mappings such as those described in *Chaos in Wonderland* will depend on the way in which the computer program is written (e.g. with regard to parentheses), the compiler used, the language used, and the computer used. This computer sensitivity has broad implications for the study of chaos and means that although the global structure of your patterns will usually look identical to the ones in this book, they may differ in tiny details.

A book should serve as the axe
for the frozen sea within us.

Franz Kafka

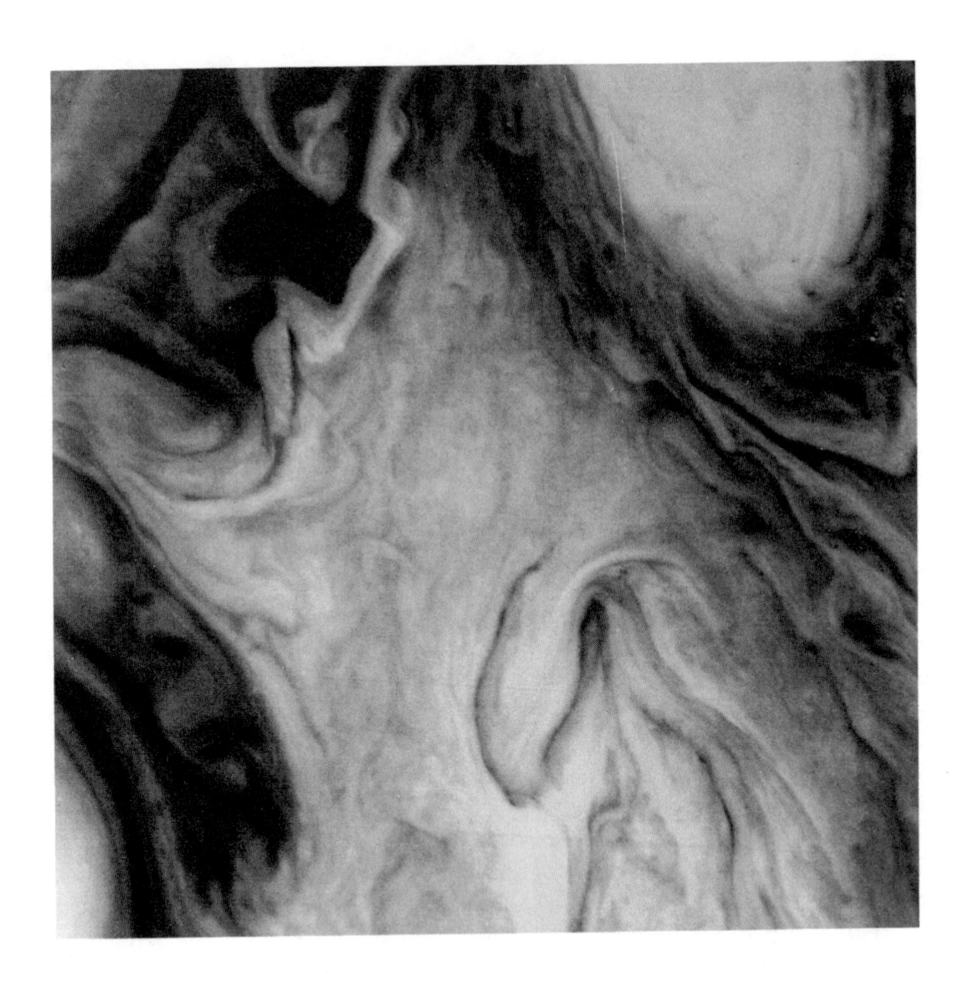

Chapter 11

The Dreams of the Lords

> *"Do we really know what the past was, what actually happened, or is history a fable not quite agreed upon? Our knowledge of any past event is always incomplete, probably inaccurate, beclouded by ambivalent evidence and biased historians..."*
> Will and Ariel Durant, *The Lessons of History*

The Lords comprise the class of Latööcarfians with highest status. Although King Yars Kotheck owns all the land on Ganymede, a hierarchy of Lords owes their allegiance to the King, and the highest of them hold land from the King. Yars Kotheck gives portions of land to the Lords in a fascinating ceremony. In this ritual, he consigns chunks of gallium arsenide which are incorporated into the receiving Lords' heads, thereby improving the speed with which they can dream new patterns. This allows the Lords to grow, increase their memory, and improve their computing power, thus challenging the King to continually devise newer and more attractive patterns. Every five years, all the Lords meet in a competition for Kingship. Like body-building contests on Earth, these competitions are lavish ceremonies where Lords display their computational prowess. Each Lord hopes to rise to Schwarzeneggerian heights beyond the imagination of any predecessors.

When not preparing for computational competitions, the Lords spend time administering their lands. They exercise a private jurisdiction over the tenants on their lands and maintain a private army for defense. They also secure from the King an immunity which excludes royal officers from their lands. In return, the King holds the Lords responsible for the public duties of all their dependents.

The Lords' functions are threefold: to supervise agriculture (the cultivation of aluminum gallium arsenide crystal gardens), to give military protection to their lands and inhabitants by organizing armies of Prohaptors, and to oversee trade and industry on his lands. The production of beautiful, chaotic pattern-dreams are the primary source of prestige, but ownership and management of land is also a source of power and wealth.

The Lords' castles are made of flawless ice. Within the walls are stables, kitchens, living quarters, and storehouses. Windows are few and small, usually made of ice lattices which keep out most of the crystal rains, but they admit much light. The ice houses keep the Latööcarfians cool and prevent overheating of their heads. Artificial light is provided by bioluminescent bacteria and also by the light-emitting gallium arsenide heads of the Latööcarfians in the house.

Legal questions are settled by the oldest members of the Latööcarfian community. The Lord and King make suggestions but do not sit in on judicial proceedings. Latööcarfian law is complex and usually severe. For example, if a serf steals gallium arsenide crystals from another Lord's crystal field, 30% of his gallium arsenide head is removed and served to the Lord for incorporation into the Lord's head. Other crimes require that the criminals be buried in the icy Ganymedean soil with their heads sticking out, and then be trampled by a silver-haired Zooz (for more on Zooz, see Part II, "The Dream-Weavers of Ganymede"). This is painful and permanently damages their memories.

As mentioned in Chapter 9, "Latööcarfian Classes," Lords are those Latööcarfians who dream of chaotic attractors almost as beautiful as the King's, as judged by a special review committee. Their patterns, like the King's, are always symmetrical, and have a center of inversion at the origin of the coordinate system.

A few of the Lords' patterns are displayed on the following pages. B.2, "Some Figure Parameters and Descriptions" lists parameters for computing many of these designs.

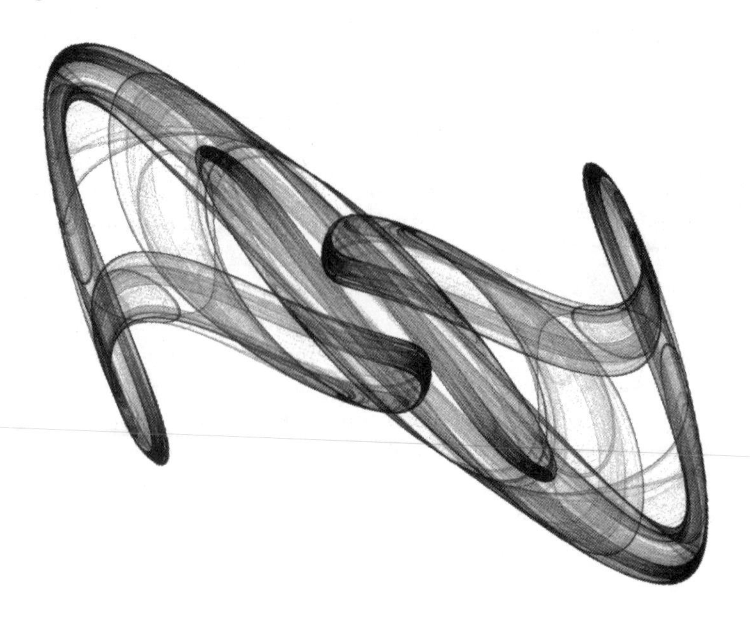

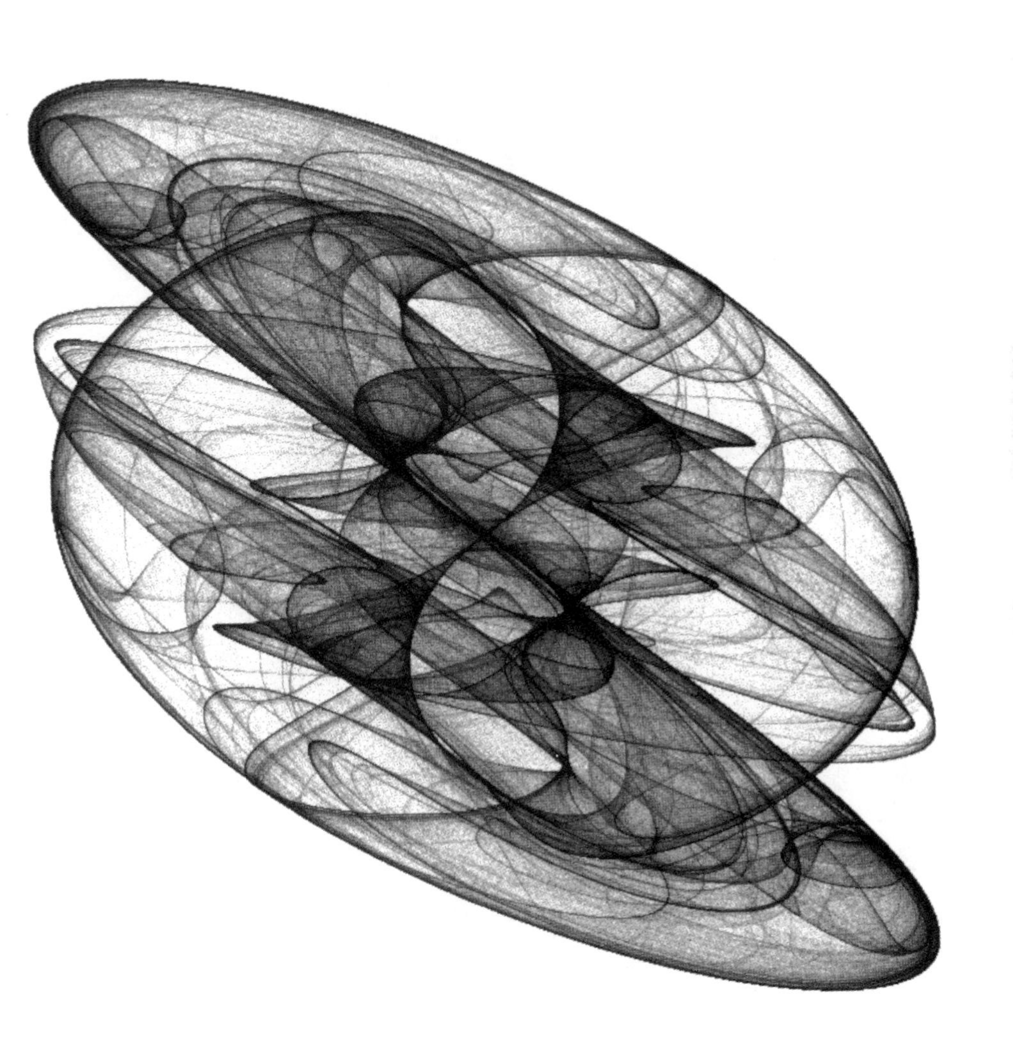

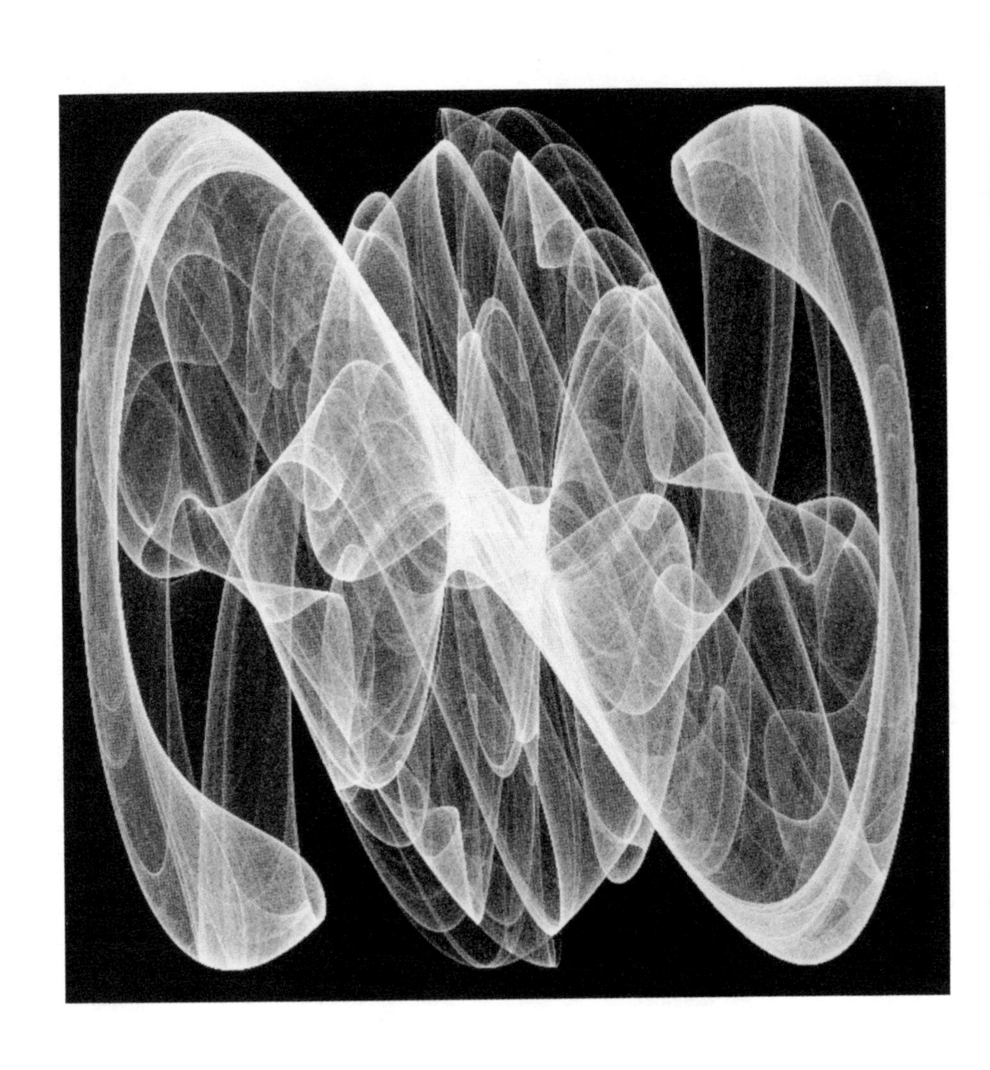

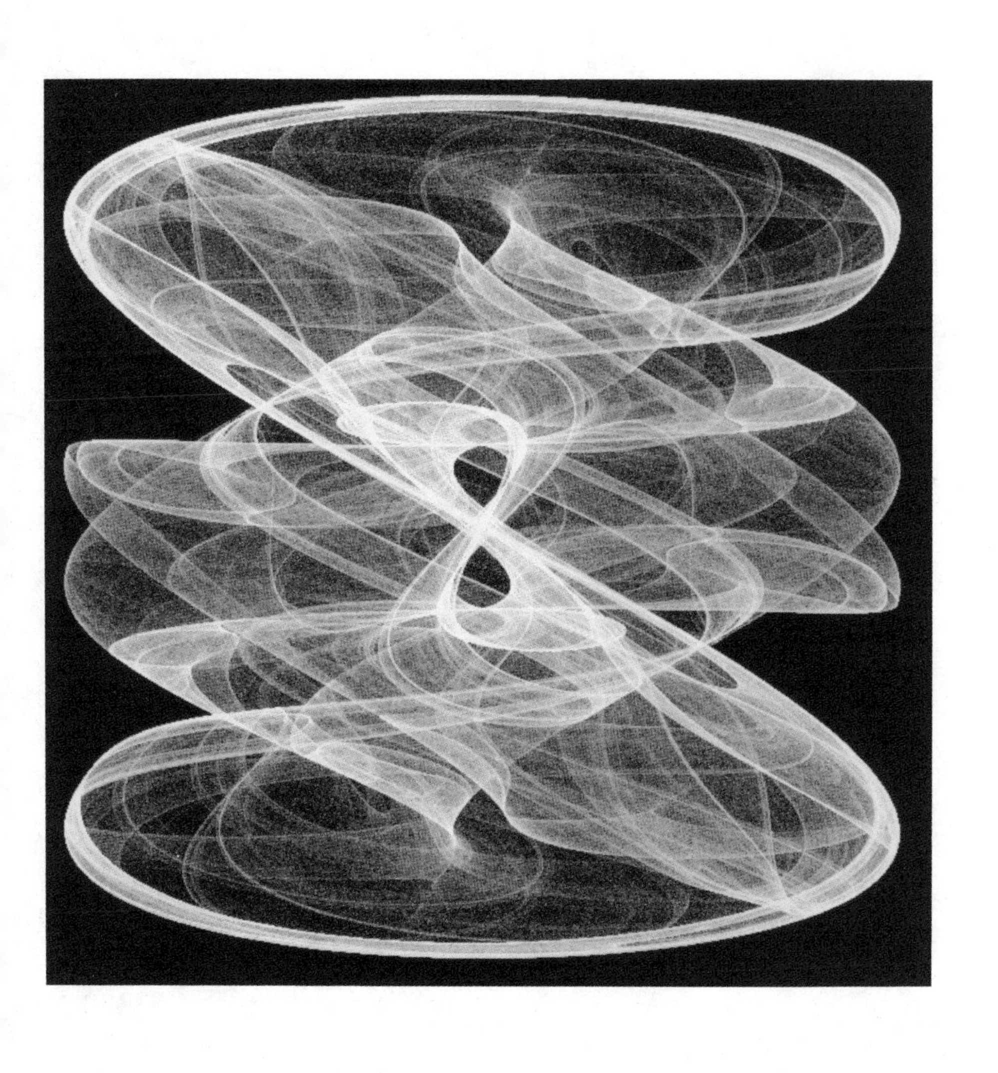

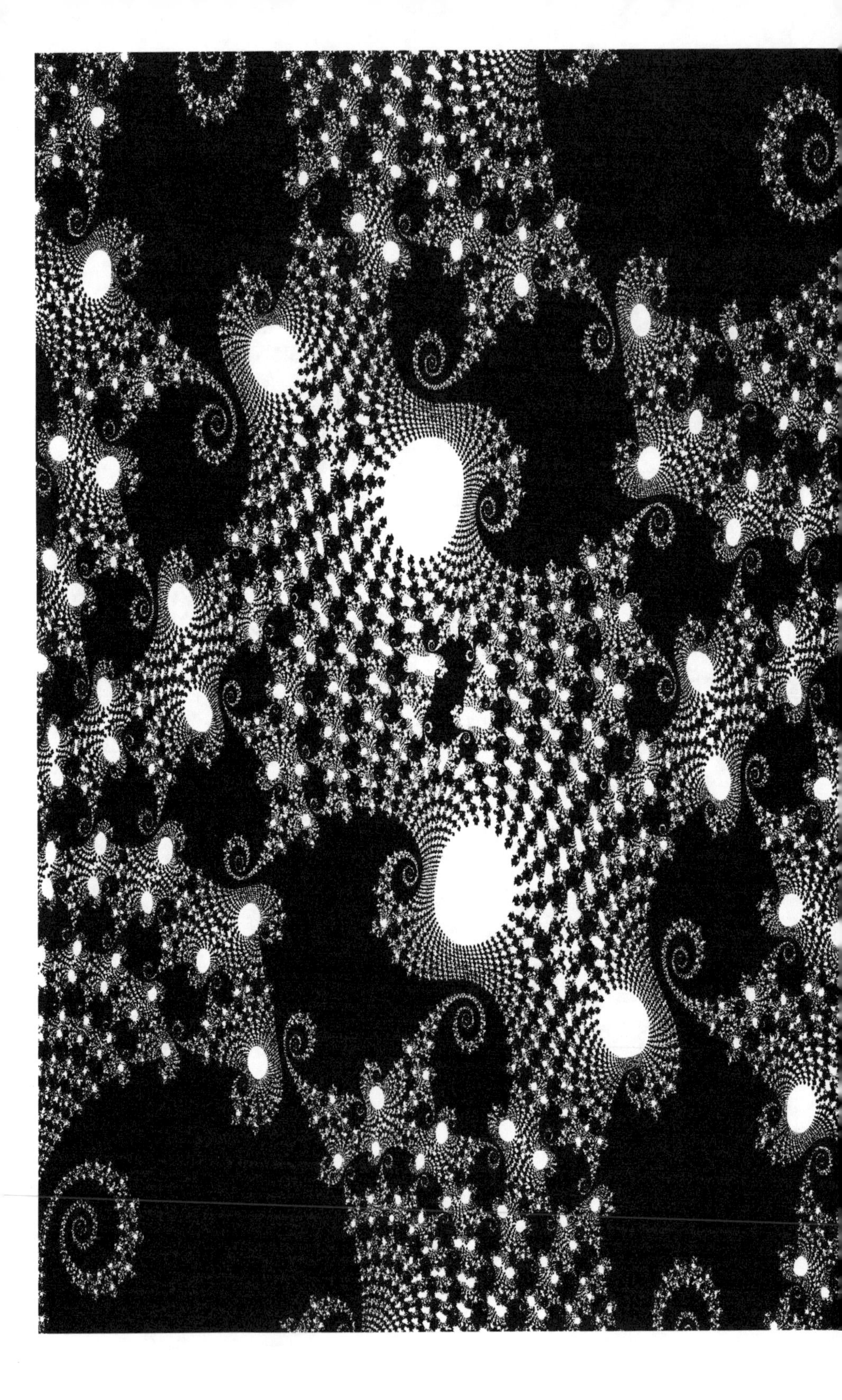

Chapter 12

Interlude: Where is the Beauty?

"The Earth is the cradle of humanity, but mankind will not stay in the cradle forever."
Konstantin Tsiolkovskii

12.1 Mathematical Observations and Tips

The line between science and art is a fuzzy one; the two are fraternal philosophies formalized by ancient Greeks like Eratosthenes and Aristophanes. Computer graphics and mathematics help reunite these philosophies by providing scientific ways to represent natural and artistic objects.

Some technical readers may wish to better understand the mathematics of the Latöócarfian's iteration process. (Most readers not interested in mathematical details may skip this chapter.) Let's start by describing the possible behaviors of the Latöócarfian equations.

As the equations

$$x_{t+1} = \sin(y_t b) + c \sin(x_t b) \tag{12.1}$$

$$y_{t+1} \sin(x_t a) + d \sin(y_t a) \tag{12.2}$$

are iterated, with some initial value of x_0 and y_0 starting with $t = 0$, successive values of x and y are determined by repeating the above equations. I arbitrarily start with $x_0 = y_0 = 0.1$. The iterates are plotted on a two-dimensional surface. After many iterations, the solution will do one of three things:

1. It will converge to a fixed point.
2. It will enter into a cycle (a repeating succession of values).
3. It will exhibit chaos and gradually fill in some complicated region of the (x,y) plane.

Many similar systems have a fourth solution: they may explode to infinity. However, the Latöocarfian set of equations cannot produce values that diverge to infinity because the values of $\sin(x)$ are between negative 1 and positive 1.

How large can the values produced by the Latöocarfian formulas grow? It turns out that the bounds for the attractors are independent of x_0, y_0, a and b, and depend only on the values of c and d. Assuming $c, d > 0$

$$|\sin(y_t b) + c\sin(x_t b)| \leq \max(|1+c|, |1-c|) \tag{12.3}$$

and

$$|\sin(x_t a) + d\sin(y_t a)| \leq \max(|1+d|, |1-d|) \tag{12.4}$$

(Note: the previous equations can be simplified using: $\max(|1+c|, |1-c|) = 1 + |c|$).)

Can we determine some values of a, b, c, d, x_0 and y_0 which do not produce beautiful patterns without even carrying out the iteration of the formulas? The answer is "yes." For example, the origin ($x_0 = y_0 = 0$) is always a fixed point, and will therefore not produce interesting behavior. We can also exclude a large range of parameters because they produce uninteresting pictures. For example, the values produced by the iteration will eventually converge to the origin if

$$1 > (|c|\ |b| + |a|) \tag{12.5}$$

$$1 > (|d|\ |a| + |b|) \tag{12.6}$$

This means that for all initial values of a, b, c, and d that satisfy these conditions, the orbits will eventually end up at the origin for any starting (x_0, y_0) values. For example, ($a = -0.5$, $b = 0.5$, $c = 0.5$, $d = 1$) will yield uninteresting patterns because of this. In addition, if $a = 0$, then the system will converge to the origin if $|b| < 1/|c|$. Similarly, if $b = 0$, then the system will converge to the origin when $|a| < 1/|d|$. These values, therefore, should be avoided if your goal is to make beautiful attractors. Finally, we can show that the Latöocarfian system produces ugly pictures (i.e., they behave as, what mathematicians call, a "global contraction") if the following conditions are met: $|b| < 1/(|c| + \alpha)$ and $|a| < 1/(|d| + 1/\alpha)$ for some α. An additional observation is that if $|bc + ad| > 2$ then the origin is not stable, and therefore these values *cannot* cause the system to converge to the origin.[17]

Several different attractors are possible for a given set of a, b, c, and d values when different starting points x_0 and y_0 are used. How often does this happen? You may wish to try different starting values and view the results.

---

[17] The results in this paragraph were derived and contributed by Michael S. Branicky, an MIT graduate student who studied the Latöocarfian equations which I presented to him. A derivation is available on request.

12.2 Ghost Siblings

Most of the Latöócarfian patterns are symmetrical. Interestingly, those patterns which are *not* symmetrical have a reflected invisible counterpart which I call a "ghost sibling." The sibling is symmetrically situated about the origin, since for any orbit (x_t, y_t) satisfying the iteration equations, the opposite orbit $(-x_t, -y_t)$ also satisfies the equations. This means that you can uncover the sibling simply by choosing any point (x_t, y_t) on the pattern, adding a single point which is its negative $(-x_t, -y_t)$, and then watching the new pattern unfold. In the designs of the crofters and cotters, these ghost siblings will not appear because a ghost point is not hit during an orbit. In the language of chaos experts, this means that if a "basin of attraction" does not contain any symmetrically situated points, the pattern will not display the ghost sibling.

If you examine values for c and d near zero you will find that most the systems behave as periodic attractors and limit cycles, and therefore small values of c and d should be avoided. Also these attractors are more likely to avoid hitting their symmetrical ghost siblings and therefore do not show their ghostly counterpart.

12.3 Other Experiments

There are many new territories for future experimentation. For example, the sine terms in the formulas may be replaced by other operations. Also, consider that computers evaluate trigonometric functions by truncating a series to a polynomial. Dr. Mike Frame suggests comparing pictures generated using polynomial approximations which can be further modified by simulating the wiggly periodicity of sine. For example, the cubic approximation to sine is $x - x^3/6$. Therefore the first Latöócarfian equation is approximated by

$$x(n+1) = by(n) - (by(n))^3/6 + cbx(n) - c(bx(n))^3/6 \tag{12.7}$$

or, with periodicity forced by adding mod 2π,

$$x(n+1) = (by(n) - (by(n))^3/6 + cbx(n) - c(bx(n))^3/6)) \bmod(2\pi) \tag{12.8}$$

Transcendental functions are hard to analyze, but polynomials aren't. If some reasonably low-degree polynomial can give good approximations to the pictures, then symmetry and periodic points might be approximated by studying the polynomials.

Of course you should exercise caution in interpreting the results of polynomial approximations to the sine. As an experiment, examine the resultant images after replacing the sine function with the first two, three, four, and five terms of the Taylor series approximation.

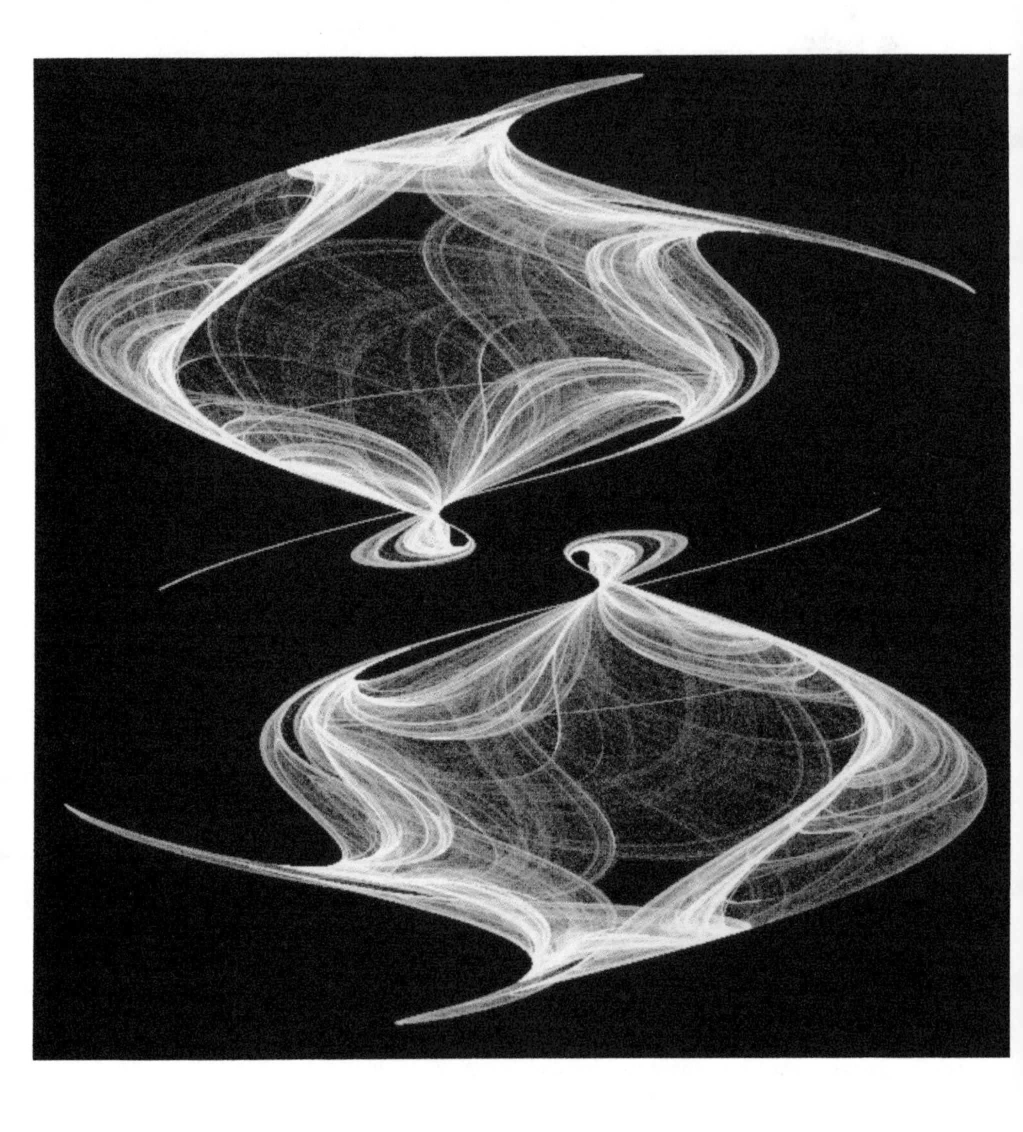

Chapter 13

The Dreams of the Villeins and Serfs

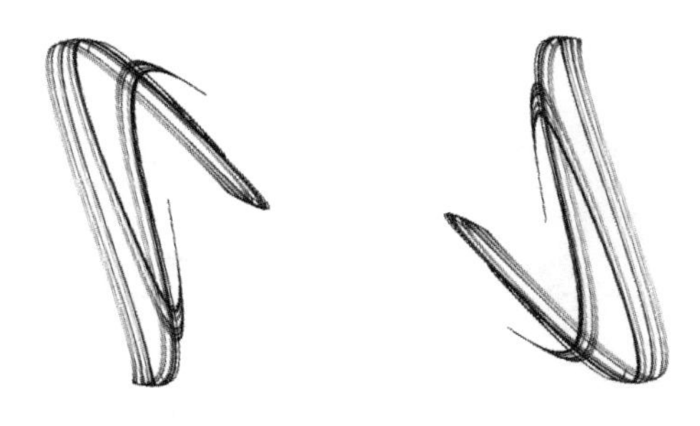

"One thing I have learned in a long life: that all our science, measured against reality, is primitive and childlike – and yet it is the most precious thing we have."
Albert Einstein

"The historian always oversimplifies, and hastily selects a manageable minority of facts and faces out of a crowd of souls and events whose multitudinous complexity he can never quite embrace or comprehend."
Will and Ariel Durant, *The Lessons of History*

Villeins and serfs are peasants under the class system of Ganymede. At night they physically attach parts of their body to the icy walls of the manor-house, symbolically showing their subjugation to the Lords. They also pay dues to the Lords and perform servile labor for them. The villeins have the power to appeal to the royal courts of justice, even against their Lord. However, they have a position somewhere between freedom and slavery. They cannot own any property, but they can make wills leaving chunks of their semiconductor heads for relatives to use. They can be bought and sold from one Lord to another. Every Latööcarfian born to villein stock belongs to their Lord and is bound to undertake service imposed by their Lord. The villeins spend the bulk of their time performing tedious computations, and therefore have little time to dream of their own patterns. A Lord who has many villeins and serfs performing menial computations for him has more time for advanced thinking and pattern formation.

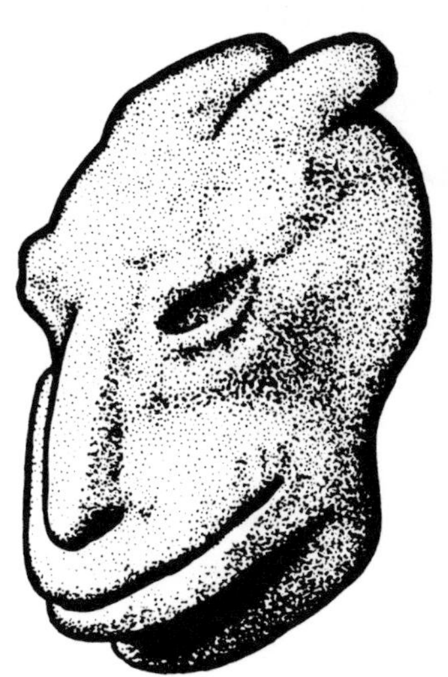

Like the Lords, villeins and serfs also dream of symmetrical attractors with a center of symmetry, but their patterns are disconnected, often separated in space by a large gap. One pattern is displayed at the end of this chapter. Another faces this chapter. B.2, "Some Figure Parameters and Descriptions" lists parameters for computing of these designs.

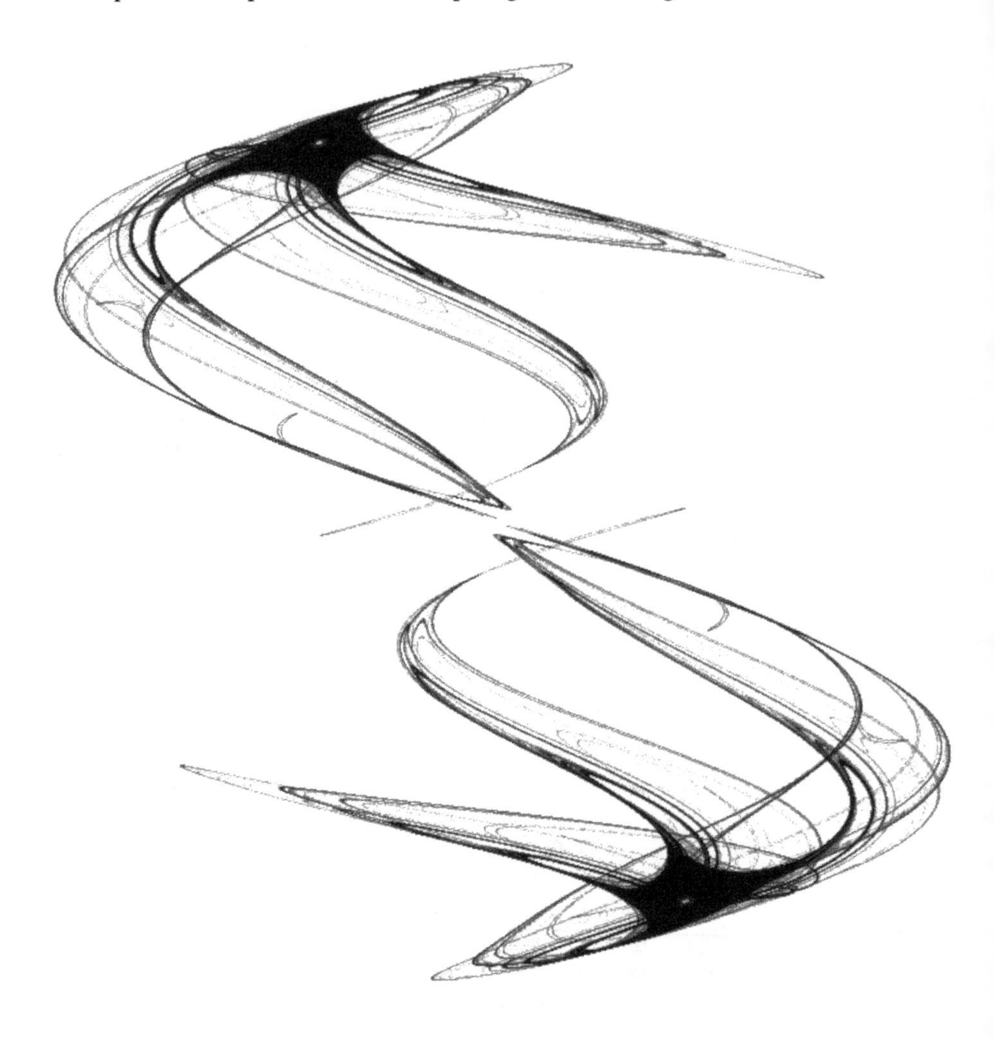

Chapter 14

Interlude: Lyapunov Logomania

"Chaos represents systems that are random in the long run but have just enough structure so that in the short run you can figure out what they're going to do."
Norman Packard

14.1 Mikhailovich Lyapunov

The visual appeal of the patterns in this book can be quantified using a single parameter known as the *Lyapunov exponent,* named after a Russian mathematician (1857-1918) who introduced this idea around the turn of the century. Alexander Mikhailovich Lyapunov was an interesting man. Both a mathematician and mechanical engineer, he produced an important doctoral dissertation in 1892, in Russia, titled *Stability of Motion* (English translation, 1966, Academic Press). In this book, he introduced the concept of the Lyapunov exponent. Sadly, Lyapunov died violently in the chaotic aftermath of the Russian Revolution, not a surprising fate for a middle-class intellectual living at that time.

This parameter called the Lyapunov exponent can be used to characterize the Latööcarfian patterns. For example, when the Lyapunov exponent is negative, the pattern is dull and boring. The greater the value of this parameter, often the more "fluffy" the pattern. I find that chaotic attractors with an exponent of around 0.5 to

```
ALGORITHM: How to Compute the Lyapunov Exponent.

Lsum = 0; n = 0; x = 0.1 y = 0.1; xe = x + 0.000001; ye=y;
REPEAT 10,000,000 TIMES
    xx = sin(y*b) + c*sin(x*b) ;  yy = sin(x*a) + d*sin(y*a) ;
    xsave = xx; ysave = yy; x = xe; y = ye; n++;
       /* "Re-Iterate" for computing Lyapunov Exponent, L */
       xx = sin(y*b) + c*sin(x*b) ; yy = sin(x*a) + d*sin(y*a) ;
       dLx = xx - xsave; dLy = yy - ysave; dL2 = dLx*dLx + dLy*dLy;
       df = 1000000000000.*dL2; rs = 1./sqrt(df);
       xe = xsave + rs*(xx - xsave);  ye = ysave + rs*(yy-ysave);
       xx = xsave; yy=ysave; Lsum=Lsum + log(df); L=0.721347*Lsum/n;
    x=xx;y=yy;   /* L is the value for the Lyapunov exponent */
END REPEAT
```

Pseudocode 14.1. *How to compute the Lyapunov exponent.* (The program coded here is in the style of the C language.) Iterate the formulas a few hundred times before starting this computation as suggested in 16.2, "Software Tips."

be the most pleasing. In my own computer programs, the Lyapunov exponent is used to automatically eliminate certain combinations of *a b*, *c* and *d*, thus speeding up the search for beauty.

Viewed in another way, visually pleasing patterns arise when the particular values of *a*, *b*, *c*, and *d* produce chaotic results, in other words, when the equations have a sensitive dependance on initial conditions. If we iterate the Latöócarfian 2-D trigonometric map with two different initial conditions that vary by a tiny amount, chaotic solutions will increase this difference with each iteration. Two trajectories staring out from nearly the same positions will ultimately end up very far away from each other as the differences are expanded during the evolution of the system. On the other hand, if the solution is nonchaotic (e.g. leading to fixed points and limit cycles), the difference between the two solutions will on average grow smaller with each iteration. The Lyapunov exponent quantifies this and can be considered the rate at which the accuracy of prediction declines as one projects further into the future.

Lyapunov maps provide an elegant way to represent the rich and complicated behavior of certain simple formulas. The mathematics of this method is somewhat technical, so I direct you to my paper for more illustrations and details.[18] A few of the details are mentioned here.

Lyapunov exponents tell you how "crazy" the behavior of your equation is. Stated more scientifically, it quantifies the average stability of the oscillatory modes. Our 2-D map actually has two exponents, since a cluster of nearby initial points may expand in one direction and contract in another. The more positive one is the one that signifies chaos, and it is the one that dominates after a few iterations (Sprott, 1993; Wolf, et. al., 1985). Our maps can be characterized by a Lyapunov

[18] Pickover, C. (1990) Visualizing chaos: Lyapunov surfaces and volumes. *IEEE Computer Graphics and Applications.* March 10(2): 15-19.

Figure 14.1. *Lyapunov plot.* Shades of gray indicates changes in the Lyapunov exponent as a and b are varied along the x and y axes, for a constant value of c and d. Dark regions correspond to values of a and b which produce non-chaotic patterns.

exponent (often symbolized by the Greek letter Λ) which is positive for chaos, zero for a marginally stable orbit, and negative for a periodic orbit:

$$\begin{aligned} &\Lambda < 0, \text{ the orbit is stable} \\ &\Lambda = 0, \text{ the orbit is neutrally stable} \\ &\Lambda > 0, \text{ the orbit is locally unstable and chaotic} \end{aligned} \tag{14.1}$$

Any system containing at least one positive Lyapunov exponent is defined formally to be chaotic, with the magnitude of the exponent reflecting the time scale on which the system's dynamics become unpredictable.

14.2 Computing Exponents

The method used in this book for computing Lyapunov exponents was presented by Sprott (1993).[19] If the two initial points are separated by a distance d_n after the nth iteration, and the separation after the next iteration is d_{n+1}, the Lyapunov exponent is determined by:

---

[19] For a background on J. Clint Sprott, the guru of strange attractors, see Appendix J, "For Further Reading."

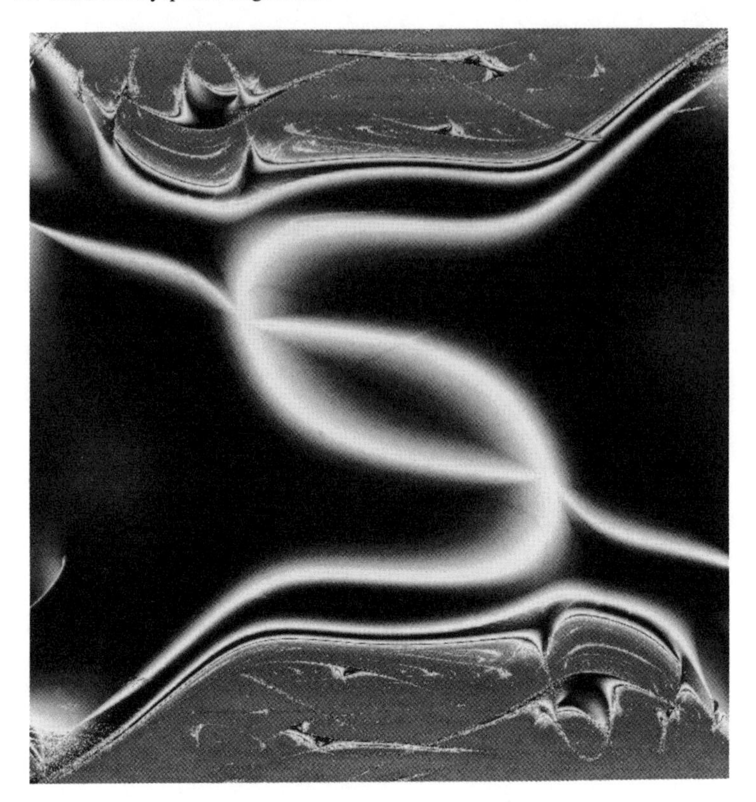

Figure 14.2. *Lyapunov plot.* (See previous figure for information.)

$$\Lambda = \sum_{n=0}^{n=N} \log_2(d_{n+1}/d_n)/N \qquad (14.2)$$

After each iteration, the value of one of the iterates is changed to make $d_{n+1} = d_n$. For my experiments, the initial separation d_0 was 10^{-6}. Also after each iteration the points are moved back to their original separation along the direction of separation. Code 14.1 is a rough outline of the process in the programming language C.

The value of Λ as a function of a and b for constant values of x_0, y_0, c, and d can be plotted to help visualize the intricate behavior of these systems. Such Lyapunov plots are shown in Figure 14.1, Figure 14.2, and Figure 14.3.

Other interesting patterns can result by plotting Julia sets for the Lyapunov exponents. To accomplish this, fix the values for a, b, c, and d, vary x_0 and y_0, and plot the Lyapunov values. These patterns have a repeating cell-like structure due to the sine functions. Each cell contains swirling patterns.

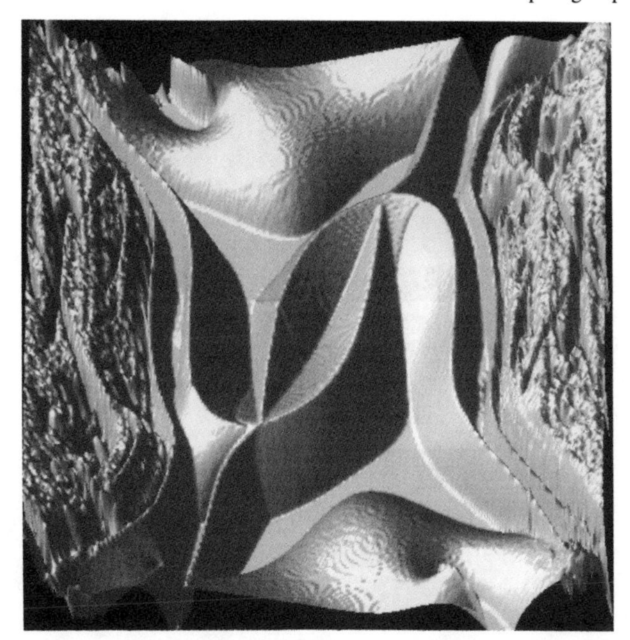

Figure 14.3. *Lyapunov plot.* 3-D representation of previous figure (rotated). Low altitude regions correspond to values of a and b which produce non-chaotic patterns.

Figure 14.4. *Lyapunov plot.* 3-D representation. Low altitude regions correspond to values of a and b which produce non-chaotic patterns.

Chapter 15

The Dreams of the Crofters and Cotters

"Einstein was once asked what he would have done if a physical experiment had contradicted his mathematical prediction, and he answered by saying that he would have felt sorry for the Lord." George Zebrowski

Crofters and cotters occupy tiny cot-houses belonging to a farm, for which they have to provide labor at a fixed rate of pay whenever required. Since they have no arms and legs (as all Latöòcarfians) their labor often consists of moving small objects. This tedious activity is accomplished by wriggling their bodies over to an object, picking it up with their piezoelectric mouths (see 4.3.4, "Body and Mouth Motions"), and moving it to a nearby location.

The cotters and crofters have no rights against their Lord who is protected from all law suits by the *exceptio cotternagii*. They may not leave the manor without permission and can be reclaimed by force if they do by the writ of *de nativo habendo*. Cotters and crofters must pay taxes and serve in the military.

As was mentioned in Chapter 9, "Latöòcarfian Classes," they often hire themselves out to the richer serfs – lending parts of their aluminum gallium arsenide heads to aid the serfs and villeins in quicker computations.

Unlike the Lords, villeins, and serfs, their patterns have no symmetry. A few of their patterns are displayed on the following pages. B.2, "Some Figure Parameters and Descriptions" lists parameters for computing many of these designs.

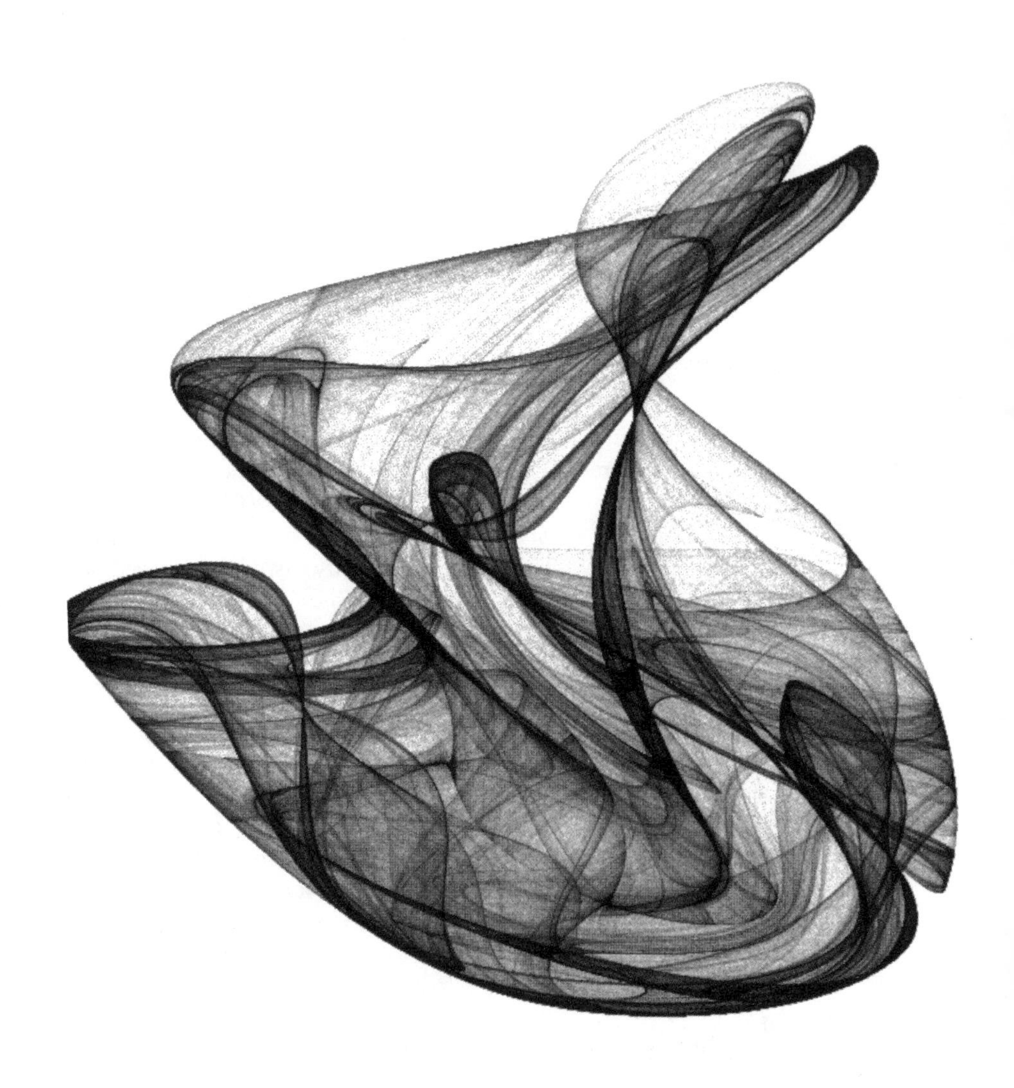

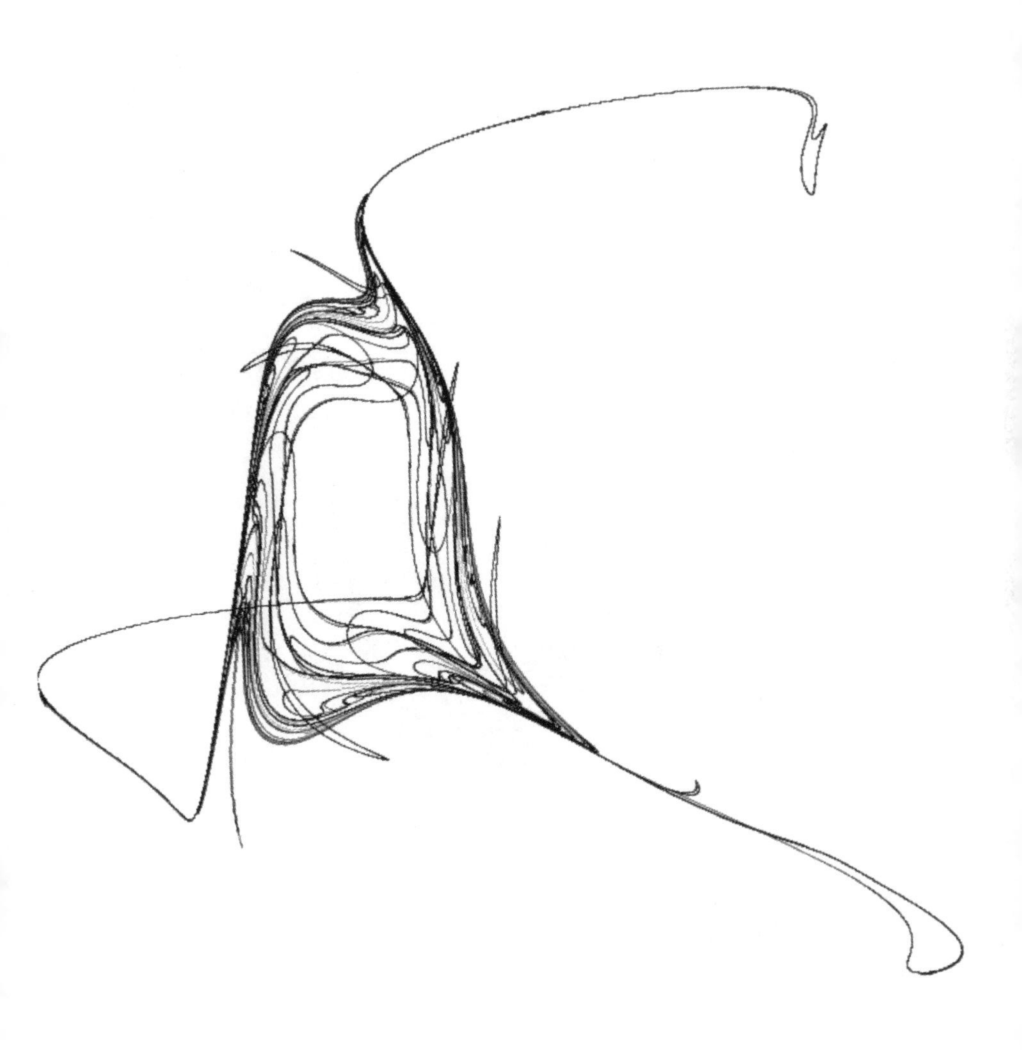

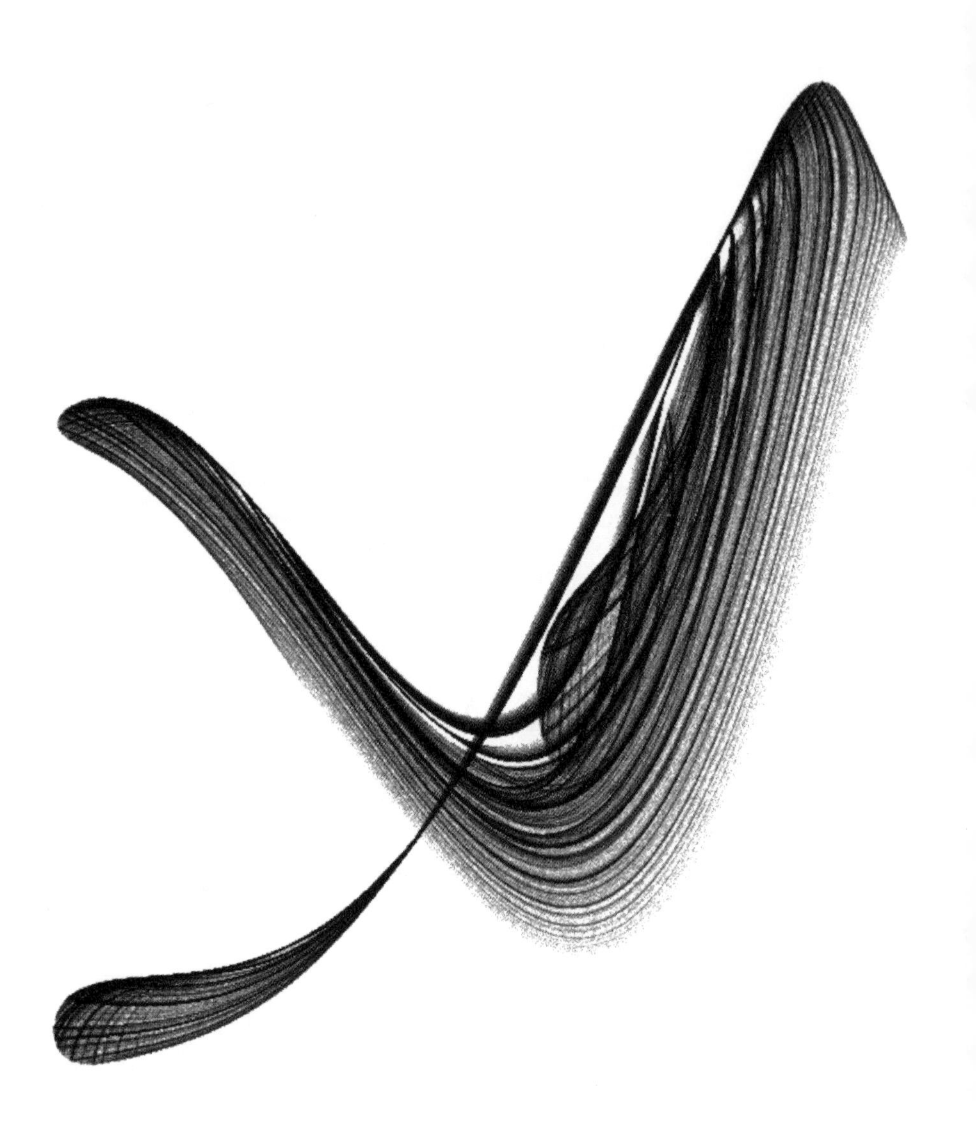

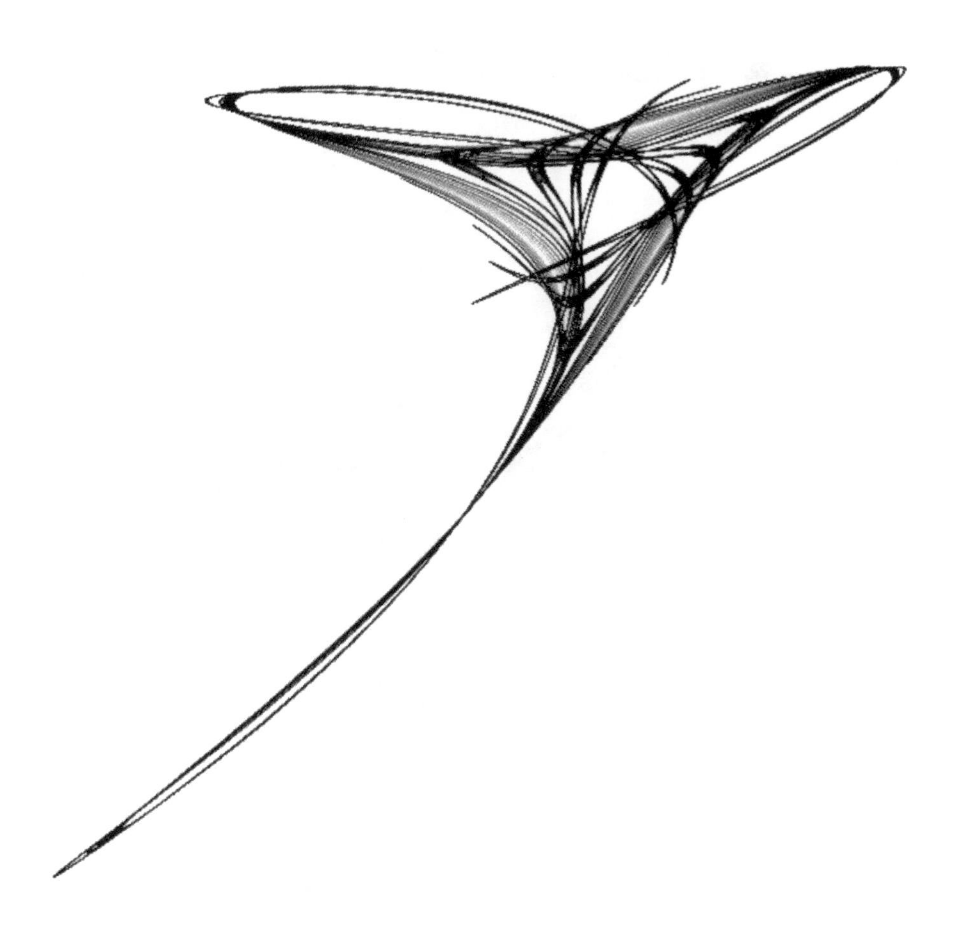

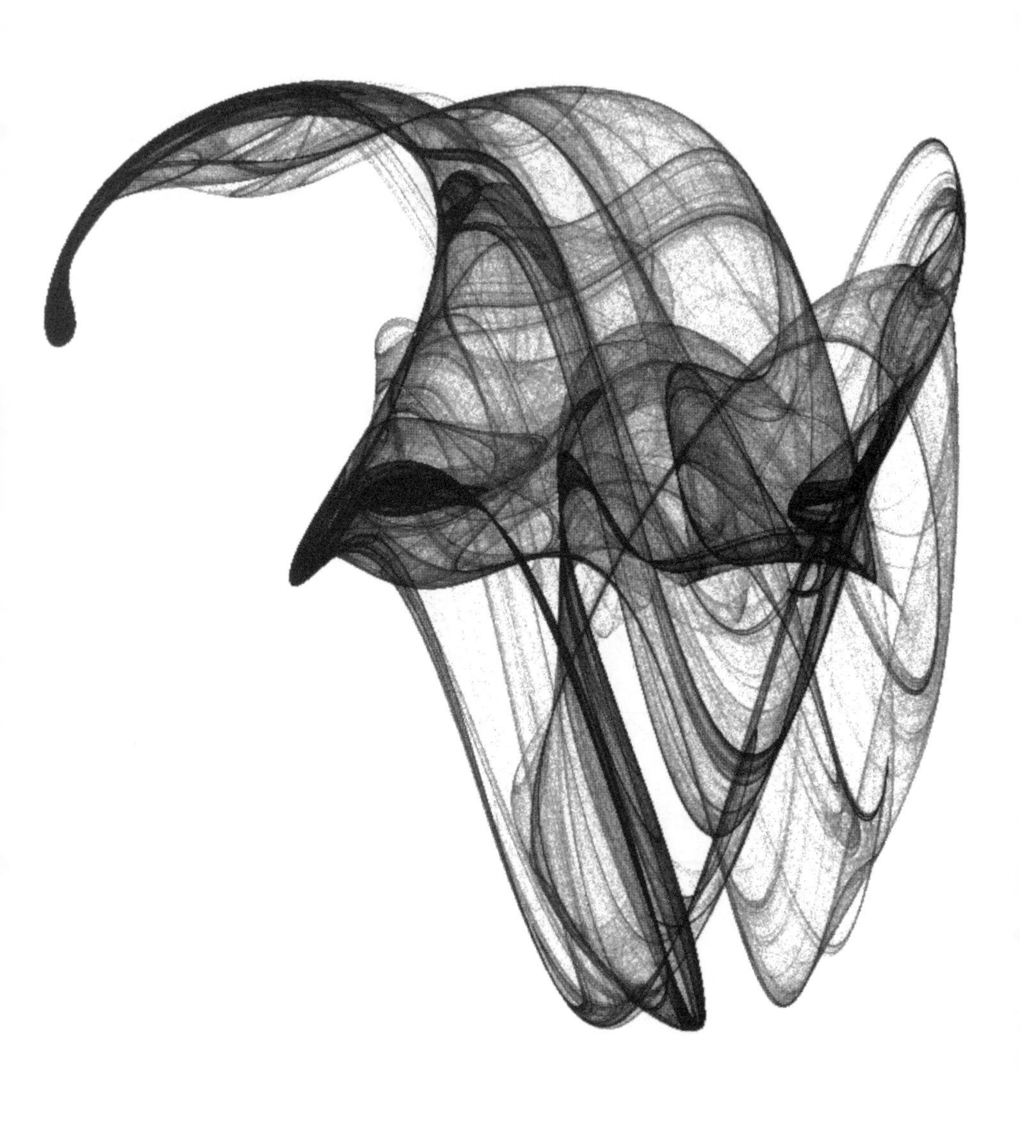

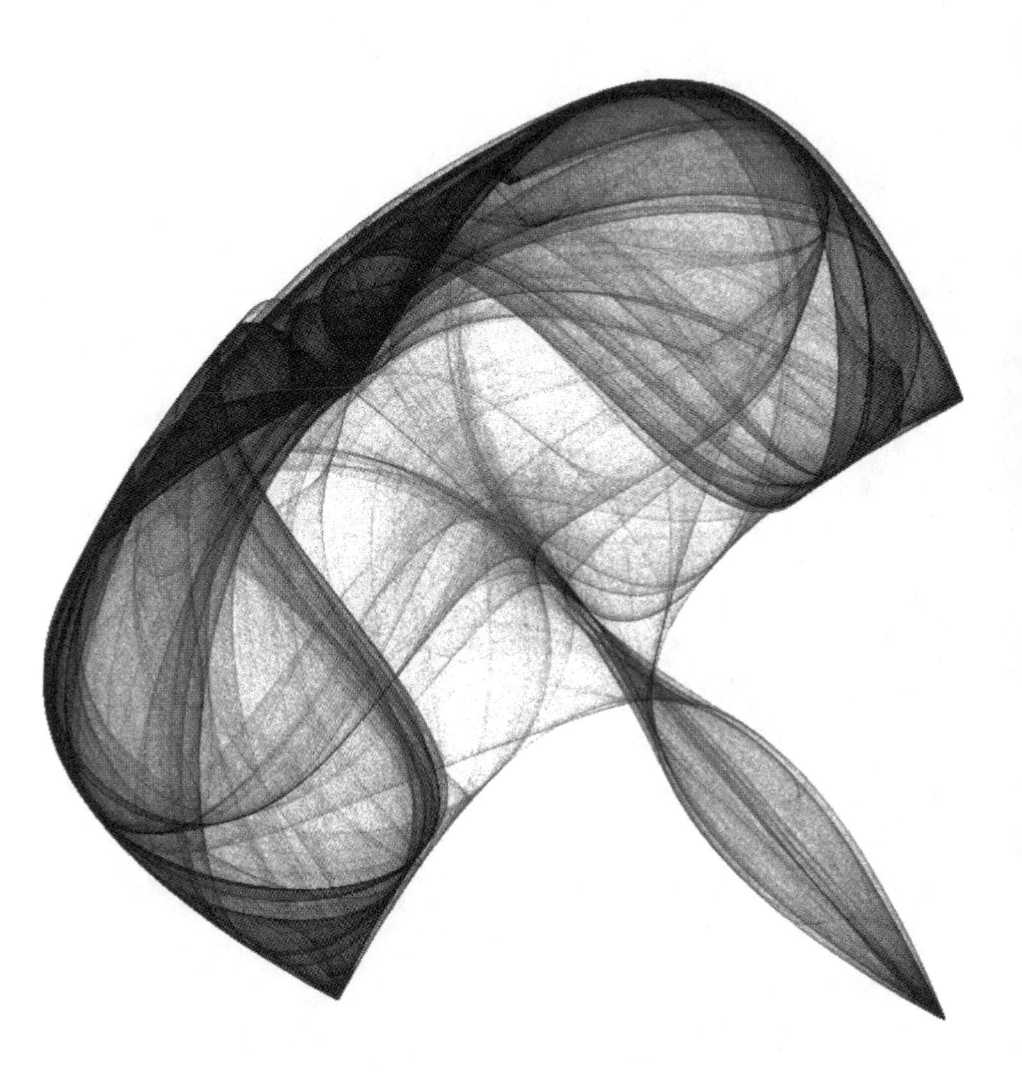

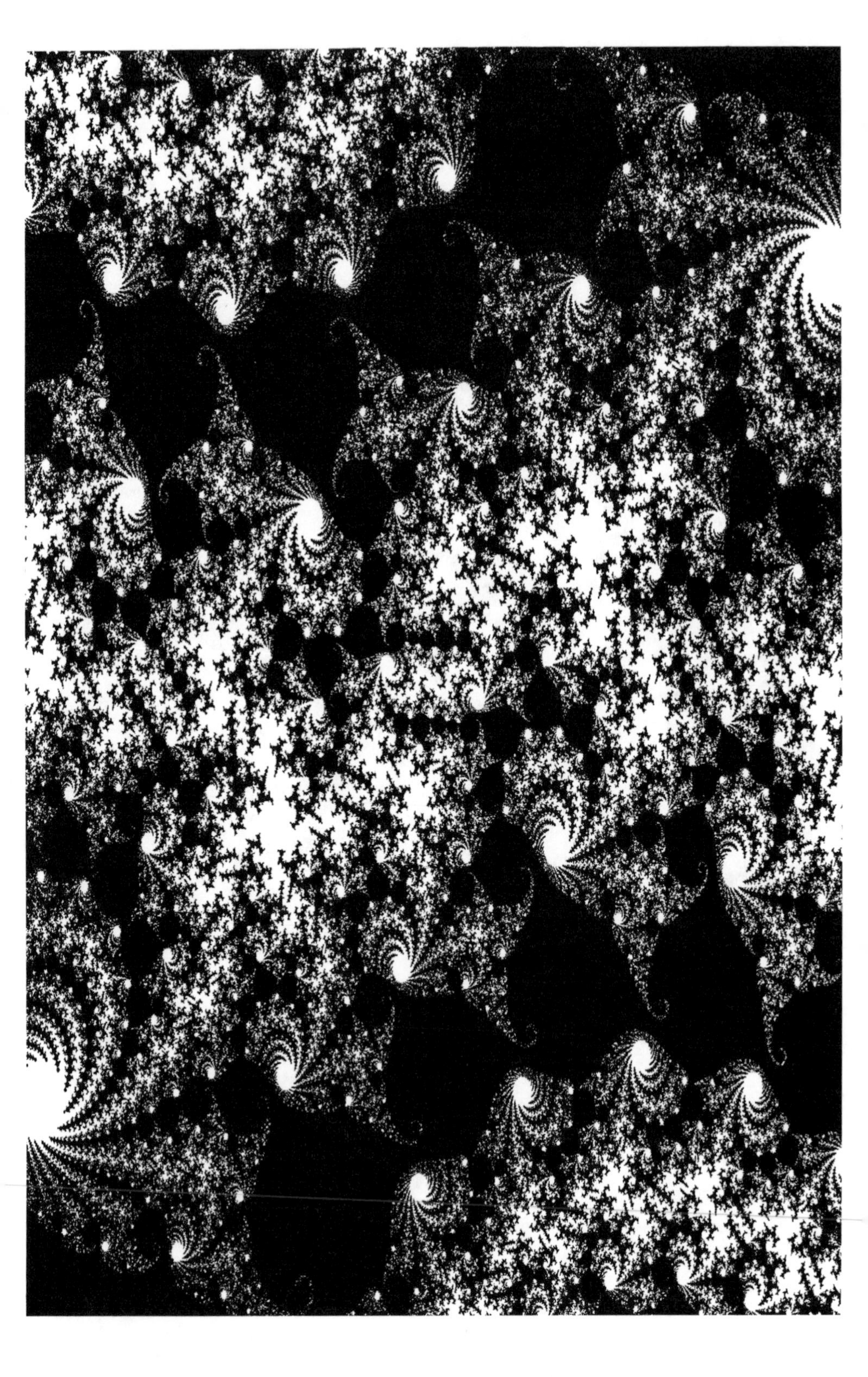

Chapter 16

Interlude: Rendering Ratiocination

"Some people can read a musical score and in their minds hear the music ... Others can see, in their mind's eye, great beauty and structure in certain mathematical functions ... Lesser folk, like me, need to hear music played and see numbers rendered to appreciate their structures." Peter B. Schroeder

"What matters in the world is not so much what is true as what is entertaining, at least so long as the truth itself is unknowable." Pierre Larousse

Obviously, I have barely scratched the surface of the subject of Latööcarfian patterns. There are endless combinations of trigonometric (and arithmetic) operators to be tried. In the process, some of you will discover worlds neither I nor anyone else has seen.

This chapter is included to help readers design computer programs that search for the kinds of patterns in this book. The following steps may be executed to compute and select chaotic patterns:

1. First, select random values for constants a, b, c and d. As mentioned earlier, for good results they should be in the ranges $(-3 < a, b < 3)$ and $(0.5 < c, d < 1.5)$.
2. Iterate the equations a few hundred times to determine maximum and minimum bounds for sizing the picture on graphics screen
3. Compute the Lyapunov exponent.
4. If the Lyapunov exponent is negative, discard the pattern and parameters immediately, and go to Step 1.
5. Iterate several million times, and plot points.

In actuality, the process I use to produce the images in this book involves several more steps, which are not really necessary for those of you attempting to implement this on a personal computer. These additional steps only increase the convenience of producing the images and also the quality of the gray-level rendi-

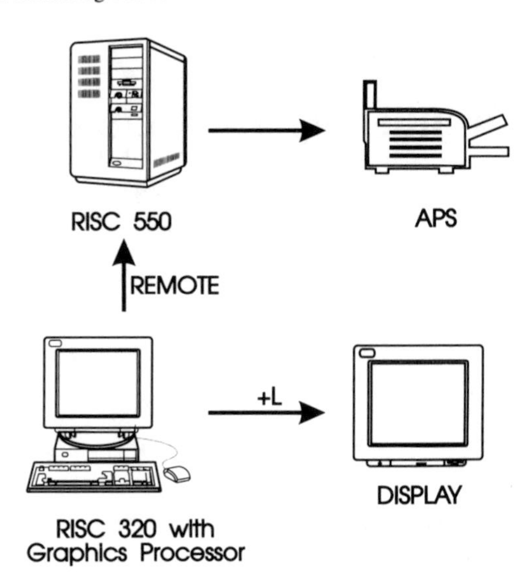

Figure 16.1. *Rendering system.* (See text for description. Shown is an IBM RISC System/6000 (model 320) with high performance graphics processor, a 19 inch display (DISPLAY), an IBM RISC System/6000 (model 550), and an APS6/108 Imagesetter. Only those patterns with a positive Lyapunov exponent (+L) are displayed.)

tions. Some of these programming accoutrements are described in the following sections.

16.1 Hardware and Software

> *"Like migrating birds, billions of atoms are flying in and out of my brain every second. It swirls with electrical waves that never form the same pattern twice in a lifetime."* Deepak Chopra, *Ageless Body, Timeless Mind*

To produce the pictures in this book, I initially perform steps 1 through 5 using an IBM RISC System/6000 (Model 320) equipped with a high performance graphics processor which allows me to rapidly view the results in low resolution for a small number of iterations. The program only presents patterns with a positive Lyapunov exponent. If I find a pattern to be attractive, the parameters (a,b,c,d) are sent to a remote rendering IBM RISC System/6000 (Model 550) which recomputes the attractor for much longer periods of time (10 million iterations). The remote renderer counts the number of times an orbit (trajectory) wanders across a pixel in a 700x700 grid. The value of this count is represented as intensity or brightness (or color). An image processing method called histogram equalization is performed (see Chapter 18, "Flavor Enhancers") to improve the contrast of the resultant attractor. (If image processing techniques are not available, you can compute a simpler representation by decreasing the number of iterations and plotting dots.)

The image, represented as a byte-map, is then converted to postscript format, rasterized on an Autologic APS-PS/PIP and imaged on an APS6/108 Imagesetter.

16.2 Software Tips

I find it useful to iterate the system a few hundred times before plotting the attractor and before computing Lyapunov exponents to eliminate transients and to make sure that the mapping has settled into the attractor. In other words, since we are interested only in the asymptotic behavior of the orbit, care must be taken to discard the first few hundred points before plotting the points.

Note that if our dynamical system producing these plots led to totally random output, then the plots would be a diffuse random scattering of points in 2-D. If the system were absolutely periodic (like a sine wave), then the figure would consist of a thin curve in 2-D. These curves, however, are delicately poised somewhere between the two extremes and have a potential infinity of values. For many cases, the starting point makes little difference; this indicates that the resulting figures are attractors for the dynamics. As long as you're somewhere near the attractor, the next few points will quickly converge to the attractor.

16.3 Spiral Julia Sets

In addition to the dusty Latööcarfian attractors scattered throughout this book, I have also graphed many examples of another kind of dynamical system,[20] the Julia set (see Glossary).[21] I computed all the spiral Julia sets in this book, such as the one below, by iterating $z_{n+1} = z_n + c$, for complex values of z and c. For a given value of c the set of initial values z_0 for which z_n are bounded form the filled in Julia set. (See B.2, "Some Figure Parameters and Descriptions" for more information.) Computer experiments have shown that different values of c can lead to a beautiful variety of Julia sets, minute changes in c often causing enormous changes in the shape of the Julia set.

[20] See Chapter 7, "Interlude: Chaos and Dynamical Systems."

[21] Latööcarfians enjoy these spiral art forms and sometimes have the Prohaptors painstakingly hand-paint the forms on their furniture and walls.

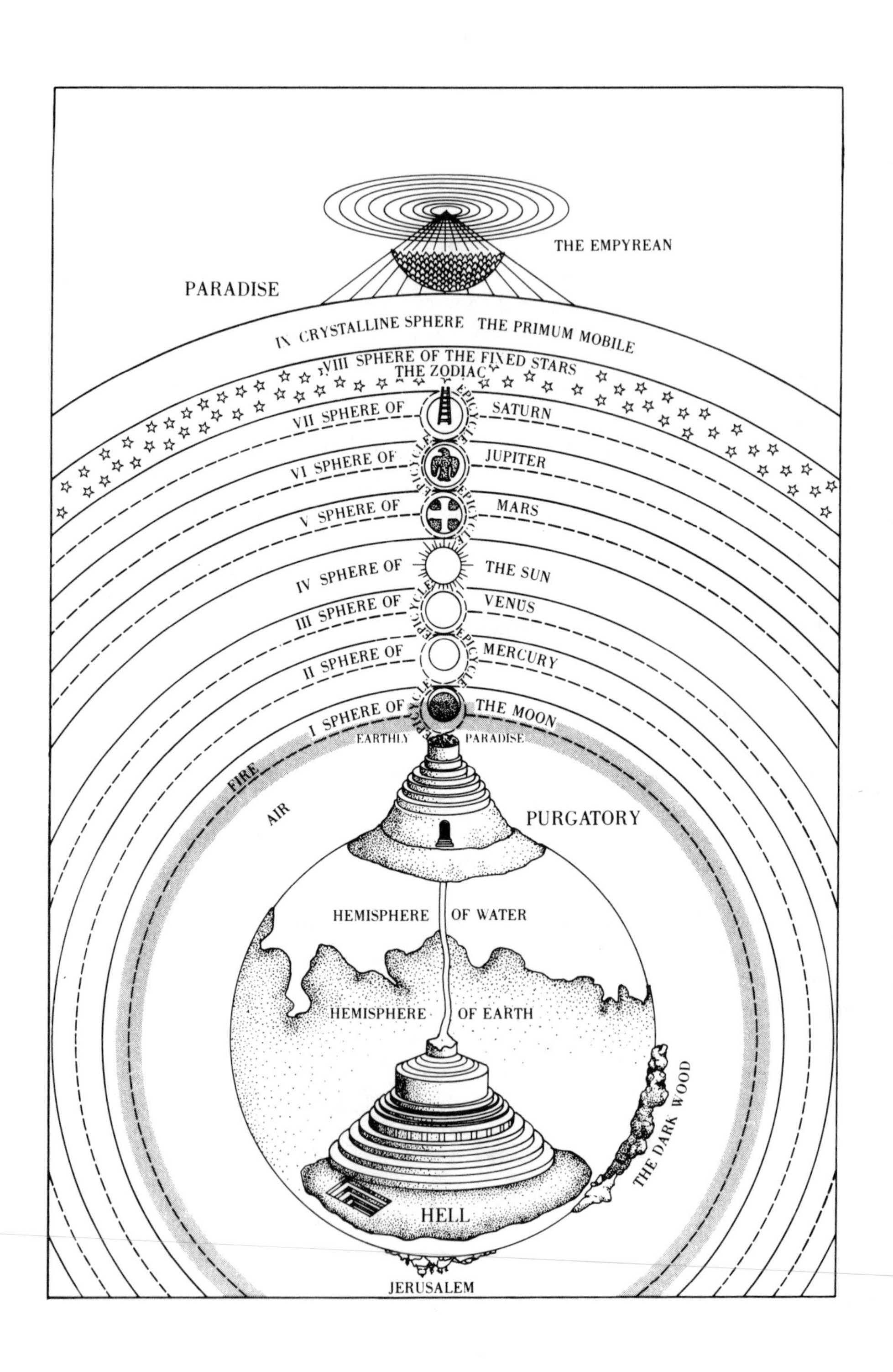
THE EMPYREAN
PARADISE
IX CRYSTALLINE SPHERE THE PRIMUM MOBILE
VIII SPHERE OF THE FIXED STARS
THE ZODIAC
VII SPHERE OF SATURN
VI SPHERE OF JUPITER
V SPHERE OF MARS
IV SPHERE OF THE SUN
III SPHERE OF VENUS
II SPHERE OF MERCURY
I SPHERE OF THE MOON
EPICYCLE
EARTHLY PARADISE
FIRE
AIR
PURGATORY
HEMISPHERE OF WATER
HEMISPHERE OF EARTH
THE DARK WOOD
HELL
JERUSALEM

Chapter 17

Latööcarfian Cosmology

"The fabric of the world has its center everywhere and its circumference nowhere." Cardinal Nicolas of Cusa, fifteenth century

"Cosmology is where science and religion meet."
George Smoot, 1992 *Scientific American*

"Infinity, curved space, the big bang, red shifts of galaxies – these are the makings of modern cosmology." Owen Gringerich, 1977, *Cosmology+1*

17.1 Aristotle and Anaximander

Cosmology is the study of the origin and structure of the universe in space and time. According to many ancient (Earth) mythologies, the universe has a complex structure but is nonetheless finite. For example, the ancient Greek philosopher Aristotle saw the universe as a system of fifty-six celestial spheres carrying heavenly bodies. The earth was a ball surrounded by these concentric crystalline spheres. The outermost sphere carried the stars. Each body was carried around the earth by the rotation of the sphere in which it was imbedded.

Another Greek philosopher, Anaximander (600 BC), believed that the world evolved from an infinite substance, which he called *apeiron*. In his cosmology, first the warm separated from the cold, which became the earth. A warm fire surrounded the cold earth and was then caught in wheel-like hoses of air. The sun, the moon, and the stars were merely holes in the hoses which made the fire visible to observers on Earth. Anaximander's wheel-like hoses were attempts at explaining the circular paths of the stars.

Figure 17.1. *Finite space.* According to Aristotelian cosmology, space was finite and had a definite edge. This idea was accepted during medieval times on Earth. The illustration here is often said to be a 16th-century German woodcut, although its origin is probably much more recent.

The finite universe of Aristotle was widely accepted in medieval Europe. For example,[22] the frontispiece for this chapter shows Dante Aligheri's structure of the universe from "Paradise" in *The Divine Comedy*. The universe of Aristotle and Dante both contained an edge which separated a finite world from the Empyrean unknown beyond. In *The Divine Comedy* Dante is guided up through the nine heavenly spheres, each sphere larger than the last, until he reaches the Primum Mobile, the ninth and largest sphere and boundary of space. Dante wishes to see the Empyrean abode of God, which finally appears to him as a blinding point of light surrounded by nine concentric spheres that represent the angelic orders responsible for the motions of the material spheres (Figure 17.5).

---

[22] Frontispiece from J. J. Callahan, "The curvature of space in a finite universe," © 1976 by Scientific American. All rights reserved.

Figure 17.2. *The world of the ancients of Earth.* The earth was believed to be a large flat disc floating on the surface of the world ocean which surrounded it. (From Gamow, see Credits.)

17.2 Fabric of Space

Only in the last twenty years have the Latööcarfians migrated to the surface of Ganymede, the bulk of their cities and civilization having formed in the Ganymedean air chamber. What strange cosmologies would creatures such as the Latööcarfians devised as they spent their lives confined in an air chamber within the ice of a moon?

As on Earth, various models for the cosmos were developed by the early Latööcarfians and Prohaptors. Scholars who have studied their ancient literature have noticed various passages relying on metaphors presenting Ganymedean time and space in terms of a pliant substance filled with knots, folds, fibers, and filaments. Both Ganymedean art and literature suggest that Ganymedeans believe their universe is bounded, and contained by long, thin, essentially supple objects of a cord-like form. The substance of these forms is not clear, but they are typically interconnected in a complex weave (Figure 17.6). It is hypothesized that their filament model of the universe arose from early observations of the wormlike öös tunnels permeating the walls and ceiling of their air chamber (see Section 4.5, "The Ice Plankton of Ganymede").

One branch of Prohaptor cosmology puts forth the idea that the infinite strands of the universe are integrated into a giant system resembling a pile of cloth, a gigantic piece of folded fabric (Figure 17.7). For example, Prohaptor manuscripts often allude to the underworld when they speak of "the folds of death." Other phrases found in ancient writings refer to the "ten folds of heaven."

Various Ganymedean cartographers have mapped their large air pocket, so that it is quite clear to them that their own world is bounded. Teams of Prohaptors and Latööcarfians have even attempted to reach the ceiling of the ice pocket using a

Figure 17.3. *The argument against the spherical shape of the Earth.* (From Gamow, see Credits.)

clever system of ropes, ice picks, and pulleys. However, the task was so formidable, and so many live were lost, that funding for the "search for the ceiling initiative" (SCI) was recently withdrawn by the Ganymedean National Science Foundation (GNSF).

Some governmental agencies have funded lateral excavations (into the ice walls) to search for other large air pockets. Some Latööcarfian scientists believe that other civilizations may have developed in remote chambers of air within the ice. The search for "extra-air-pocket" intelligence is a heavily funded scientific activity.

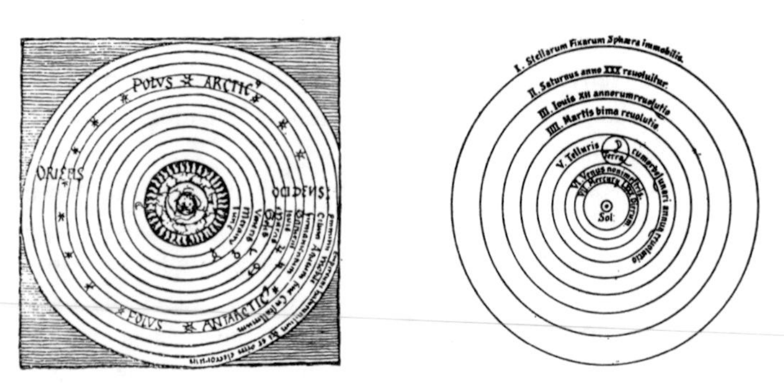

Figure 17.4. *Views of the universe.* The Aristotelian scheme is shown at left. The universe according to Copernicus (1543) is at right.

Figure 17.5. *The Empyrean.* (From Dante's *The Divine Comedy.*) Dante wishes to see the Empyrean abode of God, which finally appears to him as a blinding point of light surrounded by concentric spheres that represent the angelic orders responsible for the motions of the material spheres.

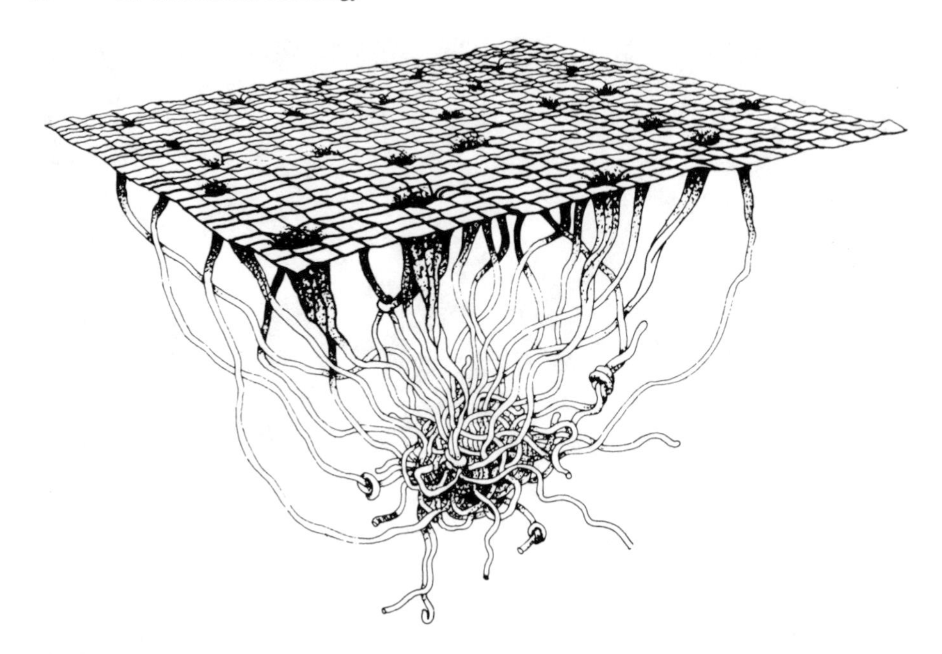

Figure 17.6. *The woven land and the tangled underworld of the Latööcarfian cosmos: a model.* (Reprinted with permission of the New York Academy of Sciences. See Credits.)

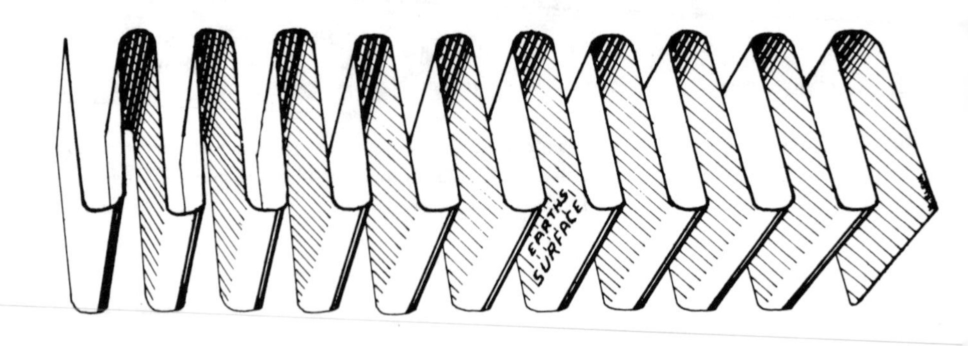

Figure 17.7. *The Latööcarfian universe as a folded cloth: a model.* (Reprinted with permission of the New York Academy of Sciences. See Credits.)

Chapter 18

Flavor Enhancers

"Why is it that the silhouette of a storm-bent leafless tree against an evening sky in winter is perceived as beautiful, but the corresponding silhouette of any multipurpose university building is not, in spite of all the efforts of the architect?"

George Zebrowski

18.1 Turning Up the Lights

How would modern humans exist without "sensory enhancers"? We rub stains on wood grains to visually bring out their subtle, hidden patterns. We add flavor enhancers like monosodium glutamate to otherwise drab foods. We use graphic equalizers in our car hi-fi systems to boost the bass and treble in Bach, Beethoven and the Beatles. In the same way that a painting hanging on a wall can be made more attractive simply by adjusting the brightness of a nearby light, computers with image-enhancement programs can improve the beauty of the Latööcarfian patterns. For many chaotic attractors, detail in the fainter regions is often imperceptible. Therefore I use computer algorithms which manipulate the picture contrast so that the final image is more suitable for human appreciation, and so that it reveals more of the detail in the fainter structures.

```
ALGORITHM: How to compute a histogram.

1. Initialize array H(Z)(0 ≥ Z ≥ L) to zero.
2. DO FOR all pixels P of the image
3.     H(f(P)) = H(f(P)) + 1
4. END
```

Pseudocode 18.1. *How to compute an array H which can be used to create a histogram.*

18.2 Image Processing

Computer graphics has become indispensable in countless areas of human activity – from colorful and lighthearted television commercials, to strange new artworks, to evolutionary biology, to processed images from the edges of the known universe. Picture processing, a subfield of computer graphics, has had a number of different goals, among them television bandwidth compression, image enhancement and restoration, pictorial pattern recognition, and contour enhancement.

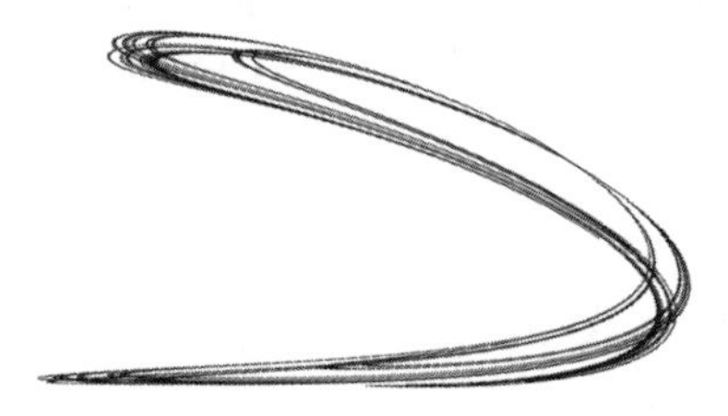

We see contours when there is a contrast, or difference, in brightness or color between two areas. Contours are so dominant in our visual perception that when we draw an object, it is almost instinctive for us to begin by sketching its outlines. In order to visually enhance the contrast of the chaotic attractors, I used a technique called *histogram equalization* to bring out low contrast and subtle features. Histogram equalization takes a raster of intensities, plots the number of times each intensity occurs, and then creates a mapping from the original intensities to a new set so that each intensity level occurs with approximately equal frequency. As indicated in the previous chapter, most of you will not use this technique to produce your own pictures, and you therefore may want to skip the remaining material in this chapter.

Let $p_1(P, Z)$ be the first order probability density function that a pixel P (specified by its location in the image) has brightness level Z. Also let's assume that Z varies between 0 (very dark) and some value $L > 0$ (very bright). We can use this probability function to resale the original image for better contrast. To estimate $p_1(P, Z)$ we can compute a *histogram* $H(Z)$ of the image. Let $f(P)$ denote the brightness level at pixel P. Then $H(Z)$ can be computed as outlined in Code 18.1. Once the histogram is computed, it can be used to enhance the image. For example, $H(Z)$ is zero for many values of Z and this means that the available gray levels are not used efficiently. Histogram equalization allows us to reassign them so that the dynamic range of the image increases. We reassign the values so that

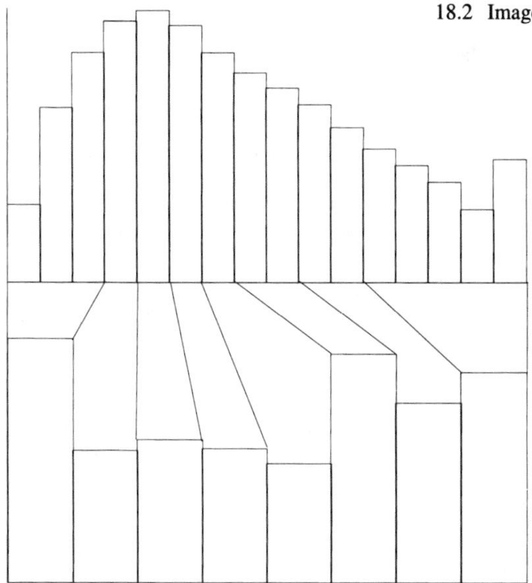

Figure 18.1. *Example histograms.* The output histogram (bottom) can be used to improve the contrast in an image. The y-axis is image amplitude for a range of gray levels before (top) and after (bottom) histogram equalization. Figure adapted from Pratt (1978).

the new resulting histogram is as flat as possible. If we let A be the image area, and N the number of available brightness levels, then for a flat histogram each level must have A/N pixels at each level. If the brightness at some level Z is k times the average, then that level must be mapped into k different levels using a one-to-many mapping. The various ways of achieving this mapping are discussed in many image-processing books, such as (Pavlidis, 1982).

One way of rescaling the image is as follows. Assume that the input image has N shades of gray, represented by N bars in a histogram bar chart (see Figure 18.1). Further assume, for example, that the final enhanced image has $N/2$ shades of gray represented as $N/2$ bars in a histogram. Each bar represents a range or band of values that are considered to be the same shade. To enhance the contrast, first compute the average value of the histogram levels. Then, starting at the lowest gray level value in the original image, combine the pixels in the bands until the sum is closest to the average. All of these pixels are then rescaled to the new intensity value at the midpoint of the bar. Repeat for the higher value gray levels. If the number of levels in the original image is large, it's possible to rescale the gray levels so that the enhanced image histogram is almost constant. See Pratt (1978) for details.

Histogram equalization is used in many practical image processing areas such as in medical diagnosis or industrial quality control where one is interested in detecting dark spots in a radiographic image. Histogram equalization can be used to make the shape of such spots visually apparent, since they must be detected against a shadowy, dark background in the original image.

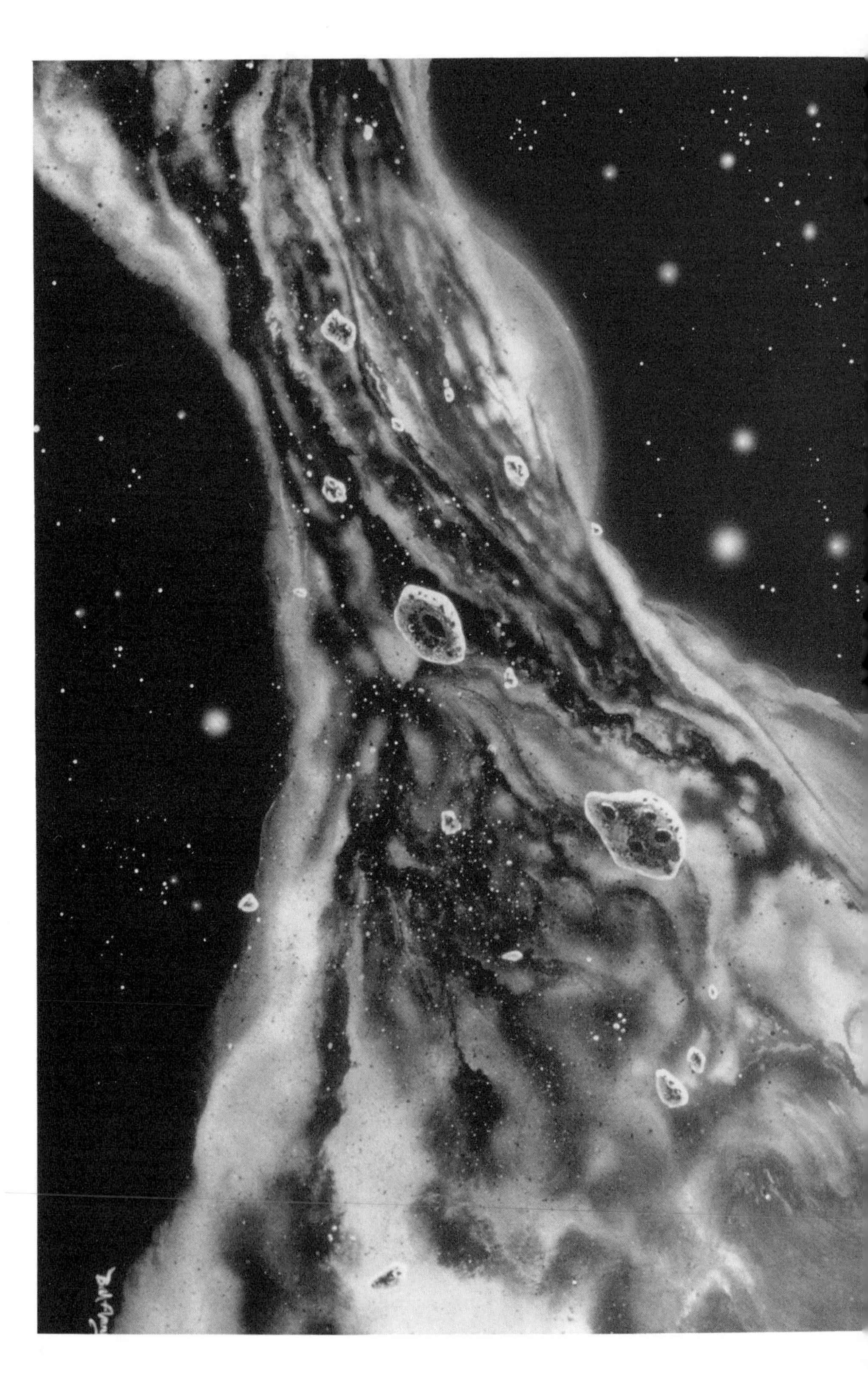

Chapter 19

Inside-Out Universe

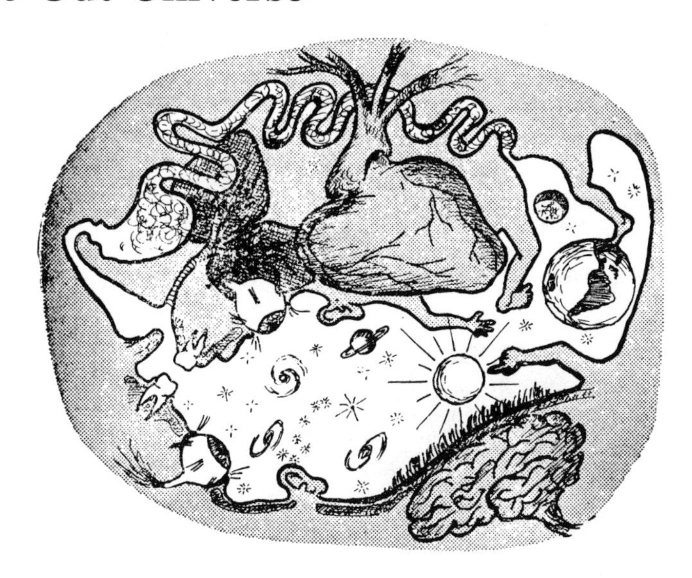

"As one goes through it, one sees that the gate one went through was the self that went through it...." R. D. Laing

"If we wish to understand the nature of the Universe we have an inner hidden advantage: we are ourselves little portions of the universe and so carry the answer within us." Jacques Boivin, *The Single Heart Field Theory*

There are few scientific illustrations that one comes across in a lifetime that really make a big impact on the psyche. Just as the airplane figure in the Introduction of this book held a particular fascination for me, the above illustration also kindled an early interest in astronomy, science fiction, and science. The figure deals with topology, the branch of mathematics concerned with the interrelatedness of geometric shapes. Two figures are topologically equivalent if one can be deformed into the other by bending, stretching, and twisting (but not cutting!). For this reason, topology is sometimes popularly called "rubber-sheet geometry." A circle is topologically equivalent to a square. Likewise, a cylinder and a sphere are equivalent. However, a doughnut is not equivalent to a sphere, because no amount of bending or twisting will deform it into a sphere.

A more imaginative exercise in topology involves deformations of the human body. Your body with its internal tube-like digestive system is topologically the same as a doughnut because it has all the geometrical properties of a doughnut. The illustration at the beginning of this chapter shows how the human body can be transformed so that it is inside-out.[23] The illustration is from George Gamow's *One, Two, Three... Infinity* (Dover), and it shows the entire universe squeezed into the inner circular channel. In particular, this surrealistic drawing represents a human walking on the surface of the Earth and looking up at the stars. Gamow gives the precise instructions for this topological transformation using "doughnut theory," and you can read his classic book for all the gory details. Notice that the earth, moon, sun, and stars are crowded in a comparatively narrow channel running through the body of the human which is surrounded by the internal organs. How many internal organs can you identify? Observe the position of the teeth at the aperture between the universe and esophagus.

"A place is nothing: not even space, unless at its heart – a figure stands."
Anonymous

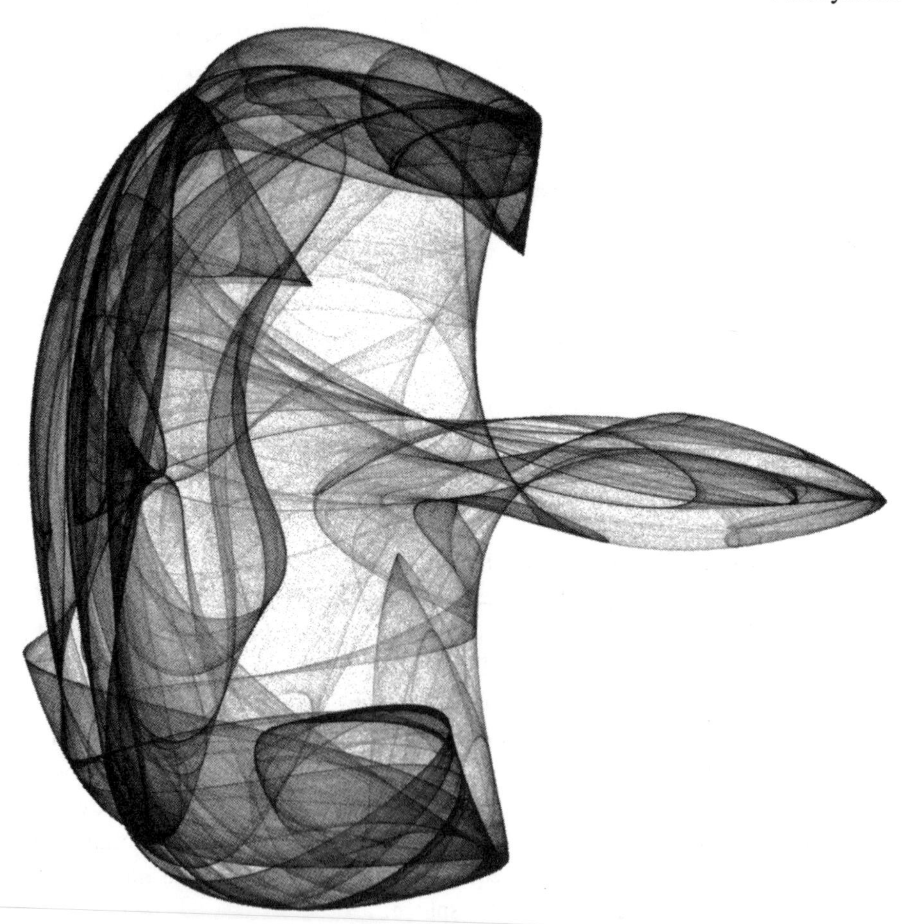

[23] Publisher's note: This topological transformation is for illustrative purposes only and should not be attempted by readers. As always, consult your personal physician before embarking on such activities.

Chapter 20

Ganymedean Blood and Biology

"Space is almost infinite. As a matter of fact we think it is infinite."

Former Vice-President Dan Quayle

20.1 Organic and Inorganic Life

Most of the chemical components of living organisms on Earth are organic compounds of carbon. Many organic biomolecules also contain nitrogen. Below is an example of thymine, a typical organic compound found in the hereditary material of all living cells. (H is hydrogen; N is nitrogen; O is oxygen; C is carbon.)

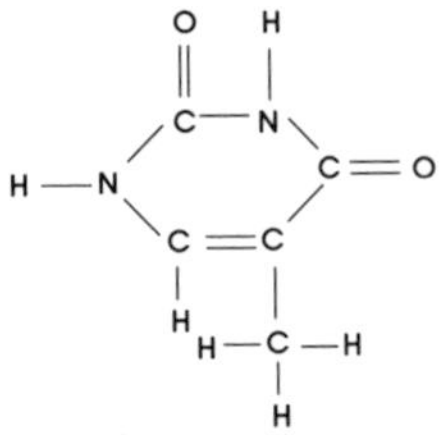

```
          O   H
          ‖   |
        / C — N \
H — N            C = O
        \ C = C /
          |   |
          H   H — C — H
                  |
                  H
```

In contrast, the elements carbon and nitrogen are rather scarce in *nonliving* matter on Earth and occur only in simple inorganic forms such as carbon dioxide or molecular nitrogen.

On Ganymede, living organisms come in three varieties: inorganic, organic, and hybrids. An example of an organic organism is the six-foot-tall Prohaptor, which has a biology closely resembling Earthly creatures. Their blood, for example, in many ways resembles the blood of invertebrates on Earth. On the other hand, inorganic creatures, such as the Latööcarfians, piezoelectric ants (Chapter 36, "Death-Fungi and Zinc Ants"), and *Kinorhyncha* worms (Chapter 37, "Starfish Soup"), are quite different possessing electrorheological blood and semiconductor brains (see 4.2, "Circulatory System"). Some creatures, such as the Navanax slug people (Chapter 29, "The Navanax People"), have hybrid bodies

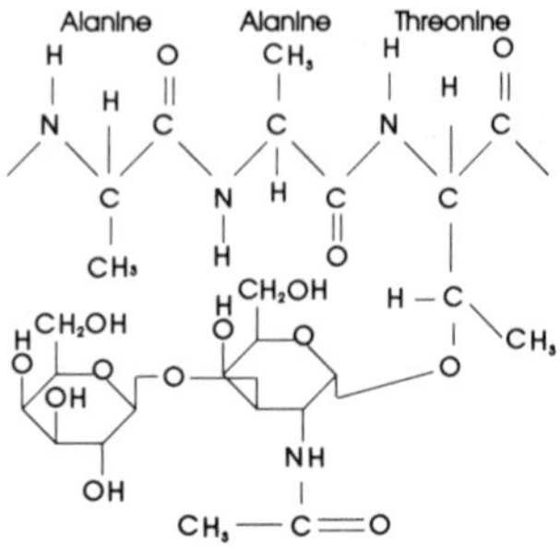

Figure 20.1. *Chemical structure of Ganymedean antifreeze.* Many of the organic lifeforms have antifreeze molecules in their blood which helps the blood to resist freezing. This is similar to the antifreeze molecules found in the notothenioid fishes of Antarctica. Shown here is the basic structural subunit of the long chain antifreeze molecules.

containing piezoelectric musculature together with tissues composed of carbon-based macromolecules. Below is an example of a typical organic compound on which the hereditary material of many organic lifeforms on Ganymede is based:

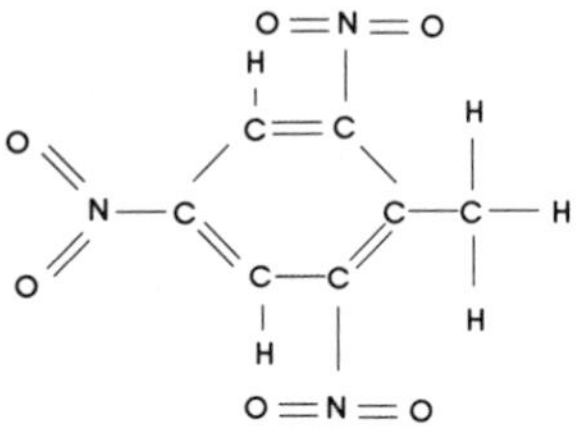

(Note the similarity with the organic structure, on the previous page, used by Earthly creatures.) Biochemist readers may recognize this as a molecule of TNT (trinitrotolulene), an explosive material. Ganymedean life actually use variants of this basic structure in their hereditary material, rather than TNT itself. However, occasionally there are metabolic diseases leading to the production of TNT, and the afflicted animal explodes.

20.2 Antifreeze

Most organic-based life on Ganymede has the ability to produce chemical compounds with powerful antifreeze properties. This adaptation is crucial to survival in the cold weather, particularly in the polar regions. These organic compounds depress the freezing point of body fluids just as antifreeze in a car's radiator prevents the fluid from freezing until much lower temperatures are reached. In fact, the survival of Ganymedean creatures rests on seven different antifreeze molecules which are found in all their body fluids, except for their urine. Those of you who are biochemists may be interested in the chemistry of these molecules. They are glycopeptides consisting of repeating units composed of a two-sugar molecule (a disaccharide) bonded to a peptide chain containing two amino acids. When these molecules absorb to tiny seeds of ice that may form in the blood, they prevent the ice crystals from growing larger. The antifreeze molecules bind to the ice through the carbonyl (-CO-) groups in the amino acid chains. The kidneys of the animal prevent the glycopeptides from entering the urine, thus eliminating the need for resynthesis of the molecules. See Figure 20.1 for the general chemical structure of the antifreeze. This chemical structure of the antifreeze molecule is similar to that of the antifreeze molecule found in the notothenioid fishes of Antarctica (Eastman, J. and DeVries, A. (1986) Antarctic Fishes. In *Life at the Edge*. Gould, J. and Gould, G., eds. Freeman: NY.)

Inorganic lifeforms, such as the Latööcarfians, are capable of living on the recently discovered surface of Ganymede. However, most of the organic forms, such as the Prohaptors, can only exist on the surface if they wear oxygen masks. They also require coverings for their external orifices to contend with the absence of pressure. The uncovered Prohaptor in a vacuum does not make for an appealing sight – its body expands and ruptures. However, most of their internal tissues are "sealed off" from the environment by a thickened, more resilient skin, making large, complex space suits unnecessary.

20.3 Hereditary Molecules

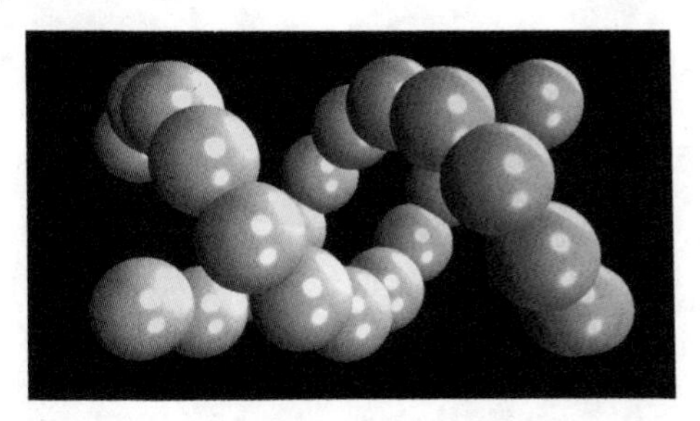

On Earth, DNA molecules contain the basic hereditary information of all living cells. If you were to view the DNA molecule under a microscope, it would resemble a twisted ladder, a twisted "double helix." Some lifeforms on Ganymede use inorganic double helices as their basic hereditary molecules. Readers familiar with inorganic chemistry may be interested in the molecular details. In particular, some lifeforms evolved from very complicated inorganic solids. These solids self-assemble from structurally simple precursors by the hydrothermal synthesis of the vanadium phosphate, $[(CH_3)_2NH_2]K_4$ $[V_{10}O_{10}(H_2O)_2(OH)_4(PO_4)_7] \bullet 4H_2O$. These DNA-like chiral double helices form interpenetrating spirals of vanadium oxo pentamers bonded together by P^{5+}. These double helices are in turn intertwined with one another in a manner that generates unusual cavities and tunnels.

Part II

THE DREAM-WEAVERS OF GANYMEDE

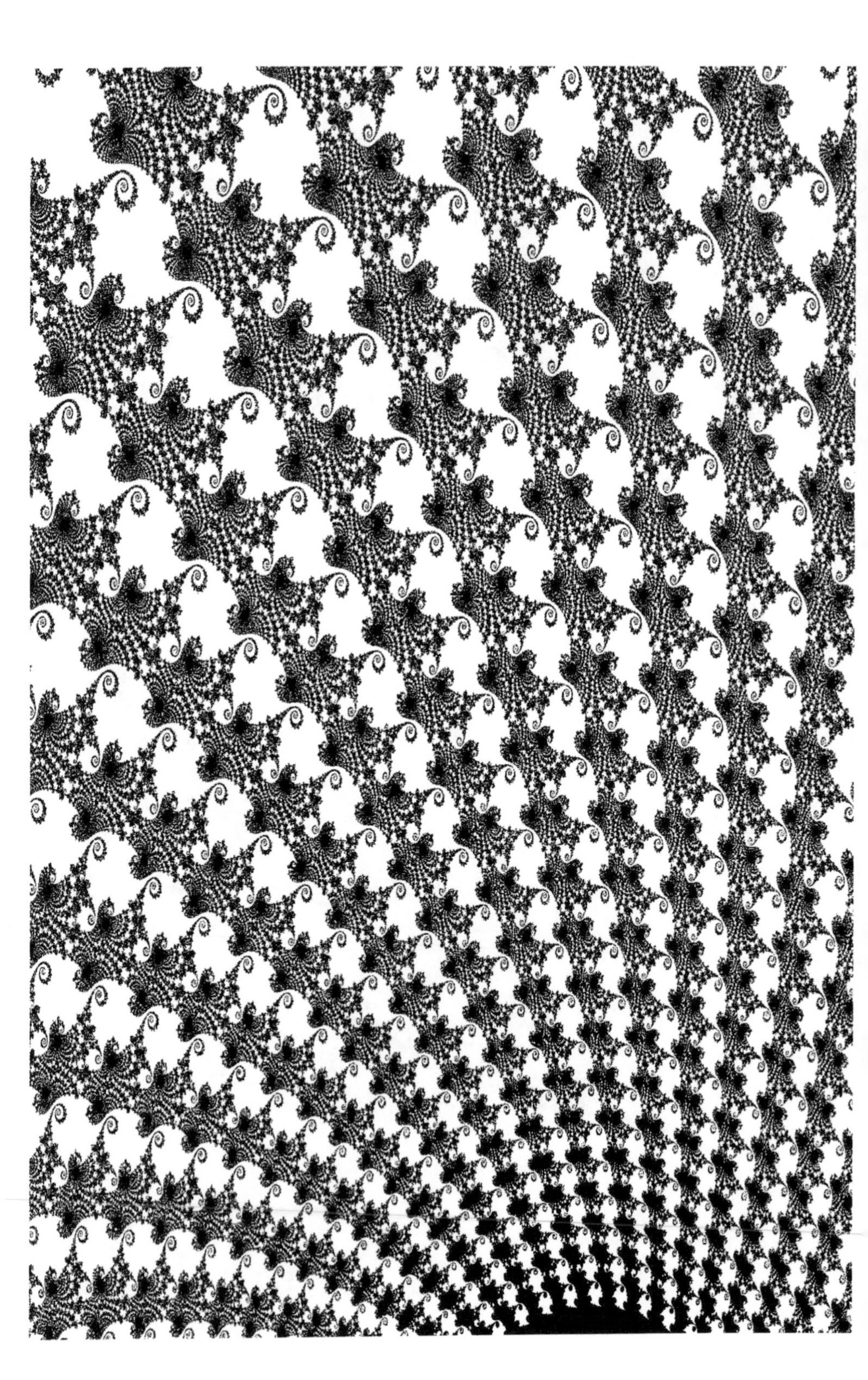

"What we call fiction is the ancient way of knowing, the total discourse that antedates all the special vocabularies.... Fiction is democratic, it reasserts the authority of the single mind to make and remake the world."

E. L. Doctorow, 1968, *Esquire*

"Did we come here to laugh or cry? Are we dying or being born?"

Carlos Fuentes, *Terra Nostra*

Late last autumn, while enjoying the brisk New England air, I took a walk with my father. Stunted trees, perhaps dogwoods, appeared at random intervals along the road. We'd been talking about this and that, and the conversation turned to books.

"Did you know how much I love your library?" I said. "Most of my favorite books are from there. I really liked that book *The Volume Library*."

In his usual, quiet voice, Dad said, "I found another strange book there the other day."

"'Found'? Don't you know what books you have?"

"Not always. Sometimes I forget. Sometimes I get gifts. It's not easy to keep track."

"What book did you find?" I said, raising my eyebrows.

"It's really old science-fiction. A guy in the 21st century explores Ganymede, one of the moons of Jupiter." Dad stopped walking and looked at me as if waiting for me to comment.

"Really?" was all I said in a neutral voice. I could never be certain about the accuracy of my father's statements. During the past ten years his stories have gotten more and more embellished, composed of myth and truth, perhaps more of the former than the latter, depending on his mood. "C'mon, show me," I said. We went back to his house.

"It's right here," he said. Dad pulled a big leather-bound volume from a dusty shelf and handed it to me. Swirly patterns decorated the binding. Stamped in large gold letters on the cover was the title *The Dream-Weavers of Ganymede.*

Chapter 21

Conflict on Callisto

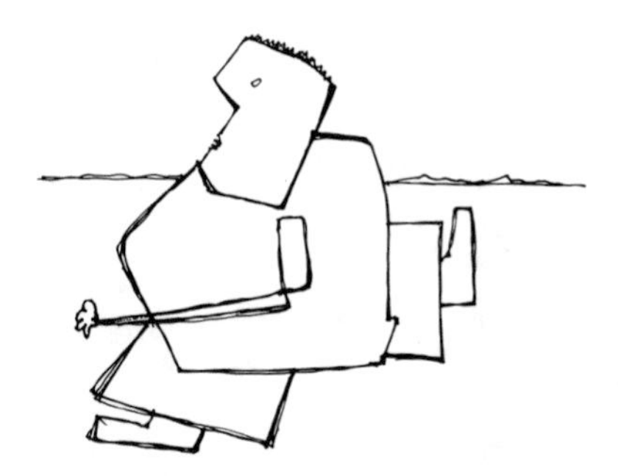

"Watch out!" I cried to Kalinda. I caught a movement out of the corner of my eye, looked across the craters of Callisto, and saw a huge bald man grab her.

"Hey, let go," she said. The man said nothing. Kalinda's beautiful eyes were widening with fright.

I jumped up and followed the splashing sounds as the man dragged her through the puddles of fungoid growths that lined Callisto's recently terraformed soil. Wispy trees occasionally blocked my path.

Thank God there are some full moons tonight, I thought to myself. Luckily, I could see Kalinda's shadow by the light reflected from Jupiter's other moons.

"Stop it! Let me go!" Kalinda shouted. I saw the man trying to drag her by the hair. A second man joined them, a wicked hooked linoleum knife in hand.

I took the second man first, hitting him with a hard side kick to the solar plexus. He fell to the ground, desperately gasping for air. In the follow through, I spun around and smashed the side of my hand into the other man's sternum. We both heard it crack. It reminded me of the sound I made as a child eating peanut brittle.

The first man rose to his feet. "Try that again," he said, as flecks of saliva sprayed from the corner of his mouth. He had the bearing of a man who was used to this. Before I could respond, his quick fists struck invisibly, like the mouths of snakes. I gasped for breath. He drew his hand back menacingly.

"What's the matter?" he said. "Can't you take it?"

"Right."

"Gonna leave the woman with me?"

"Nope." I moved swiftly and got him square in the belly, short, vicious, hard. His eyes told me his anger was controlled, but more dangerous than before.

"Not bad," he said in a flat, inflectionless voice.

Before he had a chance, I smashed my fist into his jaw and felt it give way. He dropped to the ground, unconscious.

"Let's get out of here. I could use something to eat," I said to Kalinda who stared at me with dazed eyes. Her mouth was slightly opened. "C'mon, Kalinda.

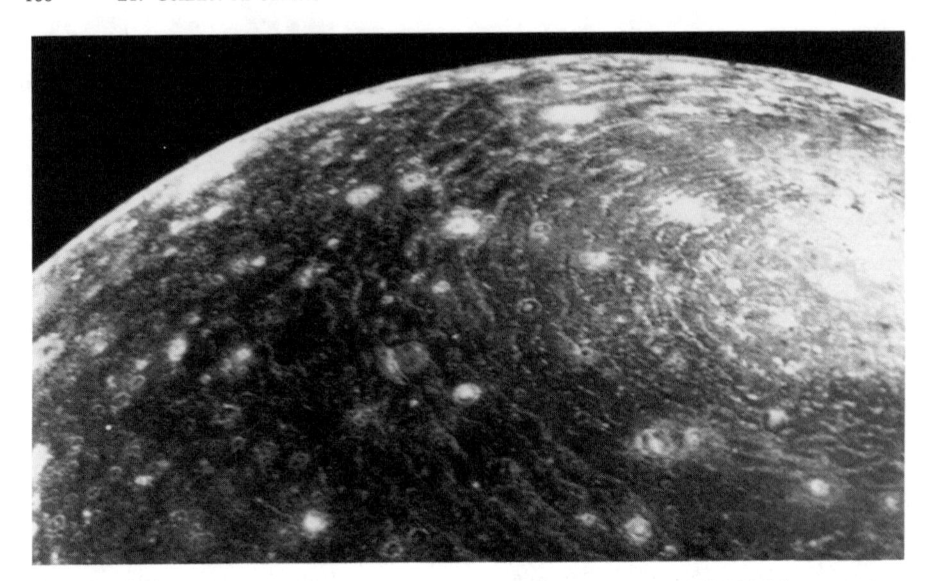

Figure 21.1. *Callisto, Jupiter's outermost moon.* (Photo taken from 350,000 kilometers.)

There might be others." I took her hand and led her out of Callisto's forest of mauve trees.

"What was *that* all about?" she said in a soft voice as she rubbed some of her bruises.

"Don't know. Maybe they wanted our scientific instruments or our weapons. Maybe they wanted you."

"Glad you were here. Thanks." She massaged her scalp.

"You OK?"

"I think I'll survive. Let's get out of here."

I thought I heard some distant voices, but I kept quiet, not wanting to scare Kalinda. "I'll get the knapsacks and specimen jars." I picked them up off the ground and also tucked several weapons into my belt. Kalinda sheathed a dagger at the side of each of her suede boots. Her cloak was the same tawny orange suede.

We wore no space suits. Callisto, the outermost of Jupiter's water-containing moons, was terraformed years ago by oxygen producing bacteria and plants introduced from Earth. It now had a breathable atmosphere.

"Did I ever tell you how great you look in that necklace?" I said while gazing at pearl-like spheres around her neck. The spheres were the egg shells of nautiloid water beings we'd discovered on a previous expedition. Kalinda just stared at me for a few seconds. This petite, yet explosive charge of a woman had spent months on a water world where I was also stationed to study the native squids. Now she was a research teuthologist, a world-renowned expert on venomous squids among other things.

"Yep, many times." She put her hands on her slim hips and fingered her necklace. "Garth, how can you think about something like that now? We nearly got killed with a linoleum knife. You're incorrigible."

"I know, too confident for my own good, right?"

"You said it."

"OK, finally finished packing," I said.

"Got the hammers and picks?"

"Right here." I patted the knapsack.

"Wish we had time to eat. I – Behind you!" Kalinda shouted as a bunch of men quietly came out of the forest and surrounded us.

"Hang on," I said. Kalinda held me around my waist and I pressed the travel button on my belt. "Here we go-o-o...." I felt the usual jarring effect that accompanies this kind of interplanetary travel. I also felt a faint nausea at the strange combined odor of vomit and rose blossoms. We were whisked away to Ganymede, another of the moons circling Jupiter.

Hartal

Chapter 22

Arrival

> *"I looked round the trees. The thin net of reality. These trees, this sun. I was infinitely far from home. The profoundest distances are never geographical."*
>
> John Fowles, *The Magus*

My name is Garth. I have only one arm. I have a black belt in tae kwon do, a Korean system of unarmed combat resembling karate. My passion is archaeology and exozoology.

A few years ago, I lost my arm to a race of nautiloid water beings who had an obsession for human flesh. The nautiloids tore it off during an underwater excavation on a small planet circling Alpha Centuri. Today the prosthetic arm serves me quite well and is only a minor impediment to my physical abilities. When I need to protect myself, my tae kwon do black belt and formidable skill with swords and fighting-sticks more than compensates for the physical loss. The worst part, these days, is not being able to scratch the phantom itches.

"Look at that," I said to Kalinda. A solitary Ganymedean seabird, wings outstretched and motionless, floated in the drizzle, inches above the limitless icy sea of orange. A pink ray of daylight reflected off its white wings and startled the multitentacular glow-worms drifting on the mist-covered ocean.

"I love this time of day," she said. We both watched as the seaworms, stimulated by the reflections of light, began to feed upon bluish grey slime which coated the ocean's surface.

"Remind me to stay away from Callisto next time," I said. "Too many crazy people running around down there."

"A little down on your own race today?"

"Humankind still has a lot of growing up to do."

"No argument there," she said rubbing her still tender scalp.

"Darn, guess what we forgot?"

"Yes?"

"Our food packs. They're still on Callisto!"

"Don't worry. There should be plenty to eat on Ganymede. I hope."

I looked at the ground. "What's this red moss?" I said, running my hands through the delicate fibers. We had arrived on a world where a rich crimson vegetation stretched for miles around us.

"Feels like velvet," Kalinda said. "Think we should take a specimen?"

"Nah, I think we'll find much more interesting life on Ganymede."

"You know exactly where we are?" Kalinda said.

"We're inside a big air pocket in the icy interior of Ganymede. Directly above us, on the surface of Ganymede, is a groovelike geological structure called the Tiamat sulcus." From a brief prior expedition, I knew that the ceilings of the air chamber were lined with phosphorescent minerals and bioluminescent bacteria, providing ample light. The bacteria derived their energy partially from the weak sunlight which penetrated the ice, and they therefore ceased their light emissions at night. The microscopic organisms produced additional energy by metabolizing ice plankton living within much of the ice of Ganymede.

"If there are lots of different air pockets, how do you know for sure?"

"There's only one pocket this large on Ganymede. I don't think we have to worry."

"Help me up," Kalinda said, extending me her hand. Slowly rising, I stretched my sore legs, and pulled her up. My bladder was painfully full.

"Not bad for a guy with a fake arm." She smiled at me.

I stand six foot three and weigh over 200 pounds, not an ounce of which is fat. Friends say my features are fairly harsh, but Kalinda seemed to like them.

"You're not so bad yourself," I smiled back and ran my natural hand through her long brown hair.

"You behave," she said wickedly.

"On second thought," I said, "why don't we take a sample of moss? I like it." I took a small trowel from my belt and used it to remove a specimen of the moss for future study. The moss had a metallic luster and shone like diamonds. "Nature must love tiny life forms, or she wouldn't have made so many of them," I said.

"Yeah, most of the worlds we've been on were simply swarming with plants and animals that were only a few millimeters," Kalinda said.

My interest in tiny creatures began in college where I researched slime molds. They resembled amoeba-like animals during one phase of life and plantlike growths during the second phase, and they felt like velvet, too.

"I think I'll also take a specimen," Kalinda said as she placed some moss in a jar. The moss varied from red through various shades of chocolate brown, according to species.

"Take a look at this sporangium," Kalinda said. "I've never seen anything like it." I watched her hold up a spore case with many bright red stalks, each stalk bearing one to multiple heads that varied from smooth to fuzzy. The fuzzy heads resembled cotton candy with pimply surfaces that looked like globs of tapioca. When Kalinda cut one with a knife, it oozed a pinkish, pasty substance.

"Garth, you're shivering."

"Just excited about this trip. Also a little cold." We had prepared for our expeditions to Jupiter's moons for almost a year. Most of our time was spent at

Mount Makalu, on Earth, to adapt our cardiovascular systems to thinner atmospheres and increase our red blood cell production. We also consumed special medicines which prepared our bodies for the tenuous, exotic atmospheres.

"Look at these rocks," Kalinda said.

I stooped down. "It's gallium arsenide." Here and there were large outcroppings of bluish-white stones which glistened in the light. "Look in the distance. See the areas of heavy mining?" I pointed.

"Yep. Are those the mines of the Latööcarfians?"

"Yes, although they seem to be abandoned."

Kalinda got on her hands and knees. "Here's some armalcolite."

I looked at the gray rock. "Know how that stuff got its name?"

"Yeah, I went to school. It's from moon rocks brought back to earth in 1969 by astronauts Armstrong, Aldrin, and Collins and named in their honor. Take the first parts of their names. Was this a quiz?"

"You're beautiful and brainy."

"I trust that's not a contradiction."

"No, not at all. Just an observation." The conversation was taking an unexpected turn. The last thing I wanted to do was to make Kalinda angry. "Hey, let's do a little exploring," I said. "The first thing we want to do is make a general survey of the air chamber we're in. Here's what we know. Principal rivers – none. Principal mountains – none. Principal towns – we don't know, except for the Latööcarfian capital city Tenochtitlãn. Recently they started to build small cities on the surface of Ganymede, but most of their civilization is below."

"Lucky for us, since the atmosphere in here is breathable!"

"Yeah, how would you like to have to wear space suits to study their civilization?"

We gazed across a landscape of pink and blue crystals as perfect as a scene inside an Easter egg. The nearby ocean glistened like orange jello. "Let's go," I said. Kalinda slipped her hand through the crook of my arm and squeezed me to her.

"Careful," I said playfully. "You don't want to break my watch."

"You and that fancy watch of yours," Kalinda said rolling her eyes.

She couldn't have broken it, of course. I was wearing the top of the line *Astrolabium Galileo Galilei* timepiece with the built-in astrolabe. It was a compact instrument that indicated due north on Ganymede, the altitude and angle of the sun and other Jovian moons, the length of day and night, the phases of the other Jovian moons, and the position of the stars. It had helped me out of some tight spots before, and I expected it might again, soon.

"Wait," said Kalinda, "Let me take a quick look at the ocean." I followed as she ran down to the shore.

The orange tides ebbed and flowed. Glittering fairylike nodes of life danced on its translucent surface and floated to its ancient rhythms. I looked closer. Within the ocean, Ganymedean jellyfish, ctenephores, and all manner of Silurian sea creatures frolicked and pranced, their diaphanous tentacles and evanescent egg sacs floating close behind in the cold liquid. Here and there were Gorgonian corals coated with algae.

"Kalinda, look. Is that a fish?" I paused, searching the nearby orange ocean. "Hmmm, I thought I saw a needlelike garfish walking across the ocean. Must have been my imagination. Garfish are on Earth. They have no business in an ocean of Ganymede."

My bladder could stand no more, and the sky was beginning to darken. I walked over to the sea, turned by back to Kalinda, and urinated into the plankton-swarming liquid. A round, bulbous fish with mucous fins came closer to investigate.

My urine was brown, a deep mahogany brown, the lingering side effects of a recent bout of malaria. I caught this case on a recent visit to Eastern Burma, which had a fresh infestation of malaria-carrying mosquitos. The culprit was a single-celled protozoa named *Plasmodium falciparum* – at least I got some good specimens. Within two weeks of being bitten by the mosquitos, my red blood cell membranes became sticky, and clumped in my liver, kidney, and brain – nearly causing permanent damage. Luckily, I got quinine and pentacycline, and was almost completely cured.

"Which way do we head?" Kalinda asked as she moistened her lips. We were right on the ocean, but the air was quite dry.

"Why don't we go northward? There are some interesting trees that way." I waved toward the huge crystalline trees in the distance.

"OK, but I can't go much further without something to eat."

"Me either."

"Wish we knew what we could eat without becoming sick."

"Let's hope we can find some intelligent creatures, like the Latöòcarfians. Maybe they can share their food with us or tell us what's edible." We walked for miles, but didn't see much evidence of life aside from the velvet moss which covered much of the ground.

Kalinda still seemed apprehensive. "I studied some of the latest maps of this place," she said. "They're not much good, but isn't it more than a day's walk to reach the Latöòcarfians? Maybe we should find some place to camp and rest first."

"That's a problem. I don't see a way to conceal ourselves here. How about we walk until we find a forest or some other cover? I don't feel to sure about the weather or any animals that might be around."

"OK."

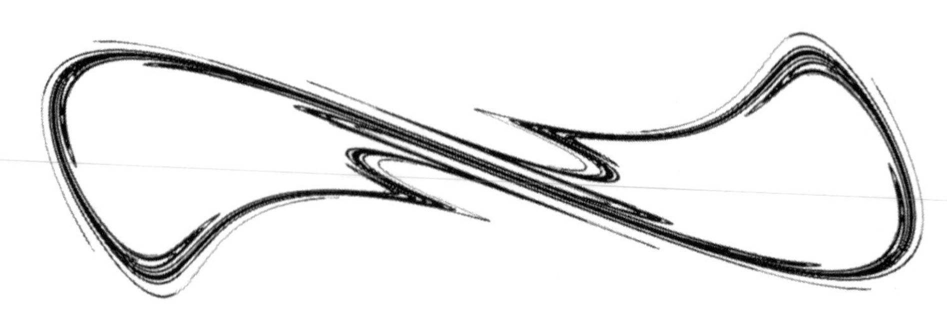

"Hope we fulfill our mission here," I said.

"Yeah, how many different animals does our employer want for his zoo?"

"Around twenty, this time."

Our ultimate personal goal was to find the Latöòcarfians, a limbless race of mathematicians. The King of the Latöòcarfians was said to live in a splendid palace constructed from intricate geometrical shapes. Most other Latöòcarfians lived in caverns, air pockets, and sulci running north to south in parallel ranges. Since they lived for many months trapped in close quarters, the creatures had worked some of the hostility out of their species by entering mathematical competitions. Individuals with beautiful mathematical thoughts, and who could create attractive patterns, had high status in their society. The aggressive ones had simply been forced into the ocean to die by the majority. Their patterns were known to mathematicians on Earth as chaotic attractors. They had a wispy, dusty appearance. We could learn a lot from them.

"Garth, when did *you* get interested in biology?"

"When I was a boy I got plastic biological models from the toy stores: the visible man, the heart, the skull, the eye, the entire human head."

"What, you never got a visible woman?" she grinned.

"No, I hate it when they turn out to be plastic. Anyway, I spent hours painting them and showing them off in my room. I even hung anatomical posters on the walls. My parents encouraged me, too. My mother hoped I'd become a medical doctor. Instead, I decided to get a Ph.D. in biology. I prefer the academic life to a clinical one."

"Don't blame you."

"What's that?" I said suddenly as I saw a dark, thin object protruding from the soil. "A fossil?" We stopped at a shrouded forest of crystalline trees and kneeled on the ground.

"I'm tired, but I can't wait to start digging. Let's go for it."

"Great."

"These should help," she said, ferreting out some metal shovels from her bundle of excavation tools.

"Get out the brushes, too. We can remove at least some of the ice and soil from whatever we find."

"Here you go."

Kalinda was a meticulous planner who tried to prepare for every eventuality. I found that she conducted her private life with the same scientific zeal that had made it possible for her to earn her doctorate in zoology, at the age of eighteen, from Harvard.

"Think we'll find something big?" Kalinda asked.

"Probably. Most scientists think that the creatures on Ganymede are large because physical bulk provides body warmth in the cold. One explorer said that some have thick fur, and others in hyperborean regions have quadruple eyelids to protect against glare and blowing snow. Most of the hearts are heavy duty organs that can sustain their big bodies in the thin atmosphere. In the warmer areas, like the one we're in now, this sort of thing isn't as important for survival."

The pale light burned away some of the morning chill, revealing a gigantic canopy of metallic creepers hanging down in a tangle from the crystal trees. At ground level were large rubbery ferns with mouths the size of a grapefruit.

"Watch out," Kalinda said as she back away from the ferns. "I'll be happier if we stay away from those."

"Luckily for us, they can't run after us."

I eased my shovel into the soil. "Whenever I start to dig I feel like Heinrich Schliemann. I guess he's always been one of my heroes."

"I never heard of him."

"When Schliemann was a boy his father told him great tales of Troy and the Trojan horse."

"Ah, I remember that one. Greek, right?"

"Yeah. Anyway, this Schliemann vowed to visit Greece and find this lost city. When he was older he went to the northwest corner of Asia Minor, where he thought Troy had been. He found a mound there that was supposed to be the home of Priamus, the king of Troy. He dug so fast that he went straight though the heart of Troy and beneath it found the ruins of another civilization at least a thousand years older. Schliemann found the remains of a mysterious race of people who in many ways were superior to the Greeks."

"And so you'd like to be the modern Schliemann?"

"My main interests are zoological, but you can always hope."

"Hey, take a look at this." Kalinda picked up a flat stone with a dusty mathematical pattern on it.

"Interesting. That's a Latööcarfian pattern. I think we'll see more of them as we get closer to their main city."

"I don't think they'll miss this one," Kalinda said as she put the flat rock in her pack.

"Right, I think we'll see hundreds of them as we get closer to where they live."

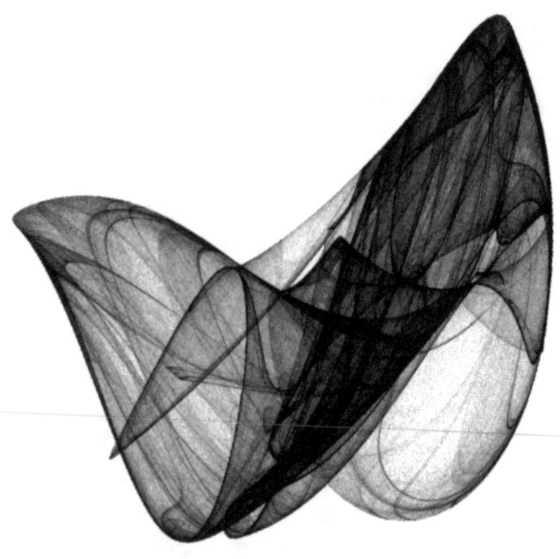

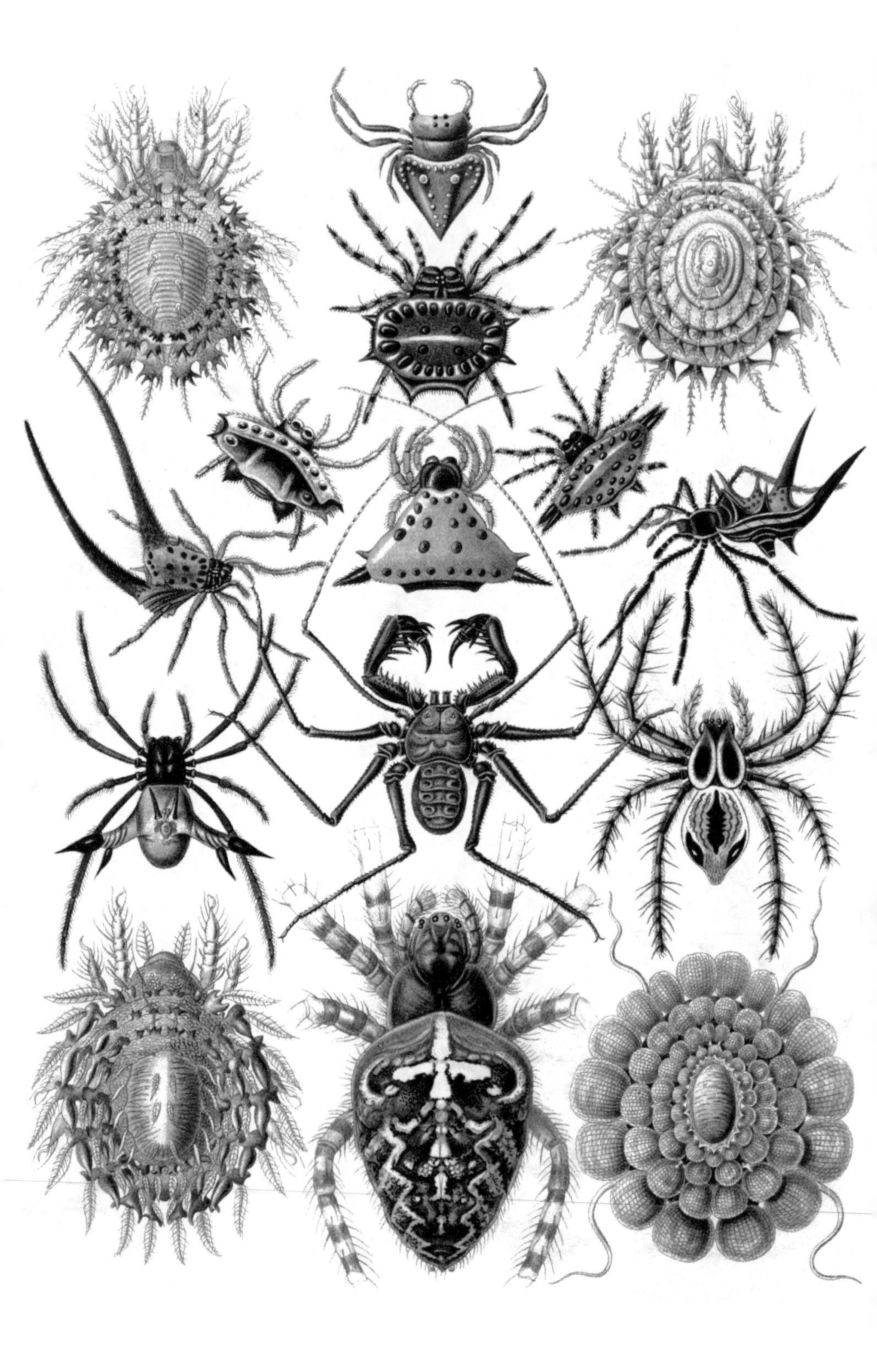

Chapter 23

Fractal Spiders

Kalinda watched as I dug and suddenly hit something hard.

"Quick, come closer." I motioned to her. "Take a look at this."

"What is it?"

"Don't know," I said bringing the dark object closer for a better look.

"It looks like a fossilized spider of some sort."

The spider's petrified legs dangled from its black body. Kalinda quickly removed a specimen jar from her pack and scooped up the spider's remains from a grave it had rested in for many millions of years. We looked closer at the ancient arachnid.

"This is a Ganymedean fractal spider!" I said as I slapped my side. "They're supposed to have legs upon legs upon legs, each set of legs becoming smaller and smaller, until the smallest legs are only 10 atoms in length." Kalinda opened the jar so that I could get a closer look. The spider had several eyes which resembled human eyes. The dangling, damaged eye balls hung from the body by shreds of red optic nerves.

"Fantastic," I cried. "This creature only lived in legends until today!"

"How old do you think it is?" Kalinda asked as she examined the spider from all directions.

"Judging from its anatomy, probably at least a million years old. Unfortunately its cephalothorax is damaged, and some of its eyes are missing. Its brain is supposed to be made of gallium arsenide." I knew that many organisms on Ganymede were thought to have inorganic components such as semiconductor brains. They were supposed to be made of materials similar to those found in computer chips.

"I wonder how its brain manipulated all the legs? You'd think it would trip all over itself!" A tart smile puckered her vivid lips.

"I bet its central nervous system is concentrated in the large cephalothorax. Tiny ganglia in the base of each leg might help regulate motions of the legs. This spider reminds me of the pycnogonids on earth."

"Pycnogonids?" Kalinda said.

"They're commonly known as sea spiders because of their long legs and spiderlike form. I'll show you if you ever come to my place for a visit."

"How do you know so much about pycnogonids?"

"I raise pycnogonids as a hobby." I smiled. "They make excellent pets. I have tank full of them back on Earth."

"I think you're pulling my leg," Kalinda said. "I never know when to take you seriously."

The fractal spider's bulging eyes seemed to gaze longingly at Kalinda. Suddenly the spider moved and jumped into Kalinda's mouth. She gagged, and I saw an eye dangle from her lips.

"Help me," she choked. There were tears of fear in her eyes. For some reason, mucous began to ooze from her nostrils. The spider had penetrated her so deeply that in seconds it would have vanished down her throat, if that had been its intention.

My reaction was instinctive. Before the spider could do anymore damage to her throat, I quickly grabbed the spider with my artificial hand and yanked it from her mouth. It started to slap at me with its sharp fractal legs. Luckily, the synthetic material of my hand could feel no pain. I threw the spider into the specimen jar and screwed the lid on tight.

Wee - Wee - Urr. The fractal spider emitted a high frequency signal and then was suddenly quiet.

"Kalinda, are you OK?" I said extending my arm to steady her.

"Just let me vomit, and I will be." She sat down.

"Damn, I wonder how the fractal spider could have been alive for all these years, buried beneath dirt? Well, at least it will make us famous." I placed my hand on Kalinda's. "That was close. I wouldn't want to lose you, Kalinda." I looked into her eyes and felt a wave of emotion.

"Yeah, a little too close."

"Maybe there are other spiders in the nearby dirt. Shall we look for some more?" I had always needed adventure, courted danger, wanted to test my own limits.

"You've got to be kidding," Kalinda snapped at me. "After what just happened to me?"

"OK, sorry. You're right. Want to move on?"

"Let's go."

We walked another mile in an attempt to find advanced life. "Look there," I said. "Spider webs."

Glowing webs were suspended from trees. I touched a few. "Feels like silk," I said. "I read about these webs in paleontology books."

"Neat," she said rubbing a few strands between her fingers.

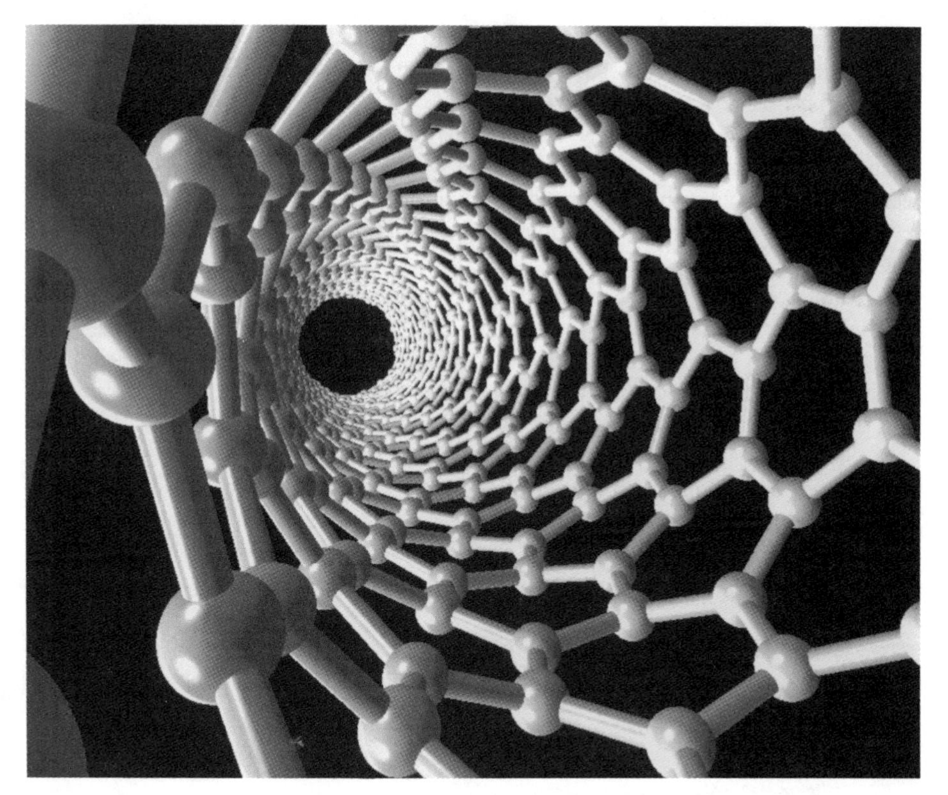

Figure 23.1. *Molecular model for the spider web (Bucky-tube) strands.* (The view is from inside the tube, as if from the vantage point of an interior bioluminescent bacterium.)

"The webs are fractal. See how the overall pattern of strands is repeated again and again, at smaller and smaller size scales? Spirals within spirals within spirals. Since the web has a large surface area, it should be more efficient at catching insects than the spider webs on Earth."

"I think I heard of them. Are these the webs that are made of long hollow cages of carbon?"

"Yep. They're made of Buckytube molecules. I once put one of these kinds of web strands under a scanning tunneling microscope."

"And?"

"The Buckytubes looked like chicken-wire. They're made of hexagonal arrays of carbon. The structure is supposed to make the web super strong. It gives the web the ability to repair itself when broken."

We continued to walk, occasionally brushing the webs away from our faces. The few glowing strands caught in our hair made us look like something from a Halloween party.

"What makes them glow?" Kalinda said.

"Lavender bioluminescent bacteria grow inside the tubes – in the small central channel."

"I bet we could make some money selling the fractal webs for decorations."

"Yeah, but only if the bacteria would survive back on Earth. Without the glow, I don't think we could market the webs."

We continued our journey through the tangle of lavender reflections more beautiful than any lilac flowers on Earth. The colorful filaments of web-light seemed to reach out at us like fingers probing the dim forest.

"I'm sure glad we're not running into anymore of those fractal spiders," Kalinda said.

"Right. They're not the friendliest of creatures. Although they could make excellent 'watch dogs'."

"I see," she said in a curiously neutral tone, as if she didn't want to insult me by pointing out how funny my idea sounded.

"Imagine if a burglar were to break into your house –"

"Garth," Kalinda said rolling her eyes. "You're a nut."

I smiled and looked at the pretty landscape. Dusk filled the hills with purple mist, and faint puffs of vapor hung over the sodden fields. We walked for an hour in full, satisfying silence.

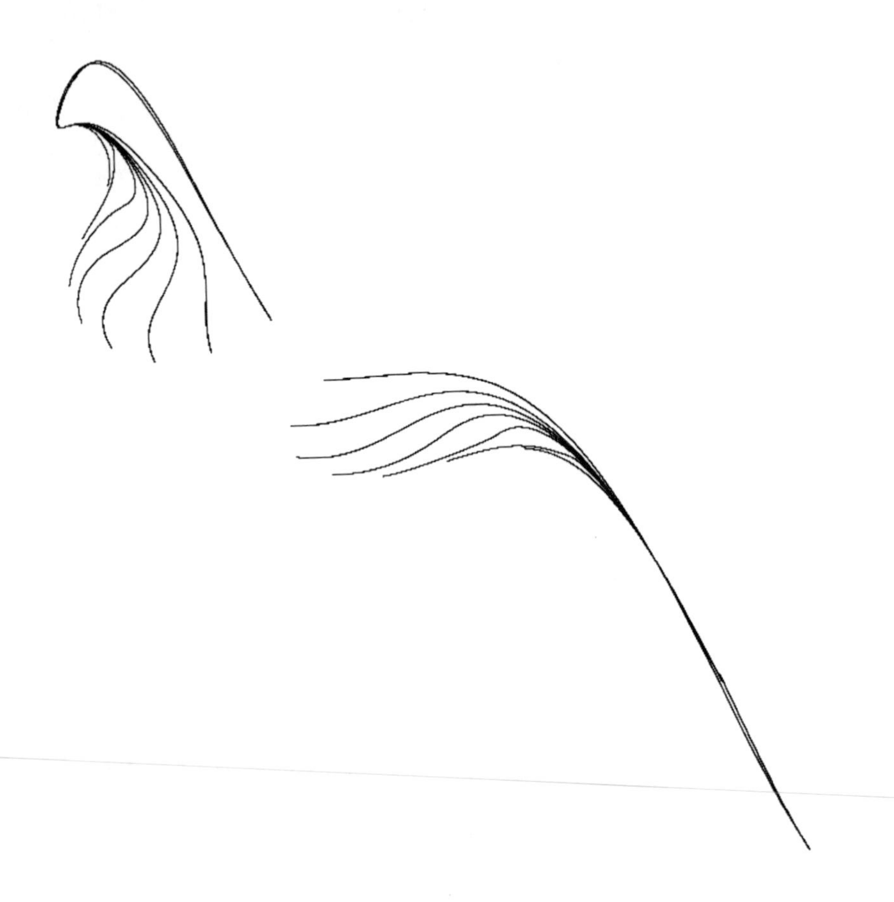

Chapter 24

Kalinda

"There is silence born of love, which expresses everything."
Conte Vittorio Alfieri

I gazed at Kalinda and took her hand. "Glad you're with me," I said.

"Me too. This wouldn't be fun alone." She gave my hand a squeeze and looked at me with a charming smile that involved her green eyes as well as her mouth. Often I'd fantasize about getting more romantically involved with her, but hesitated because of the difficulties this might create for our employer and for our scientific collaborations. Kalinda had been a great assistant, collaborator, and close friend for several missions. But I didn't think I could resist falling in love much longer!

She was a slim girl, perhaps nineteen Earth years of age, with honey-colored hair, and eyes the darkest green. She was, however, not quite human. I met her on a small planet where women lived symbiotically inside the digestive system of huge grazing animals with sheep-like muzzles. These ruminants, called "Neptune Brooms," had obscenely long bodies with hairy bases. White reproductive organs shaped like bars of Ivory soap protruded from beneath the hair. The broomstick section of the organism was actually the cartilaginous skull and torso of the creature. The highly elongated brain formed multiple connections with the base of the creature which contained hundred of poisonous tendrils. The tendrils secreted a pungent fluid which smelled like ambergris. The Broom also had huge molars and multiple stomachs.

Though the symbiotic relationship was restrictive for the woman, the Broom-woman composite had some benefits for the woman and the Broom. The women helped move the vegetable material within the Brooms' stomachs to aid digestion. In return, the women received an ample supply of nutritious food. They also received protection from the myriad dangerous animals on their planet. Bioluminescent organisms lining the Brooms' stomachs, intestines, and esophagi provided light for the women to work by. In order to protect themselves from the Broom's digestive fluids, the women coated themselves with red ointments of aromatic plants. The red coating protected the women from digestive fluids for 48 hours at which point the Broom gave the women additional ointment with which to coat themselves.

Every few weeks the Brooms allowed their women to emerge from the Brooms' mouth and spend time with the free humanoid males on the planet, and to enjoy the fresh air. I was exploring the planet at this time and met Kalinda. She confided in me her desire to escape from the dominion of the Brooms. Not only did the Brooms force the women to spend most of their lives in hellish conditions within their hot, moist innards, but occasionally they compelled the women to eat their own excrement or even their own young as punishment for minor infractions of the Brooms' commands. I helped Kalinda escape, and she has been with me ever since.

During our two years together she became proficient in English with all its idioms. At first I was fascinated by her, charmed by her. Then I started to fall hopelessly in love with her. I was desperate to teach her all of the profundities of life, to make her love books, art, classical music, and the outdoors, – to know how to fend for herself among others without doing damage or being harmed. Still she sometimes kept just beyond my emotional reach. On our Ganymedean mission, I hoped that some of the distance between us would melt away.

Chapter 25

Prelude to Battle

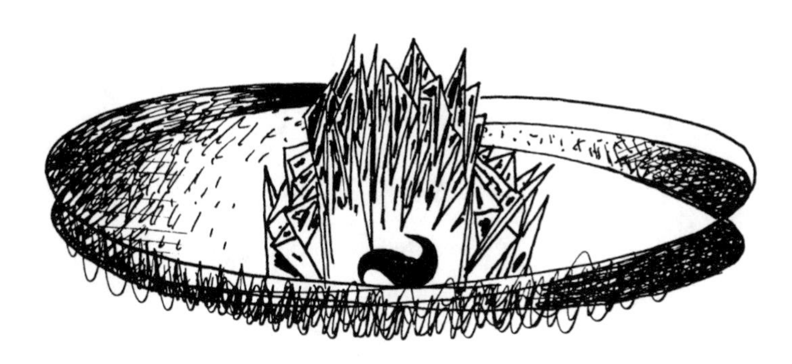

"I can show you fear in a handful of dust." T. S. Eliot

"Garth, how much longer do we have to walk?" Kalinda took off her boots and stretched her toes.

"Not much further. Look there. I think that's the Latööcarfian Royal Palace!"

"It's hard to tell with all the dark mists."

"Uh-oh. We've got company." So enthralled were we with the eerie sights that we didn't hear the riders until they were quite close.

From the corner of my eye, I spied five mounted green-skinned riders. They were clad in magnificent garments of war – golden studded harnesses and gallium arsenide encrusted helmets. At their sides dangled their traditional weapons of war, the long sword. They rode massive, black, stallion-like creatures. From their mounts hung daggers, sabers, rapiers, scimitars, wide-bladed knives called misericords, and various other instruments of mayhem, the function of which I could not quite discern.

"They're not the Latööcarfians," Kalinda whispered.

"No kidding. They're Prohaptors, another species of Ganymedean. They protect the limbless Latööcarfians."

"We haven't done anything wrong. Maybe they wont mess with us."

"Don't count on it."

"Why do they bother protecting the Latööcarfians?" Kalinda said.

"The Latööcarfians stimulate the pleasure centers of their brains." I didn't move, didn't even blink. "Those damn Prohaptors are vile. Many live in taboo-ridden, tribal societies. Some are farmers or herdsmen. Others are warriors."

"God, what do we do know?" Kalinda said. Her whole body tightened, and then she took a breath.

"Let's wait and see what happens." I felt a creeping uneasiness at the bottom of my heart.

I'd heard from exo-pologist colleagues that Prohaptors had a penchant for strong drink produced from the fermented juice of fractal fern plants. They were also known for their erotic and licentious behavior. All Prohaptor males were trained as warriors before becoming farmers or specializing in a particular craft. The primary aim of male education was to produce formidable warriors. Flamboyant military attire served as incentive. Unfortunately for us, ritual sacrifice of enemies was quite common.

When the riders were in earshot, I adapted my speech and body language to their general inflated style and manner.

"From whence comest these noble warriors?!" I cried. I tried not to show any fear.

Kalinda stared up at the riders. Waiting. Tense. "Garth, what in hell are you talking about?" she whispered.

"Play along with me. See the fractal pattern on their flags. It's the insignia of the Royal Palace of Yars Kothek." I pointed out flags bearing an intricate Mandelbrot-set emblem. The large warriors removed their helmets, and dismounted. The tallest rider came closer.

Standing before me was a six-foot tall, pea-green creature. Its eyes glowed a fierce red. From beneath its thin, drooling mouth, a hundred throat appendages quivered aperiodically. Its several feet resembled horse-shoe crabs, and its toes looked like slugs.

Two off-white snouts protruded from its multitentacular body. They began to sniff at the air. The tall creature shouted at me in something that was probably a variant of Latööcarfian speech.

"Can you understand him?" Kalinda said.

"Yeah, I studied Latööcarfian languages while on Earth."

"Inja chekahr dahreed!? What dost thou want in Our Kingdom?" the warrior said in a voice several octaves too low. Its grey-green eyelids drooped and oozed a tiny amount of liquid which smelled like absinthe.

"I seek Latööcarfian King Yars Kothek, ruler of the Third Empire, possessor of all knowledge," I shouted, and attempted to mimic his ostentatious manner of speech.

"Ha! I know not of whom ye speak, Warrior. Leave whilst thou are able!" He threatened us with his long sword. The voice of the Prohaptor was eerie, like wind whistling through a window on a lonely January night.

"And what of the pre-pubescent harlot?" taunted another as he lasciviously leered at Kalinda. I clenched my natural fist so that my fingernails dug painfully into my palm.

Suddenly the two Prohaptors removed a peculiar looking weapon from their cloaks and knocked me into a crystalline thorn bush. My *Astrolabium Galileo Galilei* timepiece shattered as it hit a rock.

Chapter 26

Battle

"Do not hurt where holding is enough; do not wound where hurting is enough; do not maim where wounding is enough; and kill not where maiming is enough. The greatest warrior is he who does not need to kill." S. R. Donaldson

Since I was not permitted to bring advanced weapons to alien worlds, the long sword was my weapon of choice on Ganymede. This was not too much of a liability since gunpowder did not exist here. In one way this restriction was a welcome one for me. I was an expert using both the English cutlass and the French cup hilt rapier. To Ganymede I brought a beautiful replica of a 17th century English cutlass with a quillon to protect my hand. I even fashioned the blade myself using bundles of iron strips, hammered together, which I cut and bent repeatedly. Subsequent carbonization of the metal improved its strength and temper.

I disentangled myself from the crystal branches and shouted back in their archaic style of speech. "Thou darest challenge me?" It was time to bring out the heavy artillery.

Whenever I'm threatened, I seem to go haywire and lose control. For some reason, I wasn't afraid. Perhaps it was my martial arts training. Perhaps I was a little bit crazy.

One of the Prohaptors kicked me in the knee. The warrior blood within me rose to fever pitch. My hand yearned for the touch of my sword, yearned for the glory of battle. I held my sword above my head. "Olagh!" I screamed to assuage my humiliation. I grinned and then lunged with the fury of seventeen raging canine bitches in heat.

"Kalinda, stay behind me," I shouted.

I lunged at all five of them, my ever-moving blade producing a wall of flashing cutting steel, through which no Prohaptor dared transgress. One Prohaptor came a little too close, and I stabbed him in the chest. Another threw his sword at me. I dodged it and then stabbed him in his arm. I grew weak, but I fought on. I've always loved a good fight.

The third Prohaptor stared at me but did not move. It appeared that he was assessing my fighting skills. He snarled something which roughly translated to, "Pretty damn good." Then he ran toward me. My sword slashed, thrusted, and parried his every attack. He was about to bring his sword onto my head when I slipped under him and smashed my fist into his groin. A second later, I ran my sword through his neck.

"Watch out!" Kalinda said. She pointed to the Prohaptor leader who wielded a fractal long sword. I had heard about these terrible instruments of mutilation and death on other worlds. One side of the blade had a sharp edge shaped like a jagged Koch snowflake curve, a fractal edge produced by superimposing triangles upon smaller triangles.

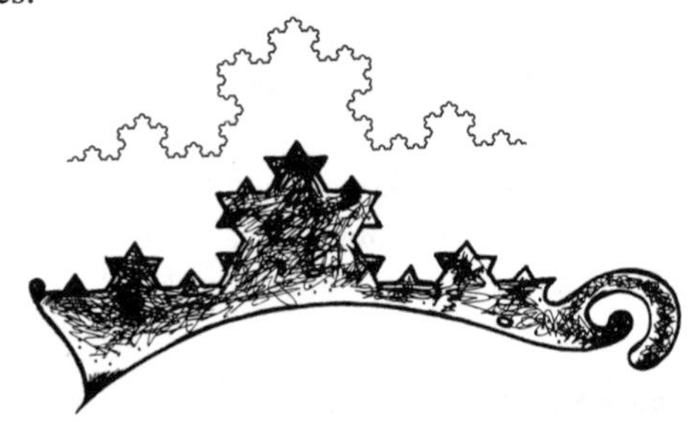

Fractal Sword Blades (Rendering by Clay Fried)

Ordinarily, when one sword hits another sword the vibration spreads out to affect the entire sword. With fractal swords, some vibrational modes are trapped within a branch of the fractal pattern, and thus the fractal edge damps vibrations – making the sword quiet and sturdy. When the Prohaptor struck my sword I heard no loud metallic ringing sound. Just silence. It was unnerving. The Prohaptor noticed by consternation and smiled. With lightening speed, I smashed his body with my prosthetic arm while at the same time throwing a piece of armalcolite at his head. He stopped smiling.

"Arrrrrr!" screamed the Prohaptor. His long nose was pinched and white with resentful rage. I screamed back at him with a furious voice. Then I ran towards him. At last, with one ruthless disengagement and thrust, I killed the leader of the five green-skinned riders.

I looked at the one remaining Prohaptor. "Prohaptor, I don't know who is uglier, you or a silver-haired Ganymedean Zooz." I laughed at my own joke with a judicious amount of false gusto. The Zooz was a huge, hairy animal, and I hoped we would never encounter one during our stay on Ganymede.

The Prohaptor roared and attacked me. His strength was so great that the strike of his sword split my sword into two useless pieces. I fell to the ground. In our large backpacks I had another sword, but there was no way I could get to this in time. I reached for the puny dagger in my belt. The Prohaptor caught hold of my wrist, its slug-like fingers forming a muscular bracelet of incredible strength. I heard Kalinda gasp and saw her run to the creature. Then she gave a swift, vicious kick to the Prohaptor's hip. The Prohaptor loosened his grip.

I was about to take a deep breath of relief when the Prohaptor's sword ripped into me. The pain was so devastating that I dropped the dagger, feeling pain like nothing I had ever known. I thought of Kalinda, and my love for her, and I told myself I had to try to protect her, and that I wasn't going to die, wasn't going to die....

The huge Prohaptor came at me again with his long sword, and I shielded my face with my prosthetic arm. I looked into his surprised eyes as he watched his sword bounce off my arm with apparently no adverse effect on my health. Taking advantage of his surprise, I launched my foot up at his throat with a powerful karate kick. He was stunned and dropped his sword. With a quick intake of breath like someone about to plunge into icy water, I quickly grabbed his sword, came forward, and removed his head.

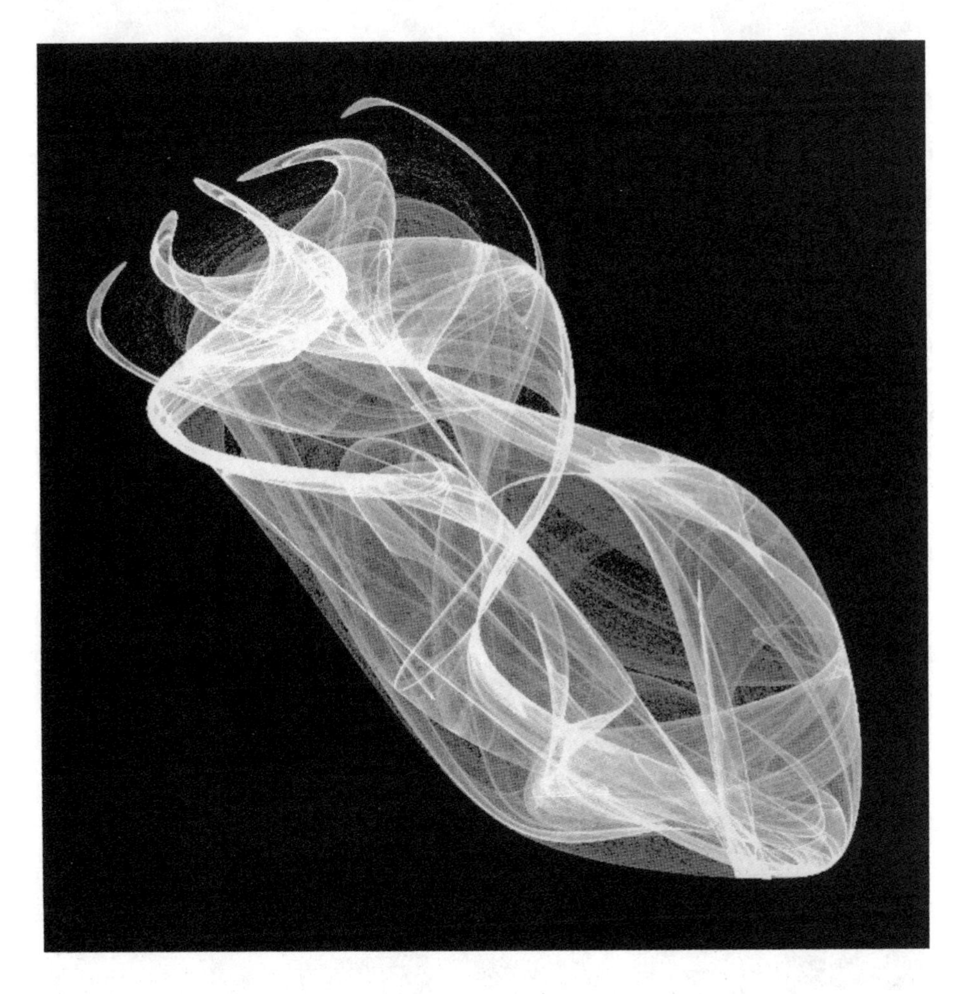

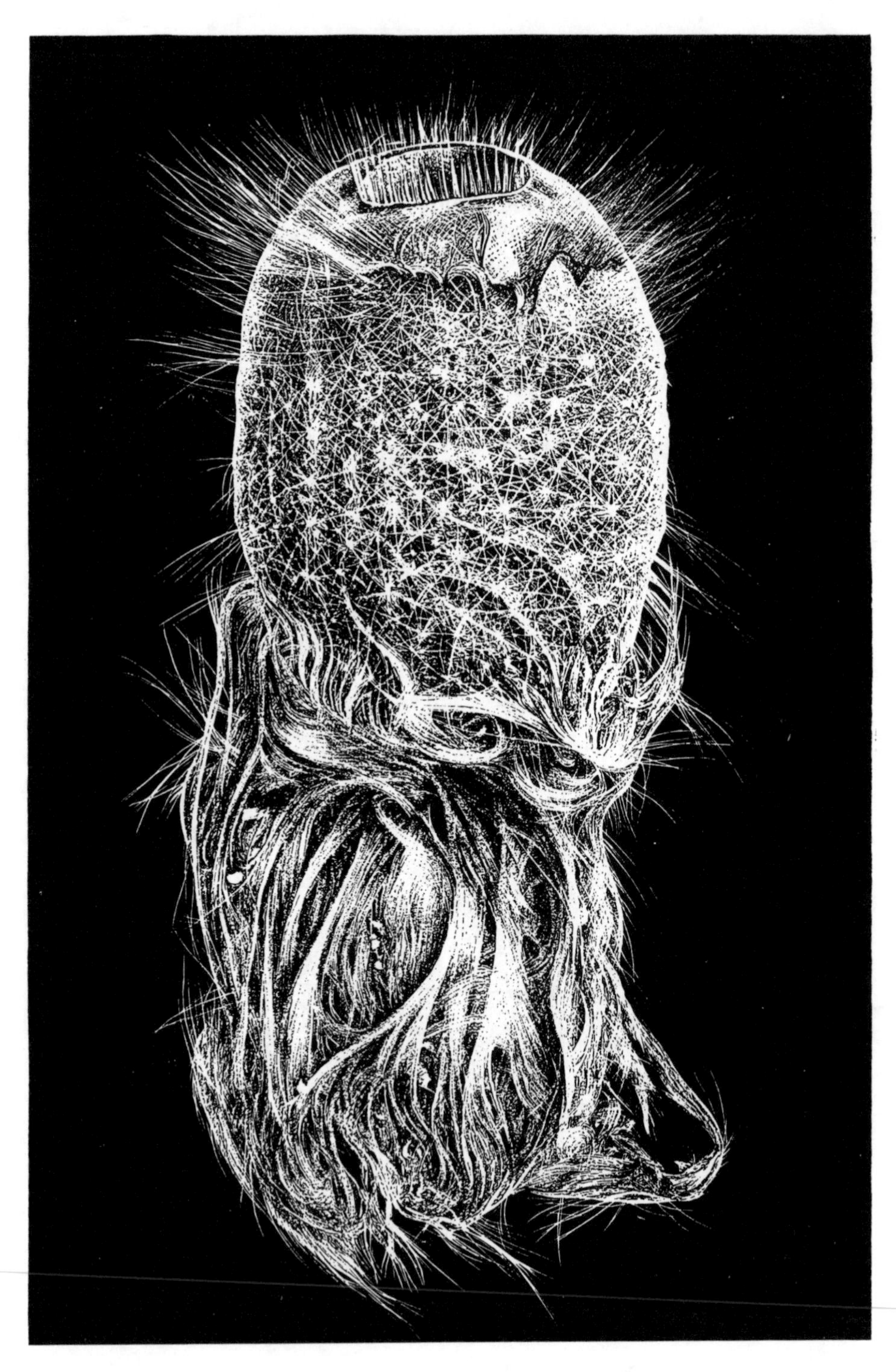

"Ka," a Prohaptor Brain Parasite

Chapter 27

Brain Parasites and Polyp Men

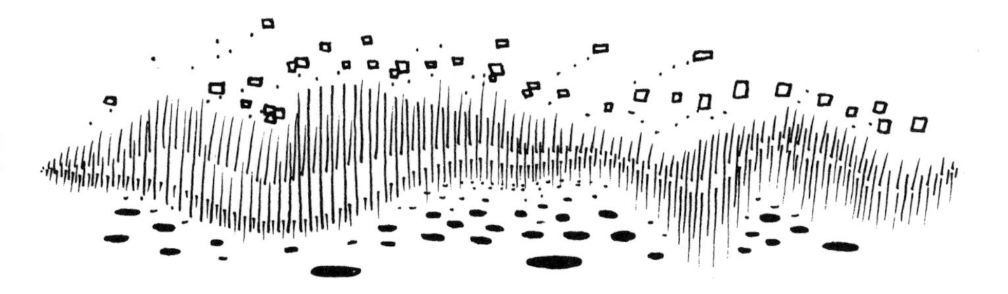

"History is subject to geology. Every day the sea encroaches somewhere upon the land, or the land upon the sea; cities disappear under the water, and sunken cathedrals ring their melancholy bells."

Will and Ariel Durant, *The Lessons of History*

"When the mind is lost, it finds solace in a new world."

Goethe, *The Walpurgis Night*

"My cuts don't seem to be bleeding much."

"Where does it hurt?" asked Kalinda as she tried to bandage my scars.

"Ouch. Everywhere, but don't fuss."

"At least let me put some medicine on those cuts." She began to apply special healing salves that would cause my torn muscles to repair themselves in days while at the same time relieving me of pain. "OK, try to take it easy," she said.

"While we have the opportunity, let's take a look at a Prohaptor brain. I'm curious about it."

"Are you up to it? I want you to rest."

"I'm all right."

She looked at me with concern in her eyes. "OK, shall I get the bone drill?"

"Yes," I said. Kalinda removed a small, battery powered drill from her pack.

"I'll place one of the Prohaptors on his side," I said. "You turn on the drill." I removed a scalpel and gently slid it across the forehead of the Prohaptor. Blood welled up in a sudden crimson thread. "Let me put on some latex gloves." I reached into a pack, snapped a pair on, and resumed cutting. I continued the incision that followed the Prohaptor's hairline across his forehead.

"Here goes," Kalinda said. The Huntington electric bone drill began to hum. She held the tip of the drill to a point on the Prohaptor's skull about two centimeters above its eyes. "Wow, this skull is thick." She pressed down harder, and the saw started to make a sick high-pitched grinding noise. Smoke, filled with tiny bone particles, rose into the cool air. After making a few holes, Kalinda replaced the drill burr with a circular cutting disk. For a second, the saw began to jump away as it cut into the bone, but Kalinda's grip held firm. In a few minutes we were able to remove a portion of the Prohaptor's cranial cap. I lifted the bone cup gently, and it made a popping sound like a cork being blown from a champagne bottle.

"Looks vaguely human," Kalinda said as she gazed at the shiny wrinkled jelly of the Prohaptor brain.

"Yes, the brain's structure is symmetrical." The small interlocking ridges on the surface of the cerebral cortex reminded me of the deep fissures and protrusions of a delicious mango – but I kept that observation to myself. "Look at the two huge optic nerve tracts carrying fibers from the eyes," I said. Here and there I saw some egg-shaped masses, perhaps thalmic in origin, centrally located at the top of the brain stem.

"Here's their equivalent of the pineal body," I said.

"We're still not too sure what the pineal body does in humans, right?"

"Yeah. It's a tiny pinecone of nerve cells near the brain stem. Descartes thought it was the seat of the human soul."

"Let's take a look at the ventricles," Kalinda said.

"If Prohaptors are anything like us, they should have four ventricles forming a network of interconnecting cavities. They should be filled with cerebral spinal fluid to cushion the brain."

"With such a hard head, maybe the Prohaptor's brain doesn't need any cushioning!"

"Let's take a look. Could you help hold apart the hemispheres as I cut?"

"Yep," Kalinda said.

"You'd make a great doctor. You're not scared by the sight of blood and gore."

I sliced into the brain, revealing the ventricle chambers. "Kalinda, what do you think of that? They're not filled with fluid. Just empty –"

"Look!" she screamed. I pried apart the cavity further and saw that inside one of the brain ventricles was a little hairy robot about the size of a dime. Occasionally the hair parted, and I saw it had a fully formed face with eyes and a mouth. A second robot in another ventricle wandered back and forth along a complex of fibers resembling cables. Both robots stopped whatever they were doing and gazed up at us from their home in the brain.

"Get the hell out of here," one of them said in the Prohaptor's language. They both stared at us. This was one of those rare occasions during which I nearly fainted.

"My God!" Kalinda said.

I felt a strange shiver go up my spine as I looked into the robots' glistening eyes. I felt a chill, an ambiguity, a creeping despair. The tiny robots were still. None of us moved. Their eyes were bright, their smiles relentless and practiced. Time seemed to stop. For a moment, the Prohaptor brain seemed to grow warm. But when I shook my head, the warmth was gone. Just my imagination. But the robots remained. Cruel. Nightmarish. I felt like I was caught in a maze and all the air were suddenly evacuated.

I dropped the brain, and the two robots scurried back inside the ventricles. As the brain hit the Ganymedean ground, I heard a pop as the cerebellum imploded. Regaining my composure, I stooped down again, reopened the brain, and asked the tiny robots the obvious, "What are you?" I held the brain as far away as I could. I didn't want them to jump out at my face.

"I am called Ka," one said in a thick, clotty voice with an uneducated accent.

"I am called Da," said the other.

I looked at Kalinda, unsure of how to proceed. She shrugged her shoulders. "What is your purpose?" I asked.

"Is this guy dumb or what?" Ka said to Da.

"All Prohaptors have a Ka and Da in their brains to help perform maintenance activities," Da said. "We help distribute oxygen, repair damage, help permanently implant memories, and regulate emotions and various hormones."

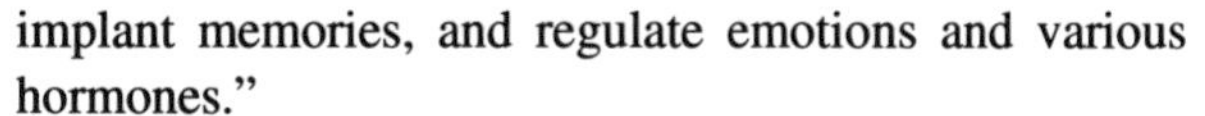

"Why?" Kalinda asked.

Da looked at her with contempt. "Madam, we evolved first as independent organisms. Later we parasitized the Prohaptor brain. Now we're an integrated living organism."

"Kalinda, this reminds me of the mitochondrial colonization of nucleated cells on earth millions of years ago."

"Aren't you forgetting I used to live in a broom?"

"Right."

"Do you think we should take them as specimens?" Kalinda said.

"'Take them as specimens'?" Da echoed Kalinda. "What in God's name is she talking about?" he said to Ka.

"I think we could learn a lot by studying them," I said. "But I'd bet they'll soon die now that their Prohaptor host is dead." I brought out some forceps and reached for them.

"Hey, buddy, what do you think you're doing?" Ka said. Both Ka and Da retreated into the depths of the dead brain.

"Garth, be careful. They might bite."

I heard a voice coming from somewhere in the wet organ. "Yeah, that's right, we might bite," it said.

I started slicing pieces off of the brain. Soon there was nowhere they could run. Nowhere they could hide.

I picked up Ka and Da with forceps as they screamed and shouted obscenities at us. "Get out a specimen jar," I said to Kalinda. "We'll put them both in the same jar." She brought a jar to me. I plopped them into the jar and stuffed it into a backpack.

"Those two are probably the weirdest little lifeforms I've ever seen. Hey, Kalinda, now you're shivering."

"It's getting cold."

"How about we make a fire? Help me look for something to start a fire with."

We watched the sky darken, watched some clouds move in, listened for thunder working its way over the glacial horizon. Lightning began to flare over the tops of far away trees. The air seemed to change, becoming electrified with infinite possibility.

"Think these will burn?" Kalinda said, pointing to the branches of a crystal bush.

"I think so. Try it."

"Give me the Prohaptor's fractal sword. I'll try to hack off some crystal branches." Kalinda was strong and handled the sword like a real pro. "Here are some branches," she said. Soon she made a small pile, and then I placed a match under them.

"Great, they seem to burn slowly. Let's get warm."

"Look at that," Kalinda pointed upward as shards of crimson froth flashed in the sky.

The fire was blazing nicely. It kept us warm and allowed us to heat some ice for drinking. The burning crystals reminded me of the time I returned with Kalinda to her home world to visit some of her relatives living in the Brooms.

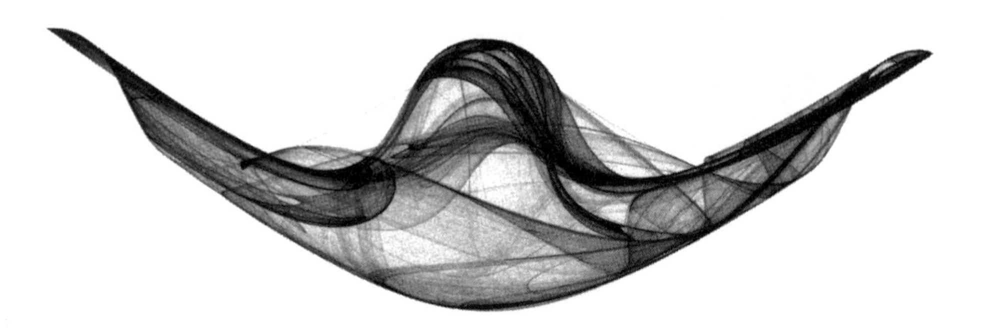

Memories of Kalinda's home opened before me as if a curtain had been ripped aside. We had entered the Neptune Broom while it slept. Kalinda lit a small crystal torch to provide us the necessary light to find our way down to the Broom's esophagi. Once in the esophagi, stomachs, and intestines, we no longer needed artificial light due to the bioluminescent bacteria lining the epithelial walls of the digestive system. Her mother lived in Stomach Seven, and several cousins lived in the large intestine. Sadly, one of Kalinda's brothers had been killed in a digestive accident when Kalinda was only seven years old. He accidentally got lost in one of the Broom's more active intestines. He was pulled in, became a bolus, and suffocated.

I especially wanted to help Kalinda's blonde-haired sister Terrie to escape. Deaf from birth, Terrie wanted more than anything to be able to hear Kalinda's mellifluous voice. On Earth there might have been an operation to correct her condition, but I couldn't convince Terrie to leave her family behind in the Broom's massive intestines and stomachs. So accustomed was her family to their secluded life, that they rebuffed my efforts to rescue them. To this day, they lived in the

Neptune Broom. It was not a bad life if the particular Broom you were in did not have a cruel disposition.... Not bad if you could get used to the constant smell of ambergris.

As we were about to say good-bye to her symbiotic family in the Broom, I saw something on the digestive wall that filled me with horror.

"Kalinda, what the hell is that?" I spoke in a voice reserved for dreaded things. Anchored to the top of the dark, damp, delequescent, digestive wall was a human-sized polyp, with eyes and a mouth. It looked like a huge punching bag suspended from the ceiling. A living, sentient uvula. The skin was pink with brown patches around the sunken eyes and at the corners of the red, suppurating lips.

It had no eyelids.

It had no nostrils.

It mumbled something in a gluey voice.

"Flectonotus is my brother," Kalinda said as she stroked the polyp with loving affection. "About a third of the males born on our world look human. The other two-thirds degenerate into polyps which remain attached to the Broom's stomachs or intestine. The Broom uses the polyp's eyes to watch us, to monitor our activities."

"How's it possible?" I gasped.

"We think the polyps' optic nerves send electrical impulses directly to the nervous system of the Broom," Kalinda said as she slumped into morose musings. "We once tried to remove a polyp, but the polyp couldn't survive as a separate organism." Flectonotus's eyes jittered for a moment and then focussed on Kalinda.

I shook my self out of the depressing reverie, and my thoughts returned to our present predicament on Ganymede.

"Penny for your thoughts?" Kalinda said to me.

"Nothing important."

"What were you thinking about? C'mon, tell me."

"The time I met your brother Flectonotus."

"Don't worry. He says he's happy." Kalinda brushed back some hair that the faint wind had caressed out of place. "Ah, my sore feet finally feel a bit better," she said changing the subject.

"Let's drink some of the melted ice and start walking again."

"Sounds good to me."

Chapter 28

Pipe World

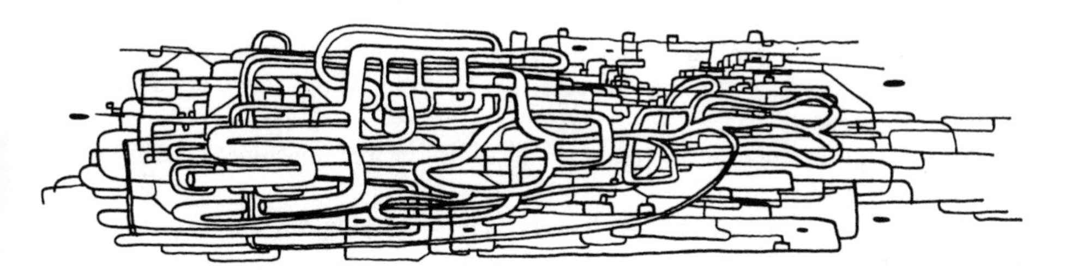

"Here's one of those chaotic patterns again." Kalinda knelt down and picked up a flat rock with a pattern containing millions of tiny dots. The designs looked like swirls, or the wisps and eddies of smoke.

"Yep, and here's another. Must mean we're heading in the right direction. The Latööcarfians can't be too far away now. Let's take a few rocks with us."

"The patterns are really beautiful."

"Yeah, they are," I said. "They're the result of simple formulas producing intricate graphs."

"How do you suppose the patterns got on the rocks?"

"Good question."

Kalinda pointed in the direction of what looked like a forest. "What do you suppose that thing is?" she said.

"Let's go find out." When we were a mile away, it became clear that it was no ordinary forest.

"Wow, look at all those.... What are they, pipes?" Kalinda said when we came to a region of interlacing cylinders.

"Looks like someone had an incredibly large box of toothpicks and blew it up with a bomb."

"Some of those 'toothpicks' are wider than our bodies," Kalinda said.

"How'd you like to call a plumber to repair a leak?"

"We don't know that the pipes are hollow," Kalinda said.

I banged on a pipe. "Sounds hollow."

"Must have been made by a plumber high on LSD." Kalinda smiled as she gazed at the chaotic structure.

"Looks like we'll have to climb through them to continue. Amazing. The pipes seem to continue in all directions for as far as I can see."

"Stay here," I said. "Let me make sure it's safe." I crouched low and began to enter the murky interior of the nest of pipes.

"The hell with that," said Kalinda. "I'm going with you."

"OK, we'll take it slowly."

"I hope we don't get lost in this maze."

"Damn!" I slipped on a pipe.

I saw that some pipes were as thin as my arm, but others were a good six feet in diameter. They were arranged in no apparent pattern. Some were vertical, others horizontal, most oriented at random angles to the ground. Kalinda reached out and touched a pipe.

"It feels cold and metallic."

"Let me see how hard the material is." I brought out my sword and started to scrape a pipe. "Nope, can't even scratch it. This pipe-world can't possibly be a natural phenomenon."

"Maybe all of this is the remains of some ancient machinery. But what purpose could it serve?"

After climbing for half an hour like monkeys on a jungle gym, the pipes completely surrounded us. Sometimes they became so densely interwoven that we couldn't pass and had to change our direction.

"From a few measurements, I can tell that the distribution of pipe-widths was hyperbolic," Kalinda said.

I raised my eyebrows. "You're quite the mathematician."

"How in the world are we going to find our way out of this?" Kalinda said.

"I don't know. Maybe we can try to find the very top of this structure and look out to see where we are. Or maybe we should try to find the bottom, where the pipes join the ground."

"I'm getting claustrophobic. Let's try to find the top and look around."

We climbed some more, and I could tell from the waning light penetrating the pipes that nighttime was approaching.

"Let's rest," I said. "If we snuggle up to some of the intersections of the larger pipes we shouldn't fall off."

"You don't expect me to go to sleep on one of these pipes do you?" Kalinda asked. Her eyes looked tired, incredibly heavy.

"Let's give it a try." We got out some blankets, and lay down. Kalinda wrapped the blanket around and around her, and curled up into an almost fetal position.

"Not very comfortable, is it?" I said, trying to make a joke of our precarious predicament. No answer. Kalinda was already asleep. I drew her close to me for warmth. My muscles ached, and I closed my eyes for some needed rest.

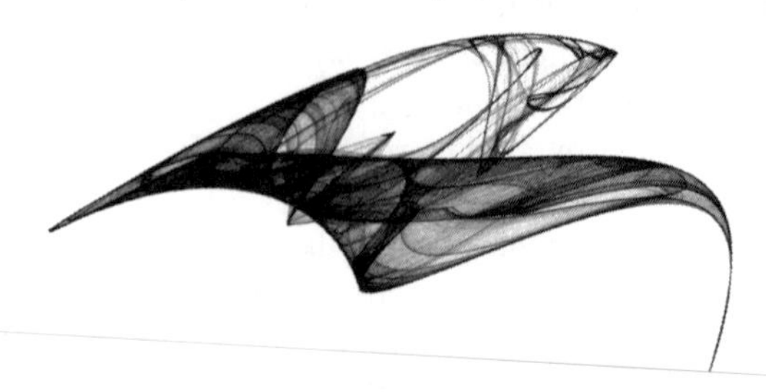

I woke abruptly, because I heard a cough. A deep cough, which could not have come from Kalinda. I looked up. In front of me was a large, handsome man with the physique of Schwarzenegger. He looked at me with his three eyes. As an instinctive result of my martial arts training, I assumed a defensive crouching position. Kalinda also assumed a similar defensive stance. He came toward us.

"Stop!" I screamed to scare him, but I did not attack. When he was only a few feet from me, I decided it was time to act. I flipped him on his back, and he fell with a sound that reminded me of a sack of potatoes landing on concrete. He laughed, as if he felt no pain, and slowly rose to his feet. It wasn't easy balancing on the pipes, but somehow he managed to do it as if he had years of tightrope walking lessons.

"My name is Thorn," he said in a Ganymedean dialect I vaguely understood. Thorn's voice was deep, colored by a vague guttural accent. His eyes probed Kalinda with an intensity that seemed to make her uncomfortable. I reached for Kalinda's hand. Thorn's intense gaze didn't exactly make *me* feel very relaxed either.

"What do you want?" Kalinda said.

"It appeared that you were lost, and I came to help escort you from pipe-world."

"How did you know we were here?" I said. "Who sent you?"

"I often travel through pipe-world. You weren't difficult to locate. I wish you no harm." Thorn paused, looked at both of us in turn, and it seemed to me that his glacial eyes suddenly seemed brighter.

"Garth, he could be telling us the truth," Kalinda whispered. I relaxed slightly but was ready to defend us at a moment's notice.

"What can you tell us about this strange place?" I said. "Who constructed this world? What is its purpose?"

Thorn sat down on a pipe. "No one knows how it was constructed. Legends describe a time when an entire civilization lived here. During the early evolution of the civilization, a few children wandered alone through the forest of pipes. Some found a hole in one of the pipes, and crawled in. The children became lost in

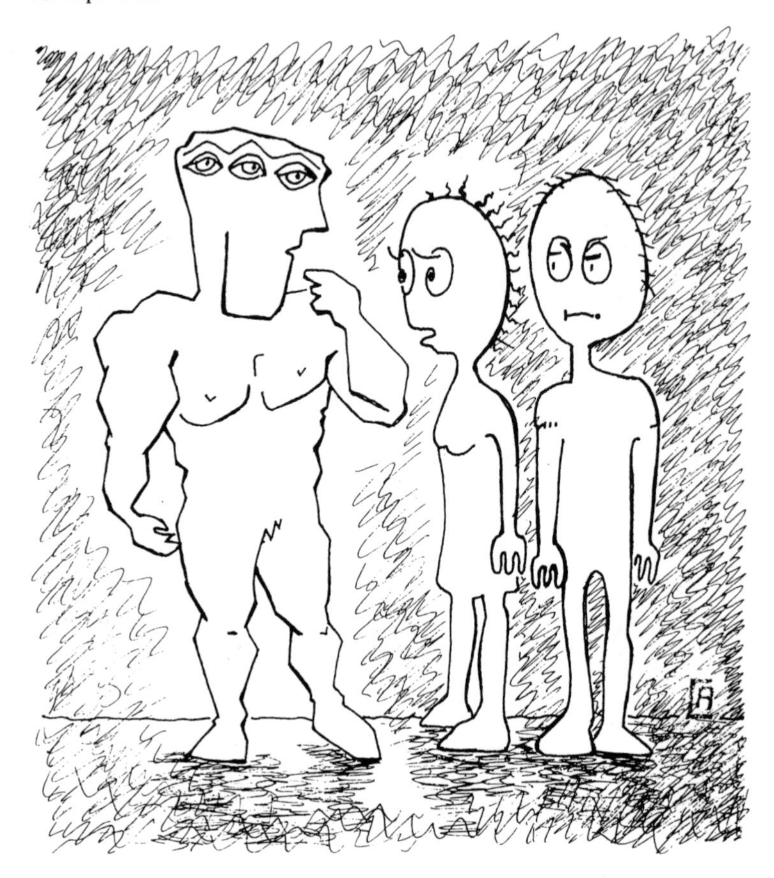

Figure 28.1. *Thorn, Kalinda, and Garth.* (Illustration courtesy of Claire Albahae.)

their inner pipe-world and couldn't find the single exit to the outside. Their parents knew of the childrens' disappearance. When the parents put their ears to the pipes they could hear the children crying. But search as they might, the adults could never find an entrance to the childrens' inner world. The childrens' prayers of hope were never answered."

Thorn paused for a few seconds to clear his throat and continued his story. "This is how two civilizations started. One on the inside listening to the outside, the other on the outside listening to the inside. The outsiders were saddened by the cries of the inner pipe-people. The inner ones lived in total darkness. After many generations, there were thousands of them crammed inside the pipes. But the outsiders could do nothing to help them, to free them."

"How do you know all this?" Kalinda asked Thorn.

"I'm one of the few living descendents of the outsider civilization."

Before I could question Thorn further and determine the authenticity of his story, we saw a movement near our feet.

Kalinda suddenly seized my arm by the wrist.

"What in hell?" she said. Her eyes grew large and glaucous as she pointed with a trembling finger to a large white thing lying on a pipe. "My God!" she said. "Get it out of here." Nestled between two pipes was a foot-long worm. Kalinda jumped and banged her head on an overhead pipe.

I understood her fear. Having encountered huge parasitic tapeworms when she lived in the Broom's intestines, Kalinda still retained a child-like horror of worms. This one was bigger than any nightcrawlers I had ever seen on earth, fatter than a garden snake...

"What is it?" I asked Thorn. I backed away.

"We call it *Aysheaia*."

The worm lifted its head which contained one large, unwinking eye.

"It's staring right at us," Kalinda said.

Aysheaia had a segmented body, with ten pairs of limbs attached to its side. Its mouth was surrounded by six or seven grublike papillae. The head had two appendages which bore spinelike branches at their tips. Its amber body limbs ended in blunt tips carrying a group of seven tiny curved claws.

"They live all through pipe-world but seem to be harmless," said Thorn. "They only eat moss and microscopic fungi that grow on the pipes." He picked the worm up and showed it to us. "See, nothing to worry about."

"May we place it in a specimen jar for further study?" I asked.

"Go ahead," said Thorn.

"Look over there," Kalinda pointed to my right. Several more Aysheaia worms were traveling along the pipes.

"Wonder where they're going," I said. "They don't seem to care about us." I walked over to one and placed it in another specimen jar.

"Thorn, I'd like to hear more about the two pipe civilizations," Kalinda said.

"Yeah, your story is intriguing, but a bit hard to swallow," I told Thorn. "How could the inner-pipe people have found food?"

"Place your ear on one of the larger pipes." Thorn's triple eyes seemed to tear slightly.

I wandered over to a pipe and put my ear on its cold surface. I heard crying! The inner pipe-people were still alive, squished in their pipe-hell, moaning in their metallic sorrow.

Thorn spoke in a mixture of sorrow, frustration, and anger. "Food? You ask about food? They eat the Aysheaia worms. Long ago the worms found the entrance to the inner-world." Thorn spoke in a quiet voice. "The worms could find the children, but we could not."

Three other worms came out to watch us. Thorn looked at me and changed the subject.

"Although you fight like a warrior, your instruments indicate you're a scientist." Thorn's voice was controlled, almost tight. "Are you interested in what our own scientists have found out about the worms?"

I nodded, and Thorn continued. "Our scientists found the worm spines to be molecular engineering marvels. Each spear is tough and resists fractures. Each is a single crystal of calcite."

"Isn't calcite very brittle?" I said.

"Yes, but a distribution of protein molecules throughout the calcite spines modifies the microstructure of the crystals, both strengthening and toughening it. We use the spines to improve the cutting edges of weapons."

Thorn proudly withdrew a dagger with an abrasive cutting edge. "The edge is made from the worms' spines."

I certainly wouldn't wanted to have engaged Thorn in battle – his massive muscles and impressive weapon made him undefeatable. Kalinda looked towards Thorn and me.

"The structure sounds similar to the spines of sea urchins on earth," she said. "Urchin spines are also made of calcite with proteins. The proteins act like fibers to prevent cracking –"

Thorn raised his heterochromous hand to put an end to the conversation. "Interesting, but enough talk of science. I believe you want to find the way out of pipe-world. Where do you want to go?" Kalinda looked hurt that she was interrupted.

"We want to meet Yars Kothek, ruler of the Latöócarfians," I said.

Thorn looked at me with his three eyes. One of the three twitched. "I don't advise this action. They're a queer bunch. Very intelligent, a race of mathematicians. They make beautiful mathematical patterns. But they are also very ruthless and difficult to understand."

"Nevertheless, we still want to find him," I said.

Thorn sighed, hesitated, and then said. "OK, follow me. I'll get you out of pipe-world." As we walked and crawled, the density of pipes dwindled. When we finally came to the last of the pipes, we looked out upon a savanna of violet grasses. A few Aysheaia worms continued to watch us but stayed on the remaining few pipes.

"I'll leave you now," Thorn said.

"Thank you for all your help," Kalinda said.

"Yeah, thanks," I concurred. "What will you do now?"

"Return to pipe-world, and resume my quest for the inner people." Thorn turned and walked into the forest of pipes.

"Interesting guy," I said to Kalinda "Interesting story. We could spend years just studying pipe-world. But now we should get going."

We took a few steps when from out of nowhere appeared two Prohaptors.

"Now you die for slaying our brethren!" one of the Prohaptors said. I felt the cold metal of a sword against my neck. Kalinda started to swing her backpack round and round and finally slammed it into the other Prohaptor. The wind came rushing out of his lungs with a "Woosh."

"Let them go," came a voice from behind a pipe. It was Thorn. The Prohaptor holding the sword at my throat did not move but seemed nervous. All of a sudden Thorn took a tremendous leap off of one of the protruding pipes and knocked the Prohaptor to the ground. The Prohaptor was unconscious.

The other Prohaptor took one look at Thorn and started to run. Unfortunately for the Prohaptor, the violet grass was moist, and even his horseshoe crab shaped feet did not have much traction. He slipped and fell on his side. With the swiftness of a cheetah, Thorn was upon him. He grabbed the Prohaptor, lifted him, and impaled him on a pipe. "Olagh," the Prohaptor cursed in a rage. The pipe went straight through his abdomen.

"Can't we put him out of his misery?" Kalinda said. Thorn went behind the Prohaptor and gave the back of his neck a quick squeeze.

"Ur - ur - ur." The Prohaptor's voice suddenly sounded like the clucking of a chicken, and then stopped entirely.

"Never did like those Prohaptor fellows," Thorn said. Kalinda and I looked at each other with wide eyes. Thorn smiled.

Pipe-World maze by master maze-maker Sampei Seki.
What's the shortest path from "S" to "G"?

Chapter 29

The Navanax People

"One of the reasons I'm not a big fan of science fiction is that none of the aliens are nearly as weird as my own invertebrates."

Janet Leonard, Marine Science Center, Oregon State University

"Maybe I should accompany you a little further after all," Thorn said, with a wink of one of his eyes. I suppose he was proud of the way he had dispatched the Prohaptors, and I had to admit we were glad he was with us. Nevertheless, I had some reservations.

"Thanks for helping, but I think we can handle ourselves now," I said.

"Garth –" Kalinda said.

"I don't like the way he keeps looking at you," I whispered to her.

"You're jealous," she said with an amused look on her face.

"Look," Thorn said, "you'll soon come to the land of the Navanax people. You may need my help in dealing with this ichthyophagous nation."

"Is he trying to impress you with big words?" I said to Kalinda, wincing inwardly as Thorn shot me a questioning look.

She whispered back to me, "I think it would be good to have him with us, if we want to get off this moon alive. The transporter on your belt won't be recharged for another week."

Inwardly, I growled. Outwardly, I was polite. I suppose I saw the practicality of Thorn's presence.

"Here's some food," Thorn said. He withdrew some brown material from a pack. I took a piece.

"Tastes like chocolate," I said. "I don't think I'll ask where it came from."

"Can you tell us more about the Navanax?" said Kalinda.

Thorn's triple eyes squinted, and he smiled at Kalinda. Again I caught him watching Kalinda with an intensity that was incongruous with the relative lightness of our conversation, as if he were judging her. "The Navanax are a race of humanoids, like you and I. They're about our size and have two arms and legs, but that's where the similarity ends," he paused as if for dramatic effect. "Around each of their bodies is a flap of sensitive skin. Most of the time the Navanax remain concealed beneath their flesh flap. Their eggs look like ellipsoids."

"Do you mean like almonds?" I asked, sketching an almond shape in the sand.

"Right," Thorn said. "Are you ready to go?"

Our feet made sloshing sounds as we travelled through some mud flats the color of burgundy.

"Aren't we making a bit too much noise if we want to avoid these creatures?" I said. Thorn seemed anxious to accompany and protect us, and therefore there must have been some danger involved. On the other hand, I was happy and excited about the idea of meeting the Navanax. As a zoologist, I'd love to bring back some of their eggs with us to Earth.

"Their skin," said Thorn, "is brown and blue with faint yellow dots. There are usually two outcomes when one Navanax meets another. If two Navanax meet face to face, they try to consume one another. If one approaches the other from behind, they have sex. They are hermaphroditic and can be male or female at their choosing. In fact, they change sex many times during a sexual encounter."

Thorn looked at Kalinda. I felt my blood pressure rise. Did he think his descriptions were too explicit for her?

"Every Navanax," said Thorn, "has a penis on the right side of its head. The penis is a few inches behind a genital slit that leads to an ovary. Each round of copulation lasts around four hours."

"Sounds like the hermaphroditic sea slugs of earth," Kalinda said, seemingly oblivious to Thorn's stares.

"Looks like we're in their land," I said. "What a sight!"

We saw hundreds of them shuffling about on the muddy ground. Their cloak of colorful flesh made them look like undulating, psychedelic jelly-rolls. "They don't seem to notice us," I said.

"Their eyes are quite weak," Thorn said. "Soon, however, they'll get a whiff of us with their phenomenal sense of smell."

"Why do some have their mouths to the ground?" I said.

"They suck food into their mouths by rapidly stretching their piezoelectric innards." Thorn said.

"Uggh," said Kalinda. "Some are cannibalizing one another." The entire scene was so unwholesome that all of us turned our heads. "Phew, smells like dead fish," she said as she gagged.

"OK, Thorn," I said. "Now that we're here, tell us what to do next. Will they bother us? Can we pass through this place without being attacked?"

"They'll treat us as one of their own, and will either attempt to consume us or...."

"Here come a few," I said. Six Navanax noticed us and shambled to within ten feet of our bodies. Thorn removed his sword from its scabbard.

"Hi," he said. "We mean you no harm. We want only to pass through your land."

The closest Navanax responded by sucking Thorn's leg into its digestive system before he had a chance to react.

"Ga... Ga..." Thorn moaned as he tried to wrestle his leg from the slobbering mouth of the Navanax. Thorn's sword dropped from his hand, and the Navanax covered it with its body.

"Garth, do something," Kalinda said. I ran and kicked the Navanax in the chest with all my might. It felt like I had kicked a bag of jello, and didn't seem to hurt the Navanax.

It continued to suck on Thorn with a slurping wet noise that reminded me of my mother's garbage disposal when it ground soft garbage.

Woosh. Woosh. Woosh.

Then I heard a flapping sound from behind me, a liquid sound, like something from an X-rated movie except much lower in frequency. More Navanax were coming.

"Ahh," I screamed as I felt liquid on my shoulder. I wiped at it with my artificial hand and saw it had turned dark sepia and crimson with syrup of some unknown composition. The sepia syrup smelled like ammonia. My human hand was beginning to sting like burning lava.

My natural arm was in the mouth of a Navanax. I pulled and pulled but could not remove my arm from its mouth which reeked of decaying fish.

If only I could get to my sword! Fear drove me to try a final attempt to dislodge myself from the creature. First I bit at it, tearing away a mouthful of horrible urine-stinking flesh. I bit down again and something crunched like the cartilage and gristle from a chicken bone. The warm, sour taste of aging bacon filled my mouth.

I tasted vomit in the back of my throat. "Kalinda," I cried.

She ran over to my knapsack and withdrew the fractal sword I'd taken from the Prohaptors.

"Let go of him," Kalinda screamed. She ran to the Navanax consuming me and cut it in two. Both halves lived, and scurried away. A few smaller pieces of flesh flapped on the ground. Then she ran and split open Thorn's Navanax. Green gel oozed from the torn Navanax flesh.

"I can't take much more of this," Kalinda screamed as she wiped some green goo from her hands onto her cloak. Within the goo were tiny skeletons, no doubt the remains of some partially digested fish. The small bones of the fish began to curl as if magically still alive. Their small, bony mouths seemed to open in a silent cry.

"Garth, are you OK?" Kalinda ran to me and held me.

"My arm looks like it has some burns, but I'm OK." I began to scrape some goo off her cloak with my artificial arm.

Thorn limped over to us. His leg was bruised but not cut.

"That's the last time I try to reason with a Navanax," Thorn said. Several Navanax responded by speaking in unison.

"You may not pass through our land. We wish to devour you."

"The Navanax are not as brave as they appear," whispered Thorn. "They're superstitious." He breathed hard and rubbed his leg. "If we could somehow scare them into thinking we have superior or magical powers they'll let us pass. I thought that the sight of my long sword would scare them, but I guess I was wrong."

Kalinda shivered from the cold and the anticipation, and her arms were goose-pimply.

"I have an idea," I said. I removed my prosthetic arm from my body and controlled it via its infrared and electromagnetic signal link, much as people from earth once controlled their TV sets with a remote control. I held the arm high above my head with my normal hand. Then I had the artificial arm wiggle its fingers as I spoke.

"Behold, the bodiless arm." I tried to sound a little like Charlton Heston in the film *The Ten Commandments*. "It will do you no good.... Tear us to bits, and still we will resist you, destroying your vital parts even as you seek to devour us."

Thorn gasped, but Kalinda, being accustomed to my prosthesis, grinned and seemed to admire the creative use of my body part.

"Urrrrp," the Navanax screamed in unison. Most of them undulated away from us.

"Looks like that did the trick," I said.

"Don't be too sure," said Kalinda. All the Navanax except one had scurried away. The bold individual came closer and swallowed my prosthetic arm. Kalinda screamed. I remained motionless. Other Navanax watched from a distance. Luckily for me its digestive fluids could not damage my artificial skin. I looked challengingly into the Navanax's eyes. Next I sent my disembodied arm a signal to continually poke its thumb into the digestive canal of the Navanax.

"Feel a little indigestion?" I said. The eyes of the Navanax bulged. It started twisting and turning like Turkish taffy in a pulling machine. Soon the anthropophagite regurgitated the arm and looked considerably more frightened.

"Thanks I needed that." I said. I reattached the arm to my body after wiping away some of the Navanax's digestive fluids. Thorn looked like he was about to drop his teeth. The Navanax creatures who had watched the entire incident were now undulating, jumping up and down, shrieking, and consuming one another, perhaps in nervous anticipation of our next action. The whole scene looked like an explosion in a sausage factory from hell.

I took a deep breath and looked around at distant crystalline forests which burned in the last glow of twilight. An unusual black tree with very low, spreading branches looked like several ghostly hands rising from an oceanic abyss. Vague patches of light gleamed coldly along the horizon. I looked into the eyes of the bold Navanax who, minutes before, had my arm in his stomach. Perhaps he was their leader. This Navanax stared back, entirely unaffected by the mayhem all around him, or so it seemed. I noticed his body was twitching slightly. His oral cavity opened and he spoke.

"You'll notice that your friends are gone."

I spun around and saw no sign of Thorn and Kalinda.

"Where are they?" I cried.

"They'll be returned to you if you follow my instructions. We're at war with our neighbors, the mole-people, but we've been unable to defeat them. However, they should be no match for you, with your great powers."

"What powers?"

"You're arm for example. Our soft bodies are not very effective against the mole people. If you destroy their queen, you and your friends will be allowed to leave unharmed."

"You little... I could kill you right where you stand."

"Don't try it. My people will eat your friends alive if you kill me."

I looked into his emotionless eyes for a few seconds. "You don't give me much choice. But if you hurt them..."

"You have my word we wont hurt them."

"Let's get this over with."

"Follow me." The Navanax led me away into a hell-black night.

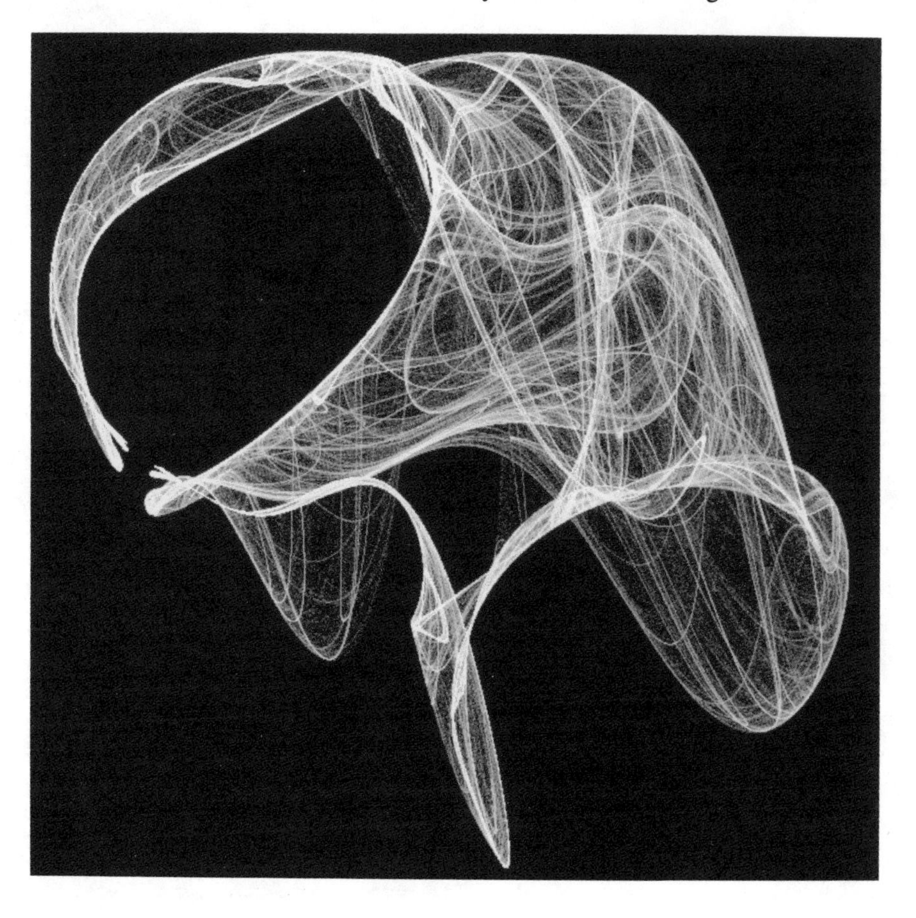

Chapter 30

The Underground Association

"I am concerned here with some animal unknown to me. The world is full of diversity and is never wanting in painful surprises. Yet it cannot be a single animal, it must be a whole swarm that has suddenly fallen upon my domain, a huge swarm of little creatures." Franz Kafka, *The Burrow*

The Navanax and I slept for a few hours and then walked to a desert of Ganymede. Occasionally I tried to engage him in conversation to learn more about his race. Any knowledge I gained might help me save Kalinda and Thorn. Of course, I still wasn't sure if the Navanax would let them go once my deed was accomplished. His "word" meant nothing to me.

I watched the Navanax stop and pour water on himself. "Why do you do that?" I said.

"Keeps my skin from drying out."

"Care for some Nivea skin care cream?"

"What?"

"Never mind."

As we walked – or I should say as I *walked* and he *undulated* – he told me about the mole people we were about to meet. These six-foot tall humanoids stood upright and spent nearly all their time underground living like social insects, part of a giant colony, part of an underground society. The mole people had white, wrinkled skin, pinhole eyes, huge colorful whiskers, and flamboyant saber-like teeth. They used their large incisors to dig complex networks of tunnels and chambers which housed as many as three hundred creatures in a single colony. To ward off the cold Ganymedean nights, they huddled together with other colony members, shivering and pressing their bodies together.

We came to a hole in the ground, about the size of a manhole cover. It looked dark in that hole, and I didn't want to get near it. Next to the hole stood a six-foot tall statue of a creature with a conical head, large teeth, and arms that reached from the shoulders to the ground.

"That doesnt't quite fit your description of a Ganymedean mole," I said as I walked closer to the strange statue.

"You're right. The moles put it there to scare off possible invaders."

"I wonder if it works."

"You can enter their borrow here," the Navanax said as he motioned to the hole in the ground. "Find the queen and kill her. If you do not, your friends will die."

"How will I know the queen?"

"She's the fat one."

"I don't want to kill anyone."

"You must."

"How do I know you'll return my friends?"

"If I don't, I'm sure you will kill me with your weapon," he said, pointing at the fractal sword.

"You're right. You'll be the first to die."

I took one last look at the bizarre statue and then slowly entered the hole.

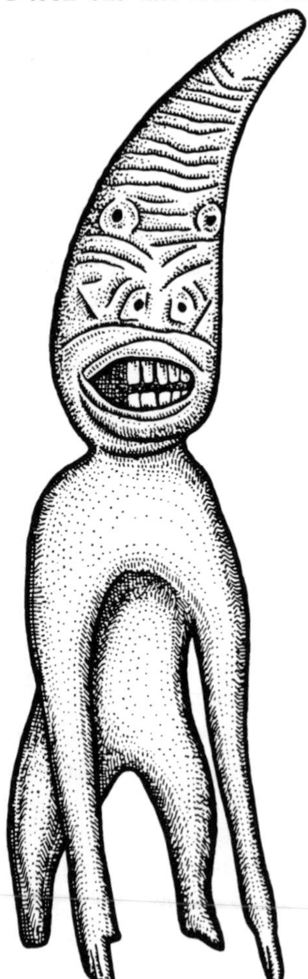

After a few steps I was surprised at how bright the interior was. Their tunnels were a marvel of artistry and engineering. The walls were painted with phosphorescent chemicals that glowed blue and violet. Many tunnel walls pulsed like a living lava lamp. A series of pastel circles moved around the walls of the caves, as in a patternless kaleidoscope. Slowly they coalesced into a beautiful blob of bathybius.

As I travelled through the maze of tunnels and chambers, I heard some voices and saw a movement in the shadows to my right.

"Get him," snarled a mole to some his colleagues in the tunnel. The moles all started to growl at me with their saber teeth.

"What is that ugly thing?" said another mole as he looked at me. It took a step in my direction.

"I wouldn't come any closer," I said. A few swings of the fractal sword convinced them that aggression was not recommended. "Now you get out of here and you wont be hurt," I said. They quickly scurried away.

They appeared to have a variety of specializations for life underground: streamlined bodies, short appendages, and loose skin. Their jaw muscles were huge. Aside from their whiskers, they had no hair at all. Perhaps hairlessness evolved as a way to minimize cover for parasites such as mites which plague most furry animals on Earth.

All their digging produced an intricate system of connected tunnels and multiple chambers as big as a house. As I observed them, I saw that digging was generally a cooperative effort. The mole people got on all fours and lined up head to butt behind a leader. The leader gnawed on the soil at the end of the developing tunnel. The ones at the end helped cart the soil away.

Suddenly I saw a few of the Navanax people. I watched them from a dark area of the tunnel feeling like an anxious child who had stumbled on something he didn't quite understand. Finally I approached one of the underground Navanax.

"What are you doing down here?" I said.

"Hey, I know about you," said the Navanax. "You're the weirdo who's supposed to kill the mole queen."

"Answer my question."

"Lots of us live down here," said the Navanax. "We have a thriving parasitic relationship with the mole people. We sometimes even eat their young."

I didn't know how to react to the Navanax's statement. Was it a boast or the truth? How would it affect my ability to get at the queen?

"Why wasn't I told about this relationship?" I said. The Navanax said nothing. "The mole people don't seem to notice you," I continued. "Don't they care that you're down here? They seem to treat you with astonishing cordiality." In fact, as we walked, I occasionally saw the mole people feeding, grooming and rearing baby Navanax as if they were the mole peoples' own brood. Finally the Navanax decided to answer me.

"Well, if you must know, we secrete a substance that attracts the mole people and camouflages our true identity. They respond to our chemical signals by grooming and caring for us. Don't we have a great deal down here?" The Navanax leaned back with excessive nonchalance, and then he looked at me with his tiny eyes. "So, I understand you are to kill the queen," he said.

"That's right. But why do you need me to do it if you're down here already?"

"They are pretty tough fighters. We don't have weapons to accomplish this task."

"Weapons –"

"We're just too soft. Their wicked teeth –"

"Why not just eat her like your people tried to do with my arm?" I said angrily.

"When you see her, you'll understand."

"Why do you want her dead?"

"Because she's heard rumors of our presence within the colony. She'll probably send a search party to find us." The Navanax's whole body tightened, and then he took a deep breath. "Follow me. I'll lead you to her."

Chromatophores ignited within the Navanax's flesh, as his color changed from maroon to red. When talking about the queen and her search for the Navanax intruders, his entire demeanor changed from one of arrogance to one of fear.

Just then, a mole person came by. The Navanax brightened. "Watch this," he said to me. The Navanax went over to the mole and then stimulated its mouth parts causing it to regurgitate its cropful of food. "We have them completely fooled and under control. We're like a whole civilization of ghosts living within their society. They don't perceive us as alien due to our camouflaging chemicals."

"Here's the queen's chamber," he said and quietly retreated.

"Hey," I whispered. "Where are you going?" But the Navanax was gone before I could ask further questions.

The queen was easy to identify by her huge oblong body, and enlarged sex organs. As an exobiologist, I was curious about her position in society, and was able to make observations without being seen for an hour. She was a terror, always shoving and biting the others to protect her position in the chamber. I could hear snippets of their speech and learned that this terrible queen nursed her young for three days, after which the children began to eat the feces of the adults. The mole people appeared to produce two kinds of fecal pellet: one variety deposited in a communal toilet chamber, the other reingested. The softer varieties were highly nutritious and contained microorganisms beneficial for digestion. In a few weeks, the children would begin to consume the same roots and multihued tubers that the adults ate. The tubers were part of perennial plants that contained water, sugar, and starch in swollen roots. The queen was the only female in the colony to engage in sexual activity. As she poked, prodded, and screamed at the other females, she induced such stressful living conditions that the other females in the colony no longer released reproductive hormones.

I also overheard the conversations of Mia, a mole with very shiny skin, who was responsible for housekeeping chores such as cleaning the phosphorescent chambers of debris. Mia's brother helped to defend the colony against Ganymedean snakes and giant mobile fractal ferns with razor-sharp teeth. The only thing in Mia's life that seemed to bother her was a minute bleb of skin on her left eyelid.

I began to worry about the time I was wasting while Kalinda and Thorn were still captives. I had to start acting, although I didn't really want to harm the mole society. Maybe killing the queen would have little effect since there were other females ready to take her place. I had no particular love for the Navanax, and didn't care who had the advantage in their quarrels with the mole people.

Without a word, I approached the queen mole. She looked at me with haughty eyes for a second.

"Sssss," she hissed at me.

Then she pounced. I struck her with my prosthetic arm. She pounced again. This time her mouth opened revealing teeth prodigious both in size and quantity. She clamped her mouth on my artificial arm and would not let go.

"Get your mouth off of me," I said. Her tongue was soft and redolent with the smell of raspberries. Suddenly her tongue grew rheopectic, as if it were hardening when subjected to mechanical agitation.

"Look," I said. "I won't have to kill you if you just leave. You can take a few of your friends and establish a colony somewhere else on Ganymede."

She responded to my offer by biting harder.

I held her as far away from my body as possible and began to spin like a whirling dervish. Her body revolved around me, lifted off the ground by centrifugal force. After a few revolutions, my arm came free of my body and together with the queen sailed away into the darkness. Although I couldn't see my arm or the queen, my arm sent remote signals back to my body permitting me to feel that she had let go of it. I sent my arm some signals so that it would grope around in the darkness in an effort to find her. Soon I felt a warm moist object and had my hand grab it. The queen let out a muffled scream. She ran at me with the speed of a striking cobra.

Holding on to her tongue was my prosthetic arm. "What's the matter? Cat got your tongue?" I asked.

"M - Mm" she mumbled. I reached for my fractal sword and cut her in two. Then I wrenched my arm from her still clenched teeth.

After a few minute's rest, I began to make my way to the surface through the complicated network of tunnels. Unfortunately I soon came to a cul-de-sac. "Damn." I turned back to find another tunnel to the surface and almost slipped on a godawful pile of mole excrement. It was not easy travelling particularly because I carried the queen's upper torso with me so I could prove to the Navanax that I carried out my promise.

A group of mole soldiers suddenly barred my way. They just stood there with mouths agape. Other moles climbed onto their backs and blocked the passageway with two tiers of clamoring, biting fury. One of the guards looked at me.

"You killed her," he hissed at me. His wrinkled skin wriggled. His teeth produced a grating noise. He smelled of excrement.

"I'm sorry." I threw the queen's upper body onto the ground in front of my feet. I saw no avenue of escape. "I had no choice. My friends were –"

"Thank you very much," said the guard. "She was a terrible bitch." Then he smiled. "Back off," he said to the other moles. Evidently the queen's popularity had slipped during the last few years.

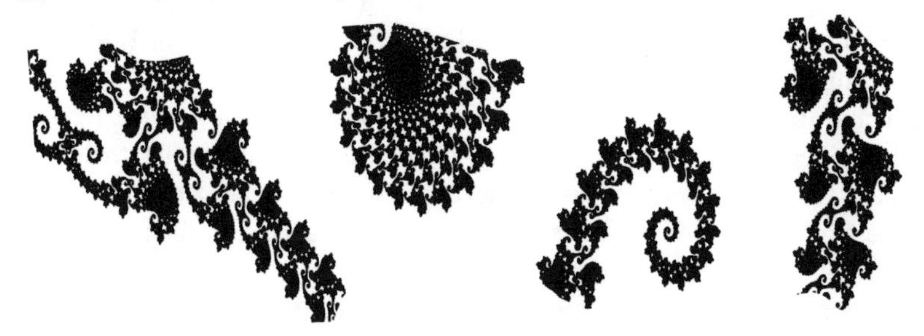

I emerged from the mole hole and saw the Navanax waiting for me. Slung over my left shoulder was the half-corpse of the mole queen. Lime green saliva dripped from her mouth.

"Did you kill her?" the Navanax said. "Is she dead?"

I threw her body at his feet. "What do you think this is? Of course she's dead."

"Good. You and your friends are free."

I turned and found that Kalinda and Thorn had been escorted to the entrance of the underground society. *God, I loved her.* The intensity of my feelings, and the loss of control over my own fate that it implied, scared me.

Kalinda ran across the ice and hugged me.

"Are you all right, Garth?"

As always, I had a sudden physiological reaction as I felt her arms around me, her small breasts pressing against me. "I was so worried," she whispered in a sweet voice that always seemed so improbable in such a strong woman.

I closed my eyes as I felt Kalinda's hugs, and the confusion of the recent experiences dropped away. I breathed in the scent of her, and wondered how I survived even the past day and a half without her. Kalinda's dark eyes bore into mine and I could see her pupils alternately dilate and contract, a sign of pleasure which not even a reserved person like Kalinda could control. Her pupil oscillations during times of great emotion were the only difference I had ever found between Kalinda and a normal human woman. "God, you're beautiful," I said.

Kalinda immediately slipped on some of the queen's large intestine. "Yuck," she said. She wiped her foot of on the grass.

"I'm just fine," I said, finally answering her question. "What happened to you and Thorn?"

"The Navanax people dragged us into some caves. They tied us down with ropes. Today, they brought us here to meet you."

The Navanax did not say good-bye but simply began to undulate away back to its people. We watched for several minutes as its sausage-shaped body grew smaller and smaller, looking more like a hotdog dancing to the sounds of Benny Goodman's "One O'Clock Jump" than a cruel, intelligent creature.

"I'm so tired," said Thorn. "It's time for me to get back to pipe-world and give my leg a rest. Those Navanax digestive fluids are hell on my complexion." He pointed to blotches on his leg.

"Can you avoid the Navanax in your trip back?" Kalinda said.

"I'll try to go around their land. It'll take a lot longer, but it'll be worth it."

"Is there anything we can give you?" I asked. Thorn just stared at Kalinda.

"No I'm fine," he said after a few seconds.

"Thanks for your help," Kalinda said. Then she turned to me and smiled. I very much liked her smile and found myself looking forward to Thorn leaving.

"Yeah, thanks," I said. Thorn placed his massive hand on mine to say farewell. His triple eyes looked into Kalinda's for a moment, and then he departed.

"Glad that's all over," I said.

"I have a surprise for you," Kalinda said to me. In her hands was an almond-shaped Navanax egg. She carefully placed it in my hands.

"Great. Good work." I put the egg in my specimen jar. "Why don't you keep the jar in your backpack?" I said. Then I kissed Kalinda. "Glad you're alive. Now I have a present for you." I handed her some tubers I'd found in the mole chambers.

"How do they taste?"

"Haven't tried them yet."

"Let's at least try to clean them off by rubbing them in some of the patches of ice." When Kalinda finally bit into one of the roots, she smiled.

"Tastes a little bit like raspberries," Kalinda said.

"Let's give some to the pipe-world worm. It's probably hungry by now." I removed the specimen jar from my pack and saw the foot-long Aysheaia in a coil at the bottom of the container. It looked like a spiral of white rope to which was attached a large spherical eye.

The worm stared back at me, its large opal eye vivid and questioning. It then swiveled its eye toward Kalinda and looked at her with a fragile expression.

"It seems so tame," Kalinda said.

"I'm going to let it out and feed it. I don't think it will try go escape if I have food in front of it."

"We have no idea it will even eat this stuff," Kalinda said.

I slowly opened the jar, and placed it on the ground. The worm smoothly glided along the icy soil to my open hand which held the roots.

"Garth are you crazy? Don't feed it from your hand. It could bite –" She stopped when she realized I was using my artificial arm. There was little damage the creature could do.

The worm began feeding. I heard little crunching sounds coming from its mouth.

"Let's see if this guy is as tame as he seems," I said. When the worm seemed satiated, I picked it up and placed it in my pocket. Occasionally it poked its head out to look around and then snuggled back inside. It seemed comfortable – or at least more comfortable than when in my specimen jar.

My eyes were sandy and my bones ached. "Let's get some sleep."

Kalinda nodded. "I'm too tired to start looking around for a sheltered area."

"Me too. Let's sleep right here and hope for the best." The shadows looked like stalking gray cats. The darkness began to press down on us.

"Goodnight, princess," I said.

"Goodnight."

That night I dreamed Kalinda and I were together, holding each other naked for the first time. Our bodies fit together like pieces of a puzzle. We moved together in breathless wonder. Her body felt smooth and soft. Her hair was silky.

"Is my artificial arm getting in the way?" I asked in my dream. "Should I take it off?"

"Whatever you like. It doesn't bother me one way or the other. I love you."

I hesitated and then decided to try something new. In my dream I had my fingers vibrate in a high-frequency sinusoidal pattern. Back and forth, my fingers traced out a wave-like pattern in the air. I held them up so she could see their sensual rhythmic undulation. Her eyes fixed hypnotically on my fingers.

"Garth, what are you planning?" she said, smiling. "Garth –" I switched the waveform from sinewave, to ramp (a slightly sharper vibration), and then to some complicated fractal motion. Kalinda's moans became chaotic. I changed to a slow Lissajous pattern. She began to gyrate in time with the low frequency waves. I slowly began to increase the frequency of vibration: 10 Hertz, 100 Hertz, 1000 Hertz. At the last moment I switched to an oscillation in the shape of a Chebyshev polynomial, and –

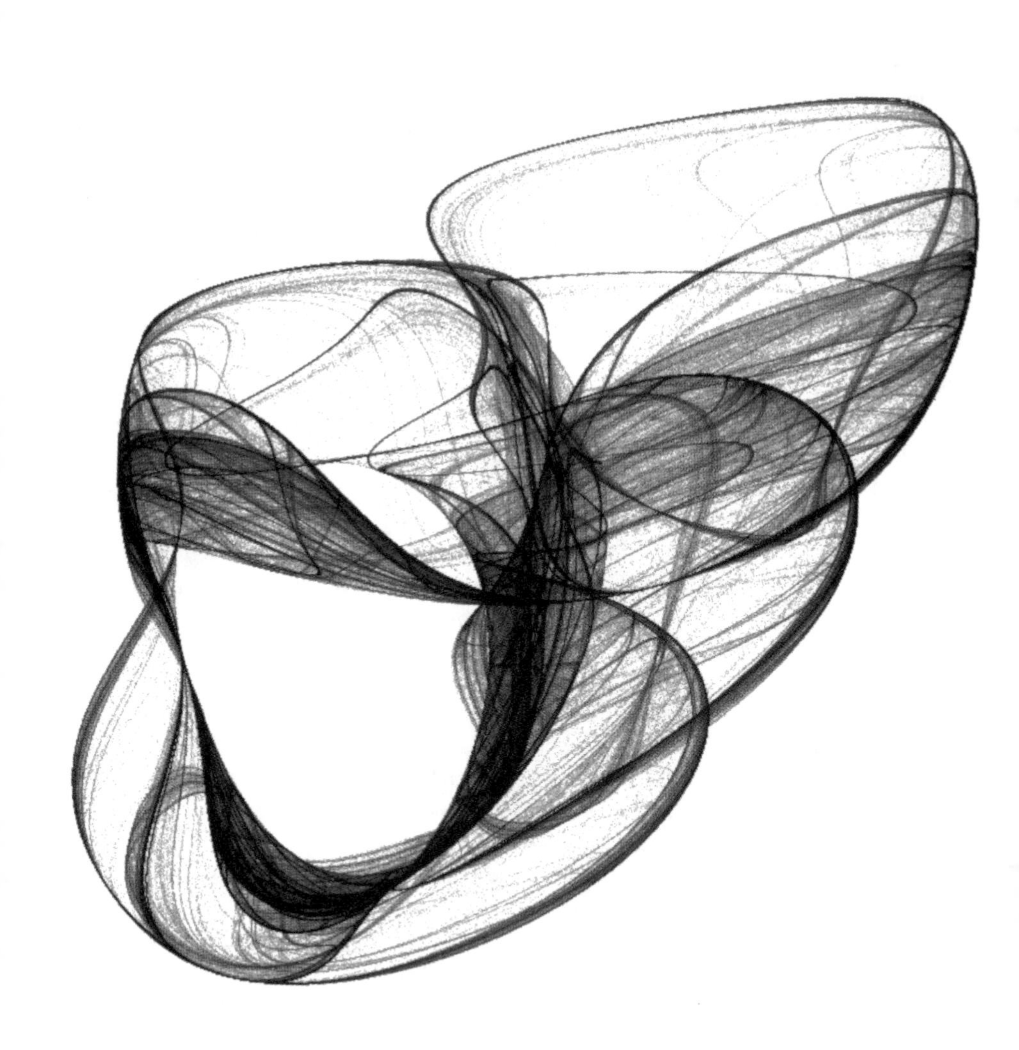

Chapter 31

The Water Beings

"Sometimes, wandering alone in the woods on a summer day, we hear or see the movement of a hundred species of flying, leaping, creeping, crawling, burrowing things. Suddenly we perceive to what a perilous minority we belong on this impartial planet, and for a moment we feel, as these varied denizens clearly do, that we are passing interlopers in their natural habitat."

Will and Ariel Durant, *The Lessons of History*

"I'm cold," Kalinda said. She shivered after awakening from an interminable, restless night.

"Yeah, and the ground isn't too soft."

"That's an understatement. I wish our blankets better insulated us from the cold soil."

"Well, I'm still so happy to be here," I said. "What an adventure. Remind me sometime to tell you about the dream I had."

"Tell me now."

"Later," I said shyly.

"Whatever you say."

"Let's go looking for the Latööcarfians."

The landscape was beautiful. The rugged plateaus formed patterns of crystalline ridges which scintillated in the light. Some crystals reflected pink beams of light into the heavens. The sky was always a source of pleasure, with cotton candy clouds, and ebony gaseous mists. Off in the western sky we saw chocolate ripples of brown and white amidst twin-clouds of flaming incandescent violence.

We came to an oasis of clear water. "Those are shrubby Suaeda plants," I said. "On Earth they grow at the sea-shore." We bent down to examine the plants.

"Can we eat these?" Kalinda asked.

"Better not. Look here." I pointed to the shallows of the oasis. Life was everywhere: beautiful naked gastropods, organisms which resembled land-snails, sea-snails, limpets, and whelks... It reminded me of the time I spent exploring the sea-ports of Galway, Limeric, and Cork on Earth....

"Care for some sushi?" Kalinda said. She picked up a clam-like animal with multiple tentacles coming from its hard shell, pried it open, and plopped the contents into her mouth.

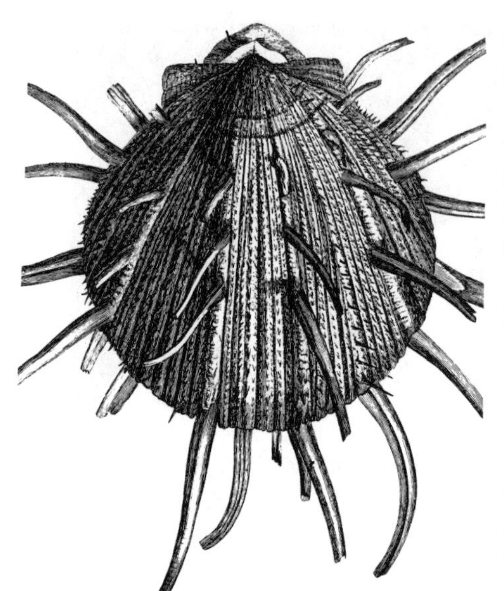

I stared at her. She stopped eating.

"I know," Kalinda said. "You want me to be more careful about what I eat. But we'll starve if we don't eat what we find here."

"You're right," I agreed reluctantly. Never did like sushi.

"Garth, do you hear something?"

"Not sure." I looked around.

"Maybe –"

"Damn," I said softly as an answer came to me with sudden clarity. "It seems to be coming from the water. But the water is clear. Could the water itself communicate?"

"Yes!" Kalinda said, looking into the water. She glanced at me sharply, and her eyes glinted with excitement. "Yes, that's it!"

"I can't quite see where –"

"I don't like this," Kalinda muttered, her excitement momentarily attenuated. "The water seems to be moving!"

We placed our ears close to the silver pool. "I think it's - it's alive," I said. "It's almost as if we can telepathically listen to the water."

Kalinda gazed into the pool of consciousness. "Wait. I think I hear some words."

We are the Leandra, said the pool.

"Garth, did you hear that?"

"Yes." I put my head near the pool and said, "What are you?" I listened for a few seconds. Silence. "This is crazy," I said. "I'm talking to a pool of water."

"Wait, I hear something," Kalinda said. "I can almost make out what they're saying."

"Maybe you're more receptive than I. I think you have to put your head closer to the pool."

"Yes. I hear it – them? They describe themselves as an energy-water intelligence. They have no physical being except for the chemicals in the liquid. Their molecules are mind. Their weak electrical fields telepathically link the Leandra with any nearby creatures. If they were allowed to dry out, they'd die."

"You mean the water intelligence is nothing more than an oscillating chemical reaction?" I said.

"Yep. The reactions form conventional transistors, just like the switches of digital computers."

"The more strange creatures we meet on our trips," I said to Kalinda, "the more I realize just how small our own mote of consciousness is in our universe."

"What do you seek on Ganymede?" the Leandra asked. Their voices had a musical quality, mostly bass but with a few xylophone notes.

"Now I can hear them too!" I said. I placed my face closer to the pool. "We're biologists looking for interesting organisms. We're also looking forward to meeting the Latööcarfian mathematicians."

The Leandra were silent.

"Do you think they're more like a chemical computer than our brain?" Kalinda asked me.

"Our brains *are* chemical computers." I placed my head even closer to the pool of consciousness to see if I could learn more.

Apparently the oscillating chemical pairs from which the water intelligence derived its being were molecules called nicotinimide adenine dinucleotide. This naturally occurring substance existed in a reduced form and an oxidized form in which it was robbed of one of its electrons.

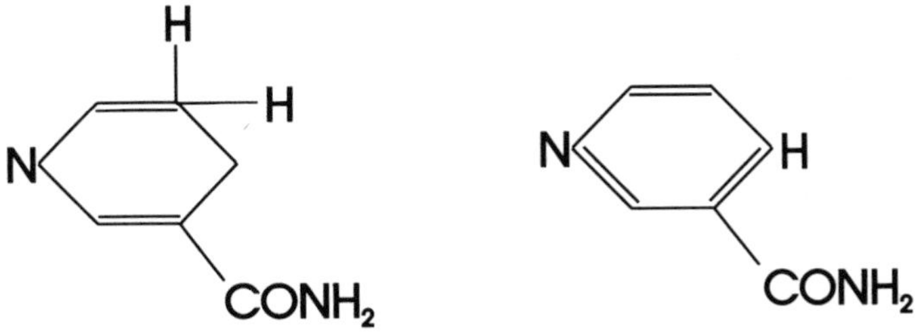

The jump between low concentrations and high concentrations of the two forms of these molecules in the water being was abrupt and worked like a tiny switch. Each chemical switch was like a neuron in our own brains, and the water beings were composed of millions, maybe billions or trillions, of networks of these switches.

"Are you seeing us right now?" I asked the Leandra.

"Our senses are not like yours, although we can visualize you now."

I dipped a specimen jar into the water to remove a pint of the liquid. I held the jar up to Kalinda.

"Garth, do you think you should do that? Maybe it's like removing one of their internal organs without asking permission."

"I don't think they're built that way. Anyway I'm sure they'd tell us if there was a problem..." I held up the jar for a closer look. "I wonder if there is any consciousness in this jar?" We placed our heads close to the glass, but couldn't tell if the portion of the water being in the jar had any sentience.

"Maybe this is too small a sample to have thoughts," Kalinda said. "Or maybe the thoughts of this small sample are so primitive that we can't communicate with them."

"What do you think would happen if we were to drink this?"

She shook her head. "Let's not try it."

As we turned to leave, the water beings spoke. "We suggest that you visit the Imaginarium. You'll find it most interesting."

The water beings' "voices" in our heads began to grate on my nerves. The pleasing xylophone notes were no longer evident in the spectrum of their thoughts. Only the ultrabass notes remained. They started to sound like a bunch of bellowing, bibulous, bassoons. I asked the obvious question.

"What's the Imaginarium?"

"It is a building managed by the Heñtriacontañes. If you go there, you will certainly have an interesting experience. These philosopher creatures will have much to show you. Aside from philosophy, they're also interested in mathematical theorems; however, unlike the Latööcarfians, they are not concerned with practical aspects of computation."

Kalinda nodded her head. "Garth, let's try the Imaginarium. At least it doesn't seem too dangerous."

As we walked among boulders and crystals, the shiny ground had just enough texture to cut the glare and give definition to the different colors of minerals. Occasionally there were clear bluish violet varieties of crystallized quartz resembling amethyst.

"Look at that," Kalinda pointed. I turned my gaze from the colorful ground to crevices in patches of rock which teemed with sinuous, ink-dark tentacles. As I went closer, the tentacles retreated into the rock.

We soon came to a pool of liquid. "Think it's conscious?" Kalinda said.

I put my head near the liquid. "Not as far as I can tell." I dipped my finger in. "It's not too cold. Would you like to wade in the water?"

"I'm going to do more than wade! I definitely need a bath."

"OK."

"The water looks clean. Think I can drink?"

"Yeah, but drink just a little at first to make sure."

"You want to bathe first?" I said. Despite my overconfidence in most situation, I was a bit nervous and excited about the idea of seeing Kalinda naked.

"You want to drink *after* I bathe? Anyway, you saw me naked in my broom," she said as she turned her back to me and took off her boots, then her knee-length cloak of a tawny orange suede. She left on her necklace of nautiloid egg shells.

"You go ahead and take your time. I'll bathe when you're done."

"Don't be silly. Pretend you're skinny dipping." As she waded into the water, she called over her back, "C'mon, it's a bit cold, but we'll survive."

I saw that Kalinda's body was even more beautiful than I had imagined. Her small breasts were alluring and looked quite human. Her brown pubic hair was neatly trimmed. I gazed at every curve of her body. *My God, do I love Kalinda.*

"You're staring, you know!" Kalinda said.

"I can't take my eyes off you. You're so beautiful."

"Thank you," she said happily.

I loved her for her body, her brain, everything...

"Now what are you thinking about?" Kalinda said as she splashed water all over her body.

"Uh...."

She swam away through the quiet pool. "C'mon in. I'm starting to get cold." We could only stay in for a few minutes, due to the cold. I stood in water up to my hips as I rubbed the clear water over my body.

After Kalinda got the caked grime loose and washed away the odor of the Navanax, she rinsed her hair. Although it remained tangled, it smelled much better. It was impossible for me not to glimpse her body in the dwindling light. I tried not to stare. Afterwards, we both shivered as we quickly got into our clothes. Kalinda kissed me lightly.

"Thank you," I said.

"My pleasure." Her long honey-colored hair was now combed straight back from her face and forehead.

"Garth, what are you thinking about right now?"

"You."

"I think a lot about you too," Kalinda said in a voice with singing in it.

How could she not be aware of the hunger within me? Sometimes I was afraid of the tremendous attraction I had for Kalinda. Over the years, I had only a few lovers, but the feelings I had for Kalinda were more emotional, more entangling. Despite a few quick kisses, I couldn't say for sure if her feelings for me were even remotely passionate. Perhaps I was afraid of an intense relationship with a beautiful nonhuman from a world where women were digestive symbionts. From a psychological standpoint, could she truly share my interests in life? Kalinda had grown up with the quiet pleasures of sleeping in a protective alimentary canal. Had her weird way of life and these creepy images forever embedded themselves in her unconscious? Just as I, who grew up by the woods, dreamed endlessly of walks in deep secluded pine forests, will she dream of the moist world of stomachs and intestines?

As we stood together near the clear water, I couldn't help hearing in my mind fragments of poems from my favorite 20th Century poet, John Celestian: *I'd like to carry this moment of time on forever... Hanging on to joys which spring out into misty airs... I must learn to stare upon your beauty without seeing... I must leave the silence... with its solitary candle... before my puzzle falls, leaving strange patterns upon my head.*

Shadows sprang up about us as if they were living creatures. The only illumination came from the green and red light emitted by the bioluminescent bacteria coating the trees. It reminded me of Christmas.

Chapter 32

The Imaginarium

"Love with care – and then what you will, do." St. Augustine

We reached the land of the Heñtriacontañes[24] by the next evening. Kalinda gazed at the winter-gloomy wilderness all around us.

"Feels kind of spooky around here," she said. I nodded. Occasionally we saw fantastically bright birds alighting upon dark aromatic bushes. Actually, I wasn't sure if "bird" was the right descriptive term to use. Their gossamer wings quivered on bodies resembling balls of twine. Some of the animals had more slender bodies. Their eyes, as well as speckles on their skin, glowed like neon lights.

"They look more like large bats than birds," I said.

"Yeah, some are hanging upside down from the branches."

The first digit in the hands of the neon bats was reduced. The other four digits were greatly elongated, with a naked membrane of skin stretched between them and attached to the side of the thorax. Some of the neon bats chewed upon the slimy black tangles of a still-pulsating creature.

"Look," Kalinda pointed. "What a neck!" One bat's neck appeared to stretch and stretch, until it was bizarrely elongated.

"Wild," I said. "Its flesh and bones must be super elastic. It's able to reach that disgusting food without moving its body." We watched as the blood blackened between its teeth.

A few of the bats seemed to have developed a fold of skin between their arms and legs. When the animals jumped off the tall bush, the skin ballooned upward like a parachute to slow down their falling speed. As one bat hit the ground, I ran

---

[24] The term "Heñtriacontañe" is a rough translation of the philosopher race's name to 1990's English. Their true name begins with the "H" sound but is rather long and difficult to pronounce. Scientists have adapted the written form "Heñtriacontañe" when referring to the philosopher race. The name derives from "hentriacontane," a hydrocarbon of the paraffin series $CH_3(CH_2)_{29}CH_3$ present in petroleum, many natural waxes – and the pheromones released by the "Heñtriacontañe" philosophers.

Figure 32.1. *Various species of Ganymedean neon bats.*

after it and caught it. Suddenly I stopped, standing perfectly still with battleship solidity. I waited for another bat to land. Waited. Finally, another bat descended. As quick and quiet as a ferret, I leaped on it and snapped its neck.

"Dinner time!" I said, holding up the bats in my hands.

Kalinda looked at me and then the dead bats. Her left eyebrow rose slightly. "OK, I'll give it a try."

I found some crystal bushes, broke off their branches, and started a small fire. "You get out some of the spices in your backback," I said. "I'll look for that old pot we carried along."

"Found the spice," she said.

"Great. You're always so organized. Add it to the pot."

Kalinda dumped some brown and red powder into the banged up metal pot I found in the pack. I added some ice which soon melted as the flames from the crystal branches heated the pot.

After removing the bats' vivid orange skin and most of their guts, I threw their carcasses into the boiling water.

"Actually smells good," said Kalinda.

"The boiling herbs give it a strong smell. Anyway, what'd you expect from the best chef on Ganymede?" I smiled. Then we got out some bowls.

"Dig in," I said.

"Mmmm. It tastes like chicken cordon bleu."

"The meat's a little on the tough side."

"Do you think this is Kosher?" Kalinda asked.

"You're broom was Jewish? We're starving."

As we visited new worlds, Kosher dietary laws became more difficult to define and uphold. Not that we cared too much. Although my parents were Jewish, both Kalinda and I were not very religious. In any case, these days 95% of the Earth and its colonies on the moon observed Baha'i religious principles. Baha'is had no Kosher dietary restrictions. We liked many aspects of this religion, for it seemed to unite much of humankind in a global community.

"Care for a leg?" Kalinda said.

"Sure, take off your boots."

"Ha, I mean a bat leg."

"Ah," I grinned. "No thanks, I'm full."

I began to clean the pot by boiling some water in it. We left the bat bones behind on the ground and packed the pot.

"Did you notice the large pectoral muscles on the neon bat?" I said. "The long abductor of the thumb –"

"Garth, do we have to discuss our meal in such detail?"

"Guess not. Ready to look for the Imaginarium?" I said.

"All set." We walked for a few minutes.

"That must be them," Kalinda said with a gasp.

"Wonder what they're smoking?" Some of the Heñtriacontañes were puffing on fronds of some kind of vermillion weed. Ganymedean weeds, in general, were beautiful – higher and bushier than those on Earth. Other Heñtriacontañes were ambling about in apparent boredom. Most circled round and round a frosted stalagmite of greenish ooze. We could tell they had been doing this for a long time as their repetitious motions were wearing a circular trench into the soil.

"They seem mentally ill," Kalinda whispered to me. I nodded. How could the Leandra have considered them as great philosophers? Finally, one of the Heñtriacontañes came forward and stared at Kalinda.

"Hello," Kalinda nervously said to the Heñtriacontañe.

"I have a small gift for you," said the Heñtriacontañe. The Heñtriacontañe displayed a vapid insectile grin and offered Kalinda a long-stemmed, luminous blossom.

"Maybe you shouldn't accept the gift," I said, remembering that the German word *Gift* meant poison. Kalinda accepted anyway.

"I'm honored by your thoughtfulness," she said to the Heñtriacontañe. The Heñtriacontañe then sprayed Kalinda with a foul-smelling substance which gushed from the pores of its body like oil from a well.

"Hey!" Kalinda screamed as the oleagionous oobleck dripped from her clothes. The Heñtriacontañe seemed startled.

"I'm sorry if I offended you, Madam. Isn't it customary for your species to spray pheromones onto new acquaintances of the opposite gender? I am male. Are you not female?"

"Our pheromones are slightly more subtle," she said.

The Heñtriacontañe had the beautiful warm eyes of Robert Redford, but unfortunately that is where the similarity ended. Above his eyes was a deep fleshy head shield, shaped like a helmet. His body consisted of seven segments diminishing in size towards his rear which was about the size of my little finger. Two very long whiplike extensions were attached to the last segment and protruded from the Heñtriacontañe's ultra-tiny rump.

We spent the next half hour talking with the Heñtriacontañe. When they weren't studying philosophy, they spent much of their days investigating a game played on a simple looking playing board. We spent a few hours studying the game ourselves.

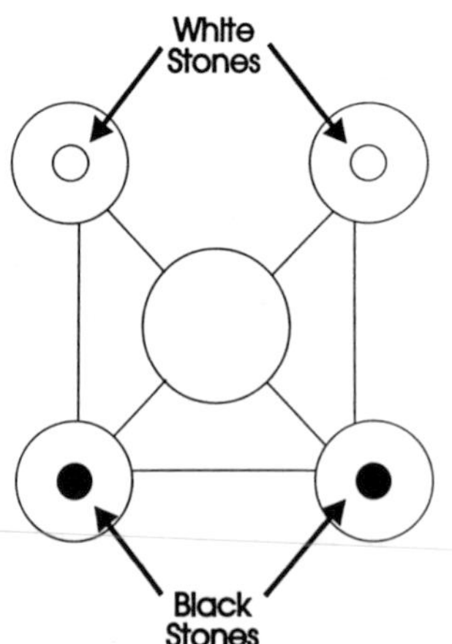

Apparently status in the henotheistic Heñtriacontañe society was based on the prowess with which an individual played the board game. The size of the two whips protruding from an individual's rump was proportional the individual's mastery of the game.

"Shall we try a game?" Kalinda asked me. Globs of oobleck still remained on her hair.

"Sure."

The Heñtriacontañe sketched out the board in the Ganymedean soil with a stick, and Kalinda and I sat on the ground at opposite sides of the diagram.

"You use the black playing pieces," I said to Kalinda. "I'll use the white pieces."

Looking at the playing board, we were easily deceived into thinking that the game was easy. But such was not the case! The Heñtriacontañes called it the "Armacolite Parasite Game," because they used little armacolite stones as playing pieces. Each stone represented a parasite moving through the body of a primitive organism. We tried the game using two black armacolite stones and two white armacolite stones. Kalinda began by placing two black stones on the top two circles. I placed two white stones on the bottom two circles. Kalinda, the owner of black stones, moved a stone along a line to an adjacent empty point. Then I moved one of my white stones. We alternately moved one stone at a time along any line to an adjacent empty space. Kalinda's aim was to block my stones so that I could not move. And my aim was to block hers.

Kalinda won a few games.

"You're a tough one to beat," I said.

"Let's hope I don't start to grow long whip tails like theirs," she replied.

"You certainly are good at it," said the Heñtriacontañe. "I find that somewhat attractive in a lady."

I scowled at the insectile philosopher. "I'm getting tired of this," I said to Kalinda. "Let's see what their Imaginarium is all about."

"May we see your famous Imaginarium?" Kalinda asked the Heñtriacontañe.

"Certainly," he responded. "I'll bring someone to lead you there. When I return with your guide, please show him the utmost respect. Bow to him. He's especially proficient at the Armacolite Parasite Game." The Heñtriacontañe walked away with his rump tightly coiled against his body.

A few minutes later, a Heñtriacontañe with two particularly long whip tails came up to us. Others bowed in his presence, and so did we. The creature was rail thin, cadaverous beneath an elegant exoskeleton which glistened like emeralds. He held out a forearm to Kalinda, all bone and hair, smiling with dozens of teeth. Kalinda shook his hand.

"My name is Hx," he said. "I'll be happy to show you the Imaginarium. Please follow me." The hippophagous Hx led us to an unusually shaped building.

As we entered the Imaginarium with Hx, his luminescing body cast a pale light on the ancient stone walls. I was surprised to see that the floor of the Imaginarium resembled a bushy, cactus-shaped object from geometry called the Mandelbrot-set. The building had a turret here and there, and some of the upper stories had windows with green panes of a material I didn't recognize. Double doors, swinging on posts made from the crystal trees of Ganymede, opened upon a shaded court. Stairways led to the upper floors.

Hx turned to Kalinda. "How do you do, Madam?"

Kalinda smiled. "Fine, thank you, except for the pheromone bath. What is this place?"

"All of our people – painters, metalworkers, potters, sculptors, and architects – have fused their skills in producing this assembly of chambers we call the Imaginarium. It's the center of all Heñtriacontañe life." Hx paused. "Would you like to attend today's show in the Imaginarium? It's called 'the four questions'." Hx gazed over my head, as if there were something or someone behind me.

"May I ask some questions first?" Kalinda said. Hx nodded his insectile head. "What material did you use to create this building?" She banged on one of the walls with her knuckles.

"The creators of the Imaginarium only had a few materials available. Ganymede doesn't have many strong rocks. We used ice and a gypsum-like material."

I noticed that the blocks of material which formed the Imaginarium walls were cut so sharply that they could be put together without mortar. Around a central court, about thirty thousand square feet in area, were spacious stairways of ice and a rambling maze of rooms.

"What's in there?" I pointed to another chamber.

"It's our knowledge producing factory," Hx said. "Here's where Heñtriacontañes add to the total knowledge of our species by proving philosophical and mathematical theorems."

"What's the incentive?" Kalinda said.

"Incentive?" said Hx.

"Why do they do it?" Kalinda said. "Do they get paid? Do they do it just for the love of knowledge?"

"Ah, I understand. The reward for every new theorem proved is sex. The duration and novelty of each sex act is in proportion to the degree of difficulty and novelty of the theorem proved."

"Interesting idea," Kalinda said.

"We determined long ago that this approach produces an extremely intellectual and educated populous. It also eliminates many of the social problems which would normally face urban populations."

I peeked into the knowledge producing factory. On one wall was a blackboard-like object with words and symbols:

91. Theorem 17.—
The locus of points equidistant from the sides of an angle is the bisector of the angle.

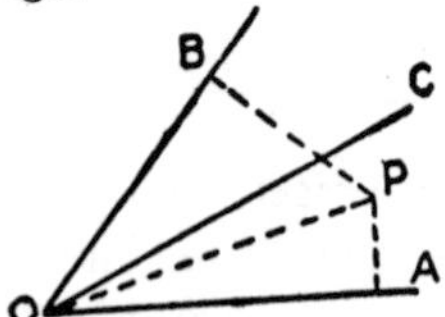

Hyp. OC is the bisector of the angle AOB.
Con. OC is the locus of all points equidistant from OA and OB.

1. Let P be any point such that its distance PB from the line OB equals its distance PA from the line OA. Draw OP.	1. Def. §90.
2. In △OAP and OPB,	2.
OP = OP,	Identical.
PA = PB,	Assumed in **1.**
∠PBO = ∠PAO = a rt. ∠.	Def. §90.

On the floor, beneath the mathematical jargon, were two Heñtriacontañes, embracing, their dual whips intertwined. I looked at another wall and saw:

152. Theorem 44.—

If in the same circle or in equal circles, two arcs, each less than a semicircle, are unequal, the chords subtending them are unequal; and the greater arc is subtended by the greater chord.

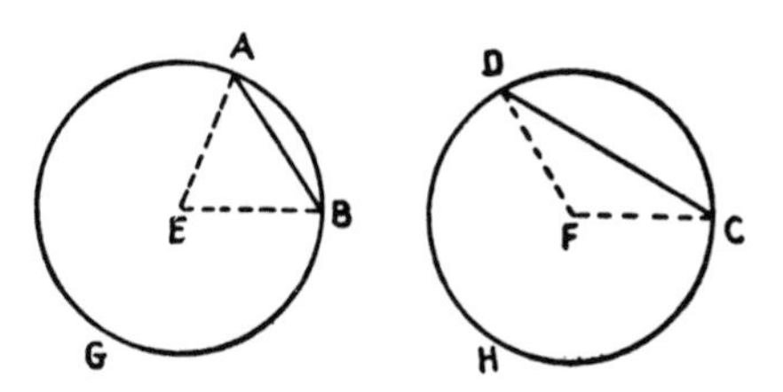

Hyp. Given the equal circles E and F with the unequal arcs AB and CD respectively, with minor arc DC greater than minor arc AB.

Con. Chord DC is greater than the chord AB.

Draw the radii, EB, EA, FC, and FD. If the two circles are made to coincide, with FC coinciding with EB, then since arc DC is greater than arc BA, angle CFD is greater than angle BEA.

I gazed up at the ceiling and noticed two Heñtriacontañes swinging from the chandeliers. They screamed some unintelligible yet erudite-sounding mumbo-jumbo concerning philosophy. Thixotrophic oobleck was everywhere.

"Ahem." Hx coughed to get our attention. "Why don't you take a seat?" He motioned us to sit in some nearby velvet chairs. He then slammed shut the door to the knowledge producing factory.

"Your knowledge producing factory is quite interesting from what I could see," I said. "Do they have a list of theorems to prove?"

"Yes, please sit down," Hx replied.

"Is it a long list?"

"Twice as long as half of it," Hx said with a serious expression.

The chairs were fairly comfortable, considering they were designed for a race with rear ends that looked like triangular tortilla chips. I looked around and felt dwarfed by the soaring pilasters, frescoed ceilings, and gilt furniture. Hx then handed us a slightly soiled slip of paper.

"Please select one of the following options." On the paper were a list of four unusual scenarios. They were:

1. Images in a mirror will no longer be reversed.
2. The color you perceive as blue will now be perceived as red.
3. An organ, selected at random, will be instantaneously removed from your body.
4. An object, selected at random during one of your dreams, will be removed from your dream and placed by your sleeping body when you awaken.

I looked up at Hx. "What is this, some kind of test?" He ignored my question.

"Please consider the four options. Now tell me: Which of the four scenarios would you want to come true? Based on your answer, you'll either be permitted to leave the Imaginarium or be kept here as our prisoners." He gazed at both of us as if judging us, coldly and calculatingly.

"Are you some kind of nut?" I said. I rose to my feet. Kalinda held me back. "Kalinda, he can't hold us as captives. And he certainly doesn't have the power to make any of the strange situations come true."

"I know," she said. "But let's go along with him. I know you're strong enough to beat him. But I'm curious to see what happens when we consider the questions. Think of it as an intriguing mental exercise."

"'Intriguing mental exercise'?"

"Give it a try. Let's start with the first scenario on his list – the question about mirrors."

"Well, OK."

"What effect would a change in the property of mirrors have on the physical fabric of the universe?"

"Kalinda, I don't have the slightest –"

"OK let's eliminate option 1. Maybe it could be dangerous. How about scenario 2?"

"I suppose a switch in colors would cause us only a minor inconvenience. It wouldn't alter other people's lives," Garth said.

"Unless the colors really did change. Think what would happen to the oceans and sky of Earth, and the absorption of harmful radiation..."

"Hx said *perception*, so the colors wouldn't really be different."

I saw Hx grin and start playing with his butt. I looked back at Kalinda. "Right. How about option 3, the one about organ removal?"

"Now that seems like the worst option," Kalinda said.

"Yeah, but remember that certain organs of the body could be removed without any harm, such as a single kidney."

"What are the odds? I can think of one organ you'd miss," Kalinda grinned wickedly. "Anyway, Hx said *at random*, remember? How many can you do without?"

Suddenly, directly in front of us was the gruesome and utterly lifelike image of an acerebrate Thorn stumbling around without a brain.

"What the hell?" I said.

Kalinda shifted uneasily in her chair. I gave Hx a suspicious, sideways glance. What had projected that image in front of us?

I pulled Kalinda close. "It seems as if the Imaginarium is somehow translating our thoughts into lifelike images," I said. "Be careful about what you're thinking."

A smirk appeared on Hx's insectile face.

"Garth, how can I stop thinking about it? I can't think about the third scenario without thinking about –"

"Yeah, I know, and when you think about the skin, the largest organ –"

Kalinda threw back her head and screamed a guttural cry of terror. Again, without warning we saw an image of Thorn, with no skin. All his musculature was intact, and he was still alive. I saw some blood vessels and nerves, but there was no bleeding.

"Uggh, could someone live without skin?" I said.

"I'm not playing this damn game anymore."

"This isn't real." I remind Kalinda. "Just an image. I guess if the skin were carefully removed, few blood vessels would be severed." The image of Thorn stood very still, as if Thorn feared that movement would cause damage to his body. Unfortunately the Thorn in our lifelike vision could not help damaging the bottoms of his feet as he stood upon them.

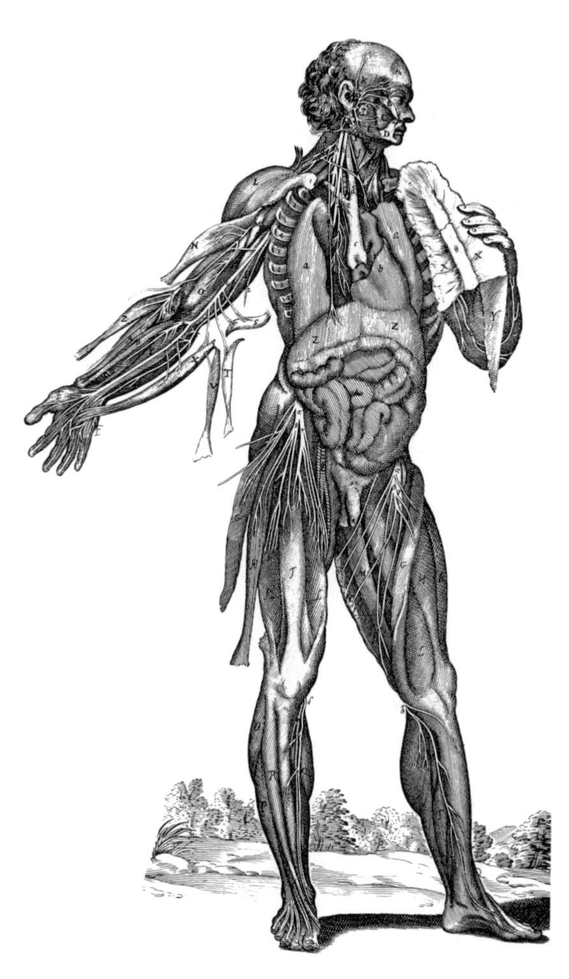

I saw all the superficial muscles of his shoulders. They extended from his trunk to the pectoral girdle, and from the trunk and girdle to the proximal portion of his humerus. I saw his thin salivary gland lying on the side of his neck, deep into the platysma. His external jugular veins quivered and his massive carotids pulsed ruby red. The loose connective tissue, the superficial fascia, could easily be torn if Thorn moved quickly.

"Please don't move," Kalinda whispered to the image of Thorn.

"Kalinda, he can't hear you. Just an illusion."

Hx smiled, as if he were delighted with the horror on our faces.

"OK, let's get this stupid game over with," Kalinda said. "Let's go to the next question."

"OK, I think we'd choose the last option. It's unlikely we would bring anything out of our dreams that would be harmful."

"How can you say that?" Kalinda said "Suppose you were having a nightmare. There could be a monster, a ravenous madman –" Kalinda said.

"But suppose it was some exotic jewel, or one of the Raquel Welch android series," I smiled at Kalinda. Kalinda did not smile back. "Imagine if you could control your dreams and bring back whatever you wanted."

"I can't. Mine are too unpredictable, especially the nuclear holocaust mutant ones."

Suddenly, there was a projection of Thorn as he might look one-hundred years in the future. His triple eyes looked weary with age, and most of the hair on his head was gone.

Hx ambled came closer. "I think you have had enough of this." he said. "My purpose was to assess the similarity of your psychological makeup to ours. Through your responses – your concern for your race and your friends, and your hesitancy about options 1 and 4 – I can see that you're a thoughtful and intelligent race. Your remarkable restraint in the face of my arrogance was also appealing."

"This is crazy," I said, both intrigued and disgusted by the whole absurd exercise. "How did you project the image of Thorn?"

"Ah, I can't give away all our tricks."

"I think we've spent enough time here," I said. "We've got to get going." I got out of my chair.

"Wait, I have a new set of questions for you," said Hx. "You may not leave yet! Here, listen to these."

1. All colors that you perceive now will be perceived as their complements.
2. Humans will now perceive sound as smell.
3. All animals on Earth will now perceive ultra-violet and infrared.
4. Florida will disappear.
5. $E = mc^3$

"Which of these would you prefer to come true?" he said. "What affect would these have on you?"

"Thank you," I said as we rose to our feet. "But that's enough for us today."

Hx's face turned red and he shouted, "Guards!" Then he turned to me and said, "Sit down, buddy."

Ten large Heñtriacontañes came and blocked the exit from the Imaginarium. Hx grabbed Kalinda and squeezed her arms together.

"Let go of me!" Kalinda screamed at Hx, a sudden declaration of war between them.

"I'll tell you what," I said to Hx. "I'm going to kick your legs out from under you if you don't let Kalinda go free. The lady wishes to leave with me, and I'm taking her with me."

"Just stay where you are," Hx said to the guards. One guard started singing a song, "They shall never leave. They shall never leave," and the other Guards took it up. Their crazy harmonies made them sound like a cross between Madonna and a dozen parakeets being vacuumed from their cages. (Don't ask me why Madonna's music was still popular so long after her suspicious death in the year 2010.)

"This is your last chance," I said.

"Guards," Hx said. "Continue singing. It'll distract the prisoners."

I aimed a kick at Hx's right foreleg. He fell backward against the icy wall of the Imaginarium. As Kalinda and I walked to the exit door, I went over to Hx and picked up the pompous, pilgarlic, popinjay by his superthin rump. I tossed him into the Imaginarium seats. Hx gasped and keeled over. The "guards" on either side of him did not try to detain us. In fact they moved sideways to make room for us to pass through the exit door. After leaving the Imaginarium, we briskly walked towards a nearby forest.

"I'm glad to get out of –" Kalinda never finished her sentence, for at this moment a loud crash shook the forest. The next moment Heñtriacontañe soldiers came running through the trees, at first in threes and fours, then five and ten together, and at last in large troops that seemed to fill the woods. They were evidently searching for us.

"Let's hide behind a tree and watch," I said.

We had never seen warriors so unsure on their feet. They were continually tripping over something or other, and whenever one Heñtriacontañe warrior went down, several more fell over him. Soon the entire sward was covered by small clumps of Heñtriacontañes. Their posterior tendrils waved like kites in the breeze.

"They don't seem to have had much practice at this," Kalinda said.

We saw Hx step out from the Imaginarium and cry, "Get them, you fools! Bring out the Nematomorphs!"

He then blew into a spiral instrument which resembled a horn.

"I remember hearing one of those blown that way at my Bar Mitzvah. Or was it on Rosh Hashanah?" I said.

But this was no religious ceremony. From a side door of the Imaginarium came animals with caterpillar-like torsos. These Nematomorphs had dark brown or black bodies and two large eyes. We tried to count the pairs of legs on each Nematomorph, and found that although the number varied, it was always a Fibonacci number like 5 or 8. Each bore a Heñtriacontañe on its back.

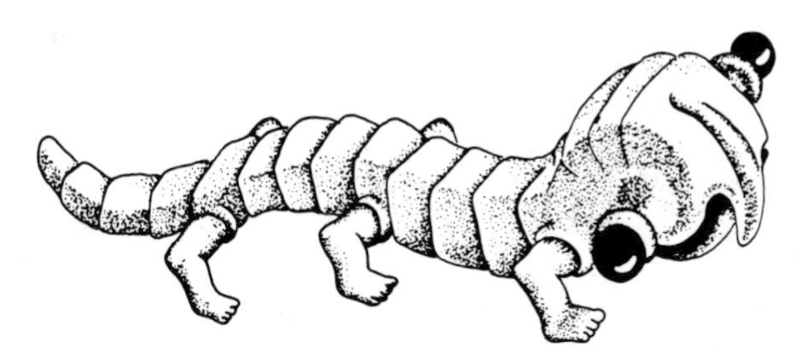

Having steeds with many feet, the morph riders managed better than the foot soldiers; but even these mounted warriors stumbled occasionally. It seemed a rule that whenever a morph stumbled, the rider fell off instantly. The confusion escalated. When a morph became riderless, it curled up into a corkscrew which could not be reopened, even by the effort of several Heñtriacontañe soldiers pulling with all their might.

"You fools!" Hx said again. His face turned red as the wild pandemonium about him grew worse every moment. Some Heñtriacontañes seemed to forget their mission and began to become sexually interested in one another. Oobleck was everywhere. By the time we decided to leave, some of the morph riders were knee-deep in the horrible goop. Others executed marvelous pirouettes atop the morph backs and snapped their whiplike tails at each other in an emotion that looked like either joy or anger. One of the creatures carried a flag, ran with his friends to the top of a hill, and slid down while yodelling. I didn't really care what was on their minds.

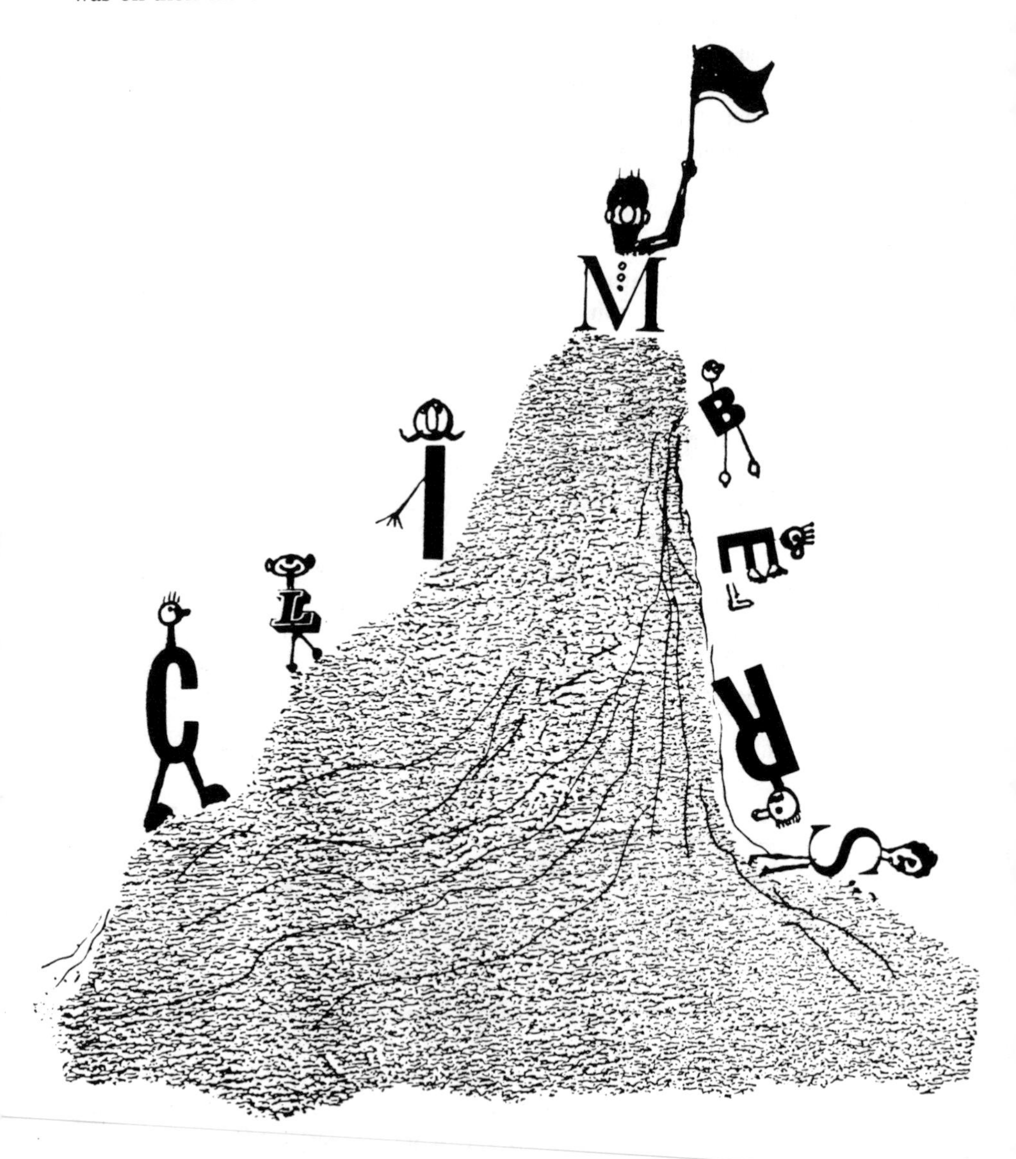

Chapter 33

The Leapers

When I was thirteen years old, a boy at school demanded my dessert every lunchtime. He was bigger than I so I always complied with his diuturnal demands. When I told my parents, my mom actually offered to make cookies with tobasco sauce so I could give them to him. I was too much of a wimp to fulfill the plan, but as an adult I often fantasized about the hypothetical outcome of his consuming the spiced up cookies. As a result of such early harassment, I learned tae kwon do, and got my black belt by the age of sixteen. This was one of the best decisions I made in life, and the defensive skills turned out to be very useful as I explored new worlds.

The turbulent sky roiled above us. Masses of clouds, painted a dozen shades of vermillion, raced from horizon to horizon. Occasionally we heard the cry of a bird in the sky, and then all was silence. We had left the Heñtriacontañes' land and had been walking for just a few minutes when Kalinda stopped.

"Look over there, Kalinda."

"A swarm of insects?" Some were flying through the air on feathery bodies, while others were hopping around the ground.

"I think I came across some of these critters while flying through the gases of Jupiter. They're like mosquitos on Earth. They suck blood from other animals. A real pain..."

"Will they hurt us?"

"I think no more than an ordinary mosquito."

"Let's get a closer look," Kalinda said while withdrawing a net from her backpack. I also took out my own net.

"Some seem to be getting curious," Kalinda said as a few started to hover near our bodies.

"'Curious' might be the wrong word for something as dumb as a mosquito."

"How in the world do they hover like that? They don't even flap their feathers!"

"I recently read an article about them in *Science* magazine," I said. "Each Leaper uses nine tiny thrusters on their abdomen to control movement. The thrusters expel high-speed bursts of warmed hydrazine gas. The gas steers the Leaper. A microprocessor-like brain distributes the gas to the appropriate thrusters."

"The thrusters must be tiny."

"About a millimeter in length, but they fire more than 100 times a second. Each thruster generates a pound of thrust in pulses of power lasting one-hundredth of a second."

"Ouch! It bit me in the leg."

"Let me catch it." I ran to her and caught the Leaper in my net. "You OK?" I placed the Leaper in a specimen jar where it continued to bang into the glass walls as it attempted to escape.

"Garth, there's more of them!" Kalinda said as she started hopping up and down. Her voice was uneasy, spiced with irritation. "Ouch, something's in my boot!" She took off her boot, and swatted at a Leaper that had taken up residence by her ankle. I rushed over to her and looked at her leg. "Looks like a small burn from the hydrazine thrusters. I guess when it lifted off, it burned you."

"Great," she screamed. "Another's inside my cloak." I slammed my hand into her ribs and saw a Leaper that was beneath her cloak fall to the ground. Its broken body vibrated slightly on the icy soil. Occasionally a small burst of fire shot out of its damaged thrusters.

"Might make a good cigarette lighter," I said.

Kalinda punched me in the arm.

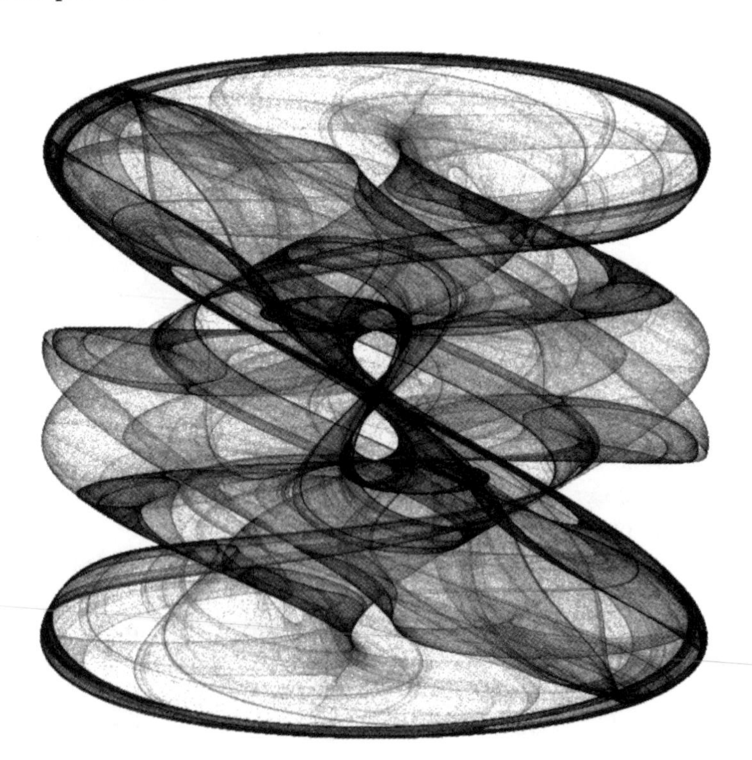

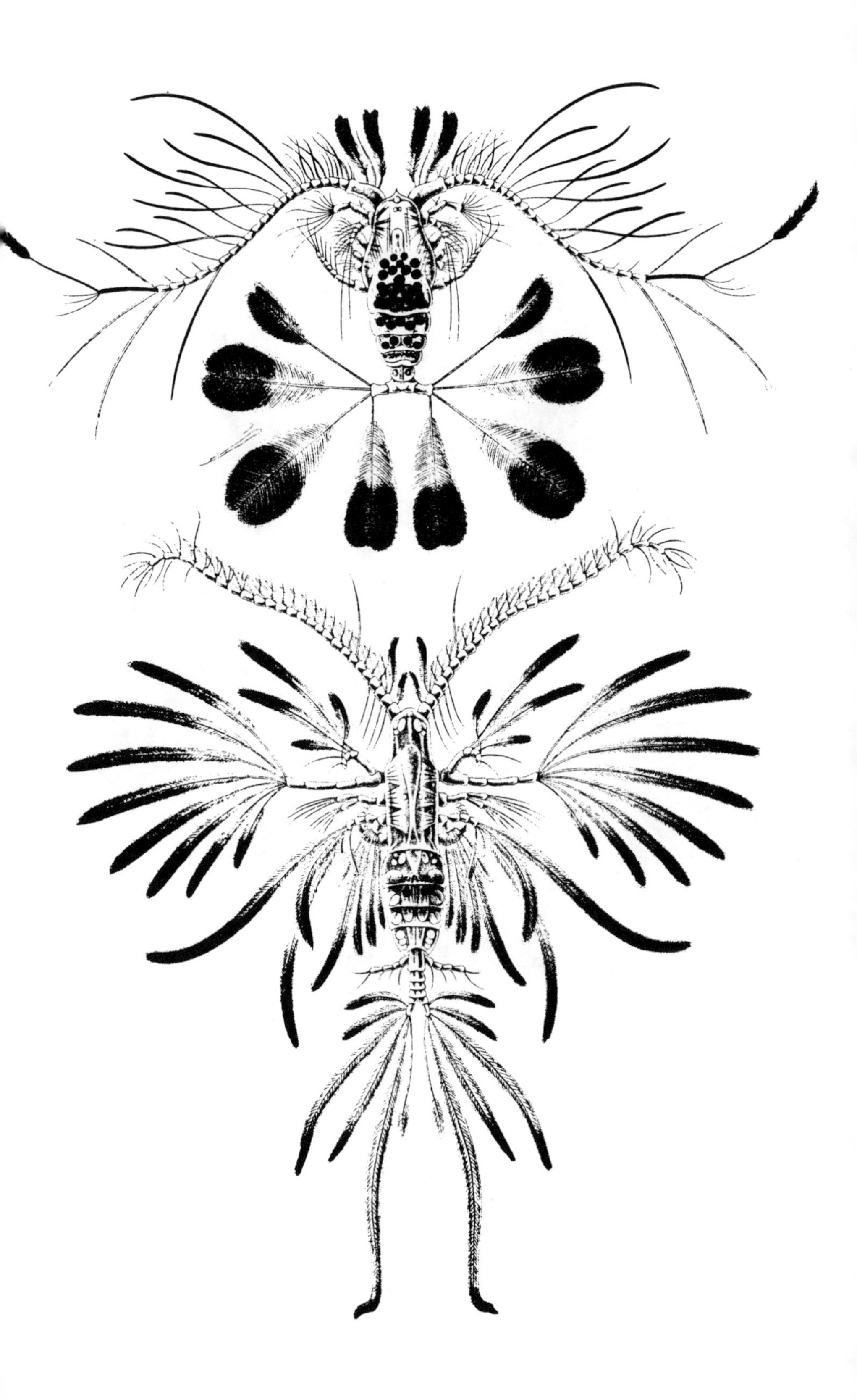

Chapter 34

The Fractal Palace of Ice

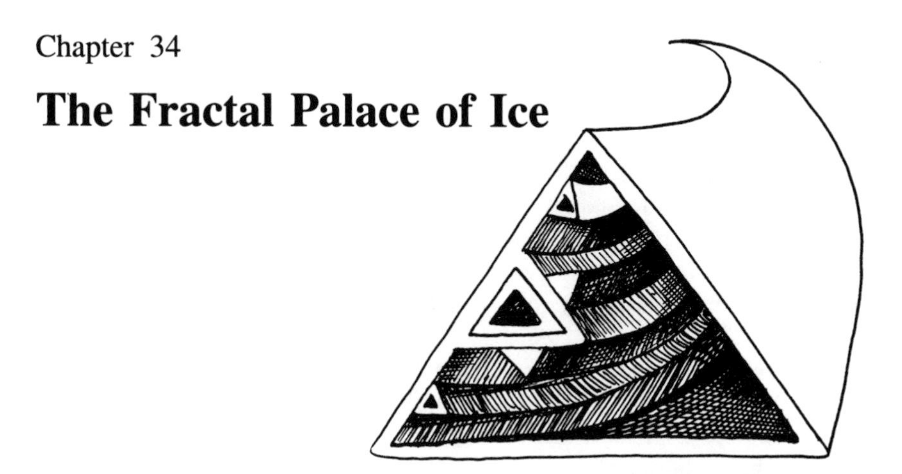

"When the universe has crushed him, man will still be nobler than that which kills him, because he knows that he is dying, and of its victory the universe know nothing." B. Pascal, *Pensees, No. 347*

I thought there was hope for Kalinda and me as a couple. We weren't all that different. She could appreciate many of my Earthly interests. For example, despite her unusual previous experience as a digestive symbiont, she immediately responded to Mozart's music I had played to her. When I played "Can't Stop the Classics," a funked-up version of classical excerpts, she moved and swayed to the music as would any human. If she could appreciate human music, speak idiomatic English, laugh at my jokes, and enjoy my company, what more could I ask for? Well, maybe the real question was: what was *she* looking for?

We left the land of the Leapers and resumed our travels along a purple sward.

"What a landscape!" I said. The treetops stirred with the whisper of a cool breeze. Occasionally I heard the muted crack of icicles.

"Wonderful!" Her smile lighted her up from the inside, like candles in a pumpkin.

"Smell the air," I said as I reveled in the zesty citrus smell which permeated the air.

"Maybe the scents are coming from the trees." She pointed to some trees which resembled French horns.

I felt blissfully happy, fully alive. Looking up, we saw a beautiful turquoise sky filled with a gold radiance. Again I heard the words of poet John Celestian echoing in my mind: *I only wanted to feel your closeness, so close as to inherit a sandy spot upon your lips. Your beauty is your knight's defense, your face his shield, see my eyes; they speak truths.*

Suddenly Kalinda stopped walking. "Garth, something's moving in my pocket."

"What –"

"Aaah," she screamed.

"Here let me take a look. Hold still." I reached into her pocket with my prosthetic arm, and grabbed onto something soft. "It's - it's a baby Navanax. The egg you found earlier must have hatched. Congratulations, you're the proud mother of a baby boy. At least I think it's a boy."

Kalinda smiled and relaxed. "Remember, they're hermaphroditic. Boy and girl in one." The Navanax turned its tiny eyes towards me and cooed softly. "Garth, I think it likes you." The foot-long *Aysheaia* in my pocket poked it head out to watch. "What do we feed it?"

"Let's try one of the raspberry roots I found in the mole hole."

"Got any left?"

"Here's a few." I reached into my pocket and found a root. "It's worth a try. You hold the Navanax, and I'll hold the root in front of its cute slug face."

"Garth, you have to break it into small pieces first."

"Break the baby Navanax?"

"No! The root."

"Oh, right." I tore the root into a small assortment of tasty morsels. The Navanax baby sniffed them, and then began to eat. Its piezoelectric mouth sucked them in whole, like a vacuum cleaner nozzle.

"He likes it," I said.

"Let's give it a name."

"A name? For an ugly Navanax?"

"Sure, why not? Anyway, the baby's not ugly."

"OK, how about Terrie, after your sister?"

"Name a slug after my sister?"

"You're the one who said the baby wasn't ugly."

"How about Nina?"

"Are you sure you want it to have a girl's name?"

"Why not? It's part girl, part boy."

"Alright, Nina, it is," I said. After Nina fed, I placed her on my shoulder where she was content to perch like a parrot on a pirate. She grasped my shoulder with a musclelike prominence on her posterior.

"OK, let's keep walking," I said.

"Garth, do you think our sexual organs are compatible?" Kalinda asked. I stopped suddenly. Her question was unexpected and shocking, but amusing. I coughed.

"What did you ask?" I said.

"I've been thinking. I'm from a totally different planet than yours. Our reproductive biologies could be very different, making us incompatible," she said.

I was pleased that Kalinda would consider the question of compatibility. It suggested more than a friendly or collegial interest. But it was totally beyond my ken to imagine that we had significant physically differences. Surely she was just kidding.

"I have a feeling that we're compatible." I said. "You looked quite, um, humanoid when I saw you bathe. But it's true that I've been thinking about your biology."

"My biology?"

"Haven't you wondered how you could be so humanlike if your species evolved as digestive symbionts. Do you think it's possible that your people descended from human explorers who came to your planet a long time ago?"

"It's possible. Why don't we examine one another to determine our compatibility."

"Kalinda, is this your way of saying you want –"

"I just want to make sure."

I stared at her for a second, trying to read her expression. I couldn't tell how serious she was.

"OK, if you insist."

Within seconds, Kalinda was naked, wearing nothing but her Nautaloid egg necklace. I thought her skin was perfect. Her curves were perfect.

"Yes, Kalinda, you – you're perfectly humanoid."

Now she smiled. "Are you sure?" She sat down on her cloak. "Take a closer look to be sure."

"You're just fine. Wonderful."

"Then there's only one thing left to do to make absolutely sure."

"Do you mean –"

"Shut up and take off your clothes. I want to closely examine you too."

I clumsily removed my shirt and pants. She came closer.

"Ooh, what an examination!" I said. "Wh – Where did you learn to move like that?" I reached a thumb down the nape of her neck. Then down her smooth stomach. She reached behind me and rubbed my thighs. I felt a feathery touch at the base of my spine, as if she had stroked me there too. From the corner of my eye I could see Nina and the Aysheaia worm staring at us.

"Uh, we have a couple of voyeurs watching us," I said.

"Who cares. Lie on your back," Kalinda said. I looked into her sparkling eyes. Her gaze was so intense that her eyes looked like they could ignite anything she stared at for a few seconds. "Oh God," I said. Kalinda made little cat-like crying sounds. I felt the cold world of Ganymede below me and a warm, wonderful Kalinda above.

"I love you, Kalinda."

"I love you too."

Without disturbing Kalinda's balance, I removed my artificial arm. I set the hand to vibrate at 500 cycles per second in a perfect sinewave with occasional phase shifts.

"Garth, I –" Every second I had a random number generator in the arm's central processor generate a random spike causing the index finger to momentarily oscillate at a random frequency before returning to the preassigned frequency.

"Garth –"

I groaned and saw Kalinda's pupils alternately dilate and contract. For few seconds I felt her whole body oscillating at rates near submultiples of 500 Hertz, and then we both were still. There were tears in both of our eyes. We held each other for a long time – oblivious to the cold, the Navanax, and the worm from the world of pipes. I held her hand tightly and promised myself I would never let go, never leave her alone and unprotected.

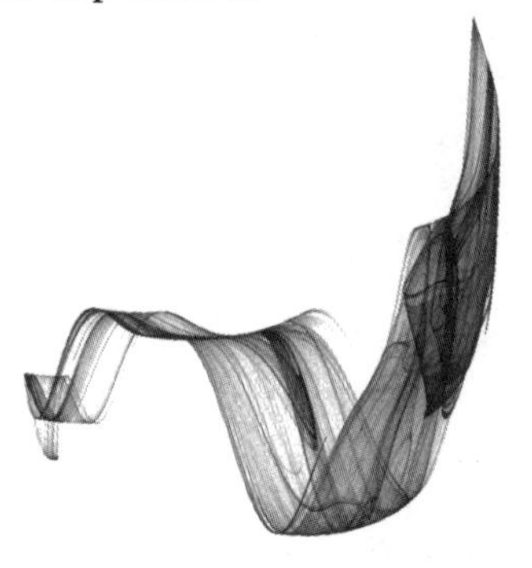

After dressing, we walked the next few miles in silence. I squeezed Kalinda's hand and smiled at her.

"Wonder what lives in those holes," I said and pointed to fist-size holes in the soil.

"They're too small for the mole people to live in."

"Let's not bother with the holes. I want to find Yars Kotheck."

"I wonder if we're travelling in the wrong direction," Kalinda said. She stooped down to massage her sore feet.

"Wanna rest?" As I asked, I saw some ruby-colored towers in the distance. "That must be it! The land of the Latööcarfians." Kalinda followed my gaze and quickly rose to her feet.

"Hey, no way I'm stopping now. Let's go." She took my hand and we walked for a few minutes. "How can you be sure it's them?"

"Take a closer look at the ground."

"My God. Look at all of them."

Scattered in haphazard piles for as far as we could see were flat pieces of stone engraved with the beautiful swirling patterns of the Latööcarfians.

"It's as if they're advertising themselves to anyone wandering nearby."

"Possibly, but who knows? This could just be their discarded mathematical experiments – a mathematical garbage dump."

"I could just see them slinging their mathematical mess over the city walls." Kalinda laughed. "Pretty weird."

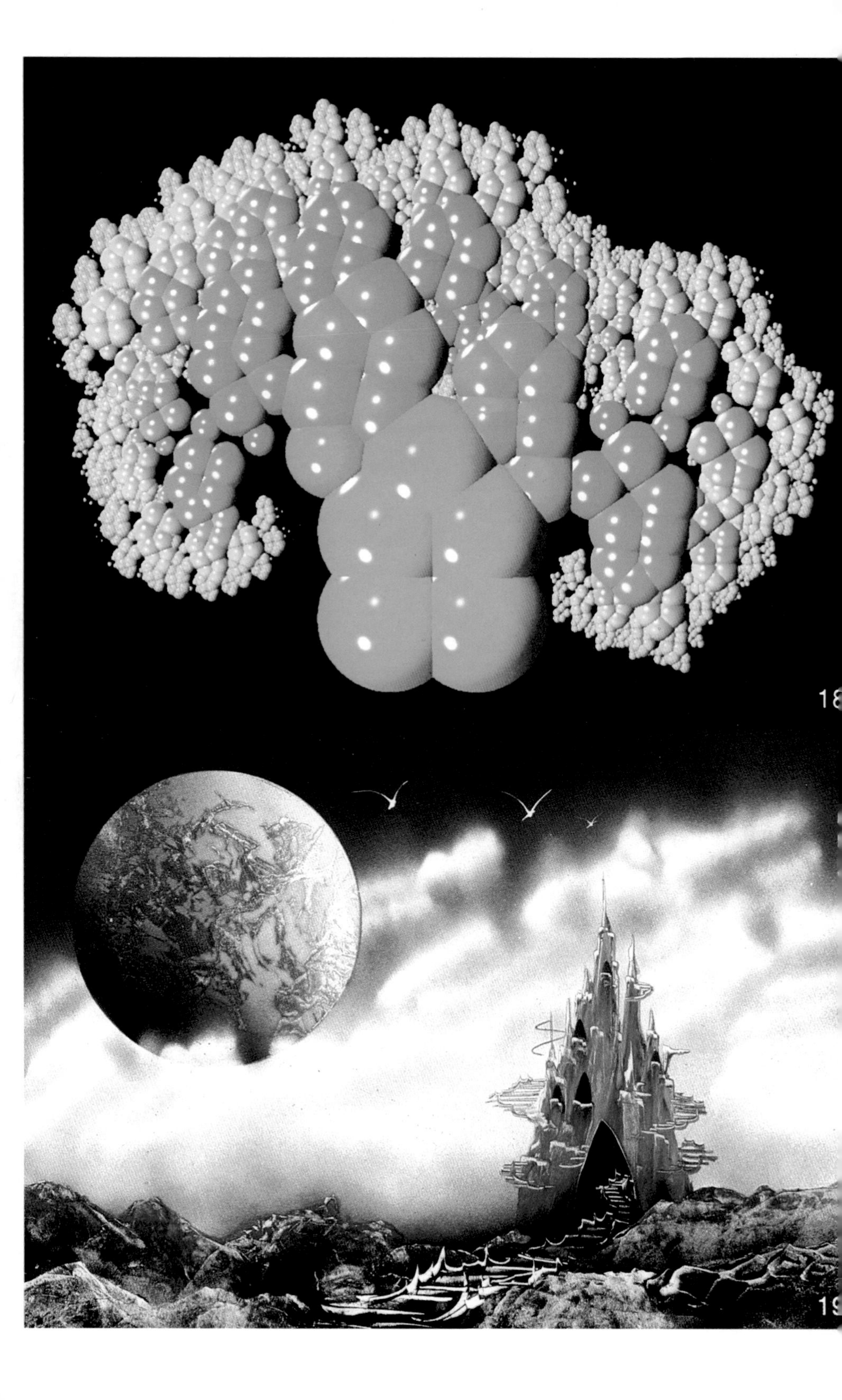

18

19

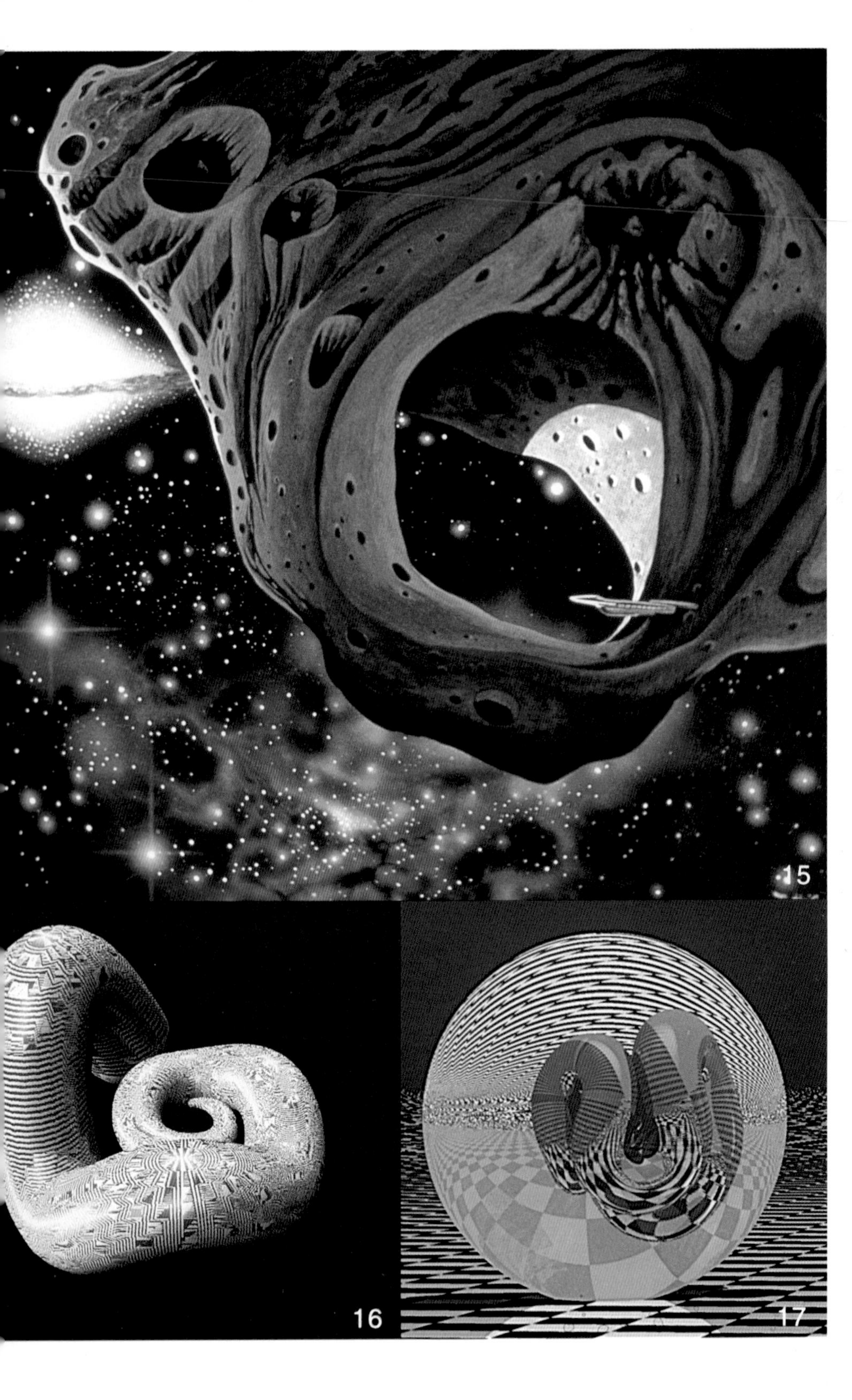
15
16
17

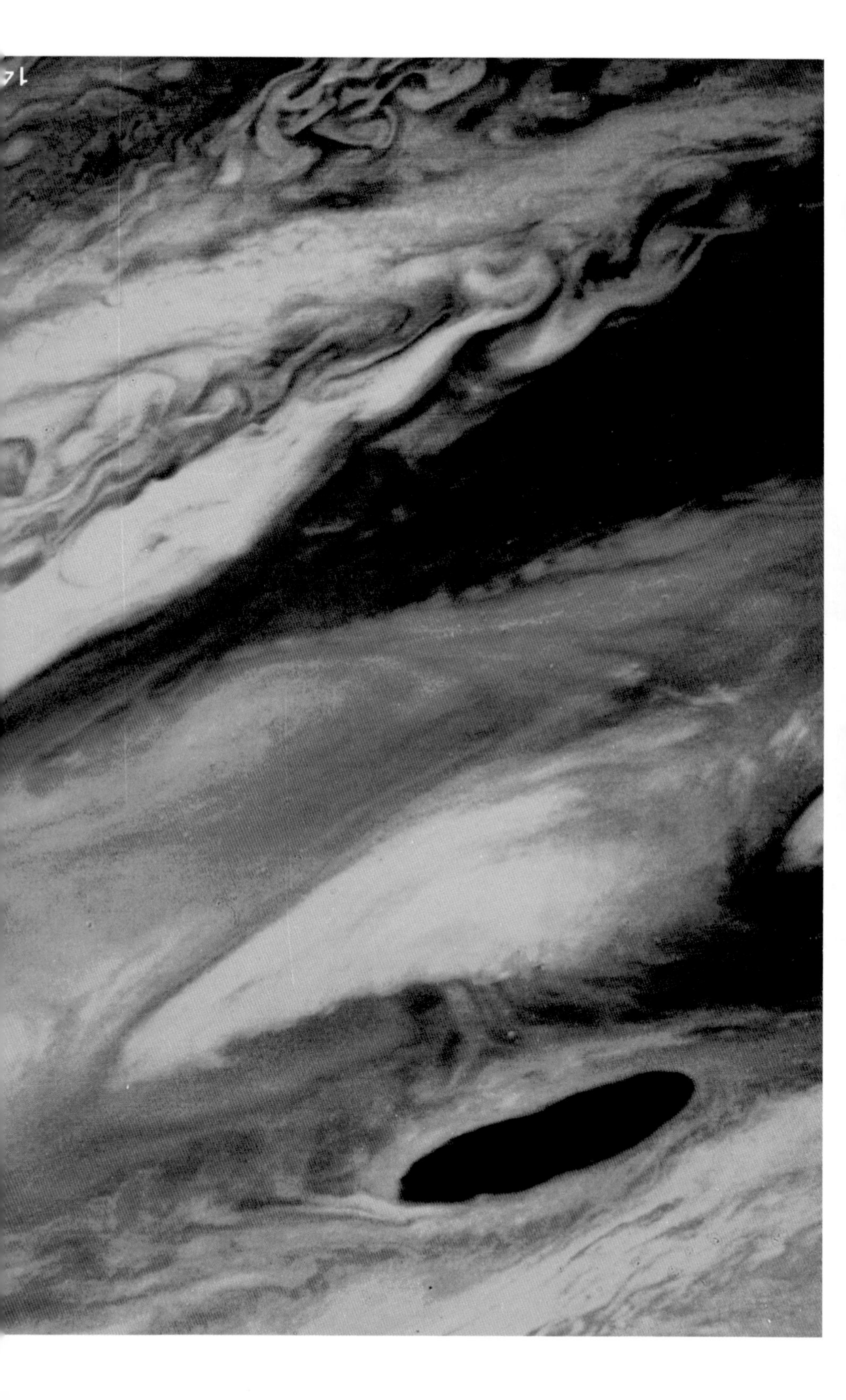

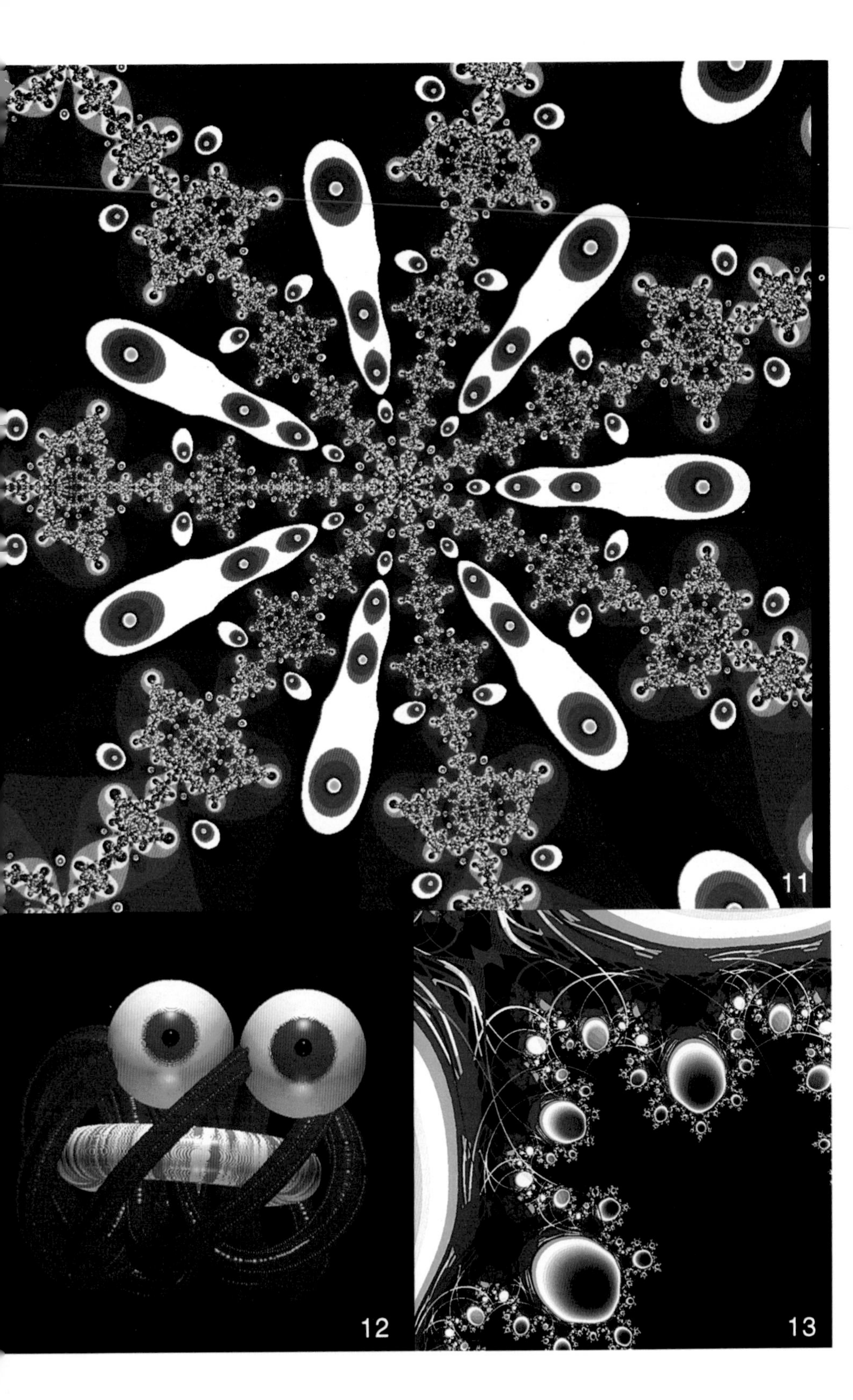
11
12
13

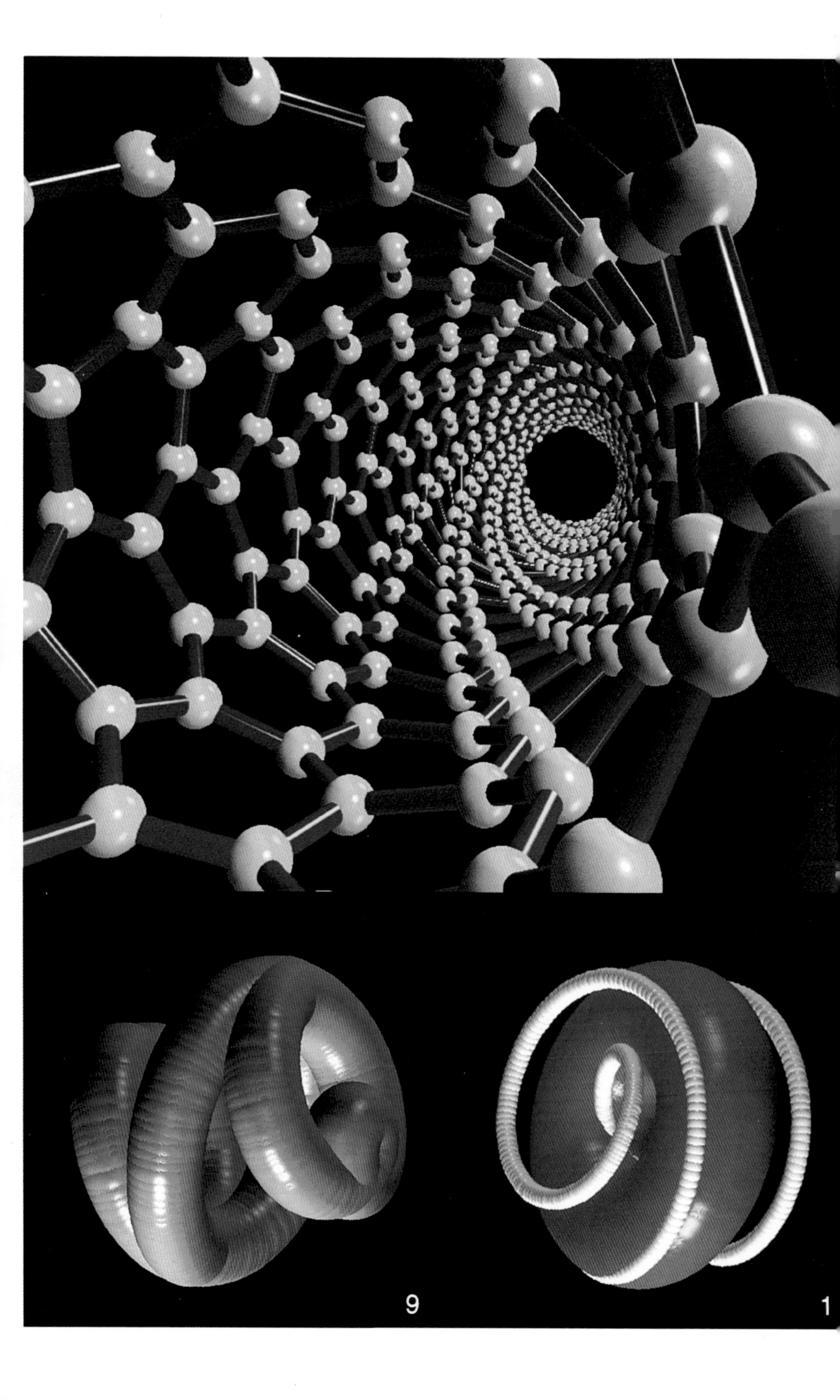
9
1

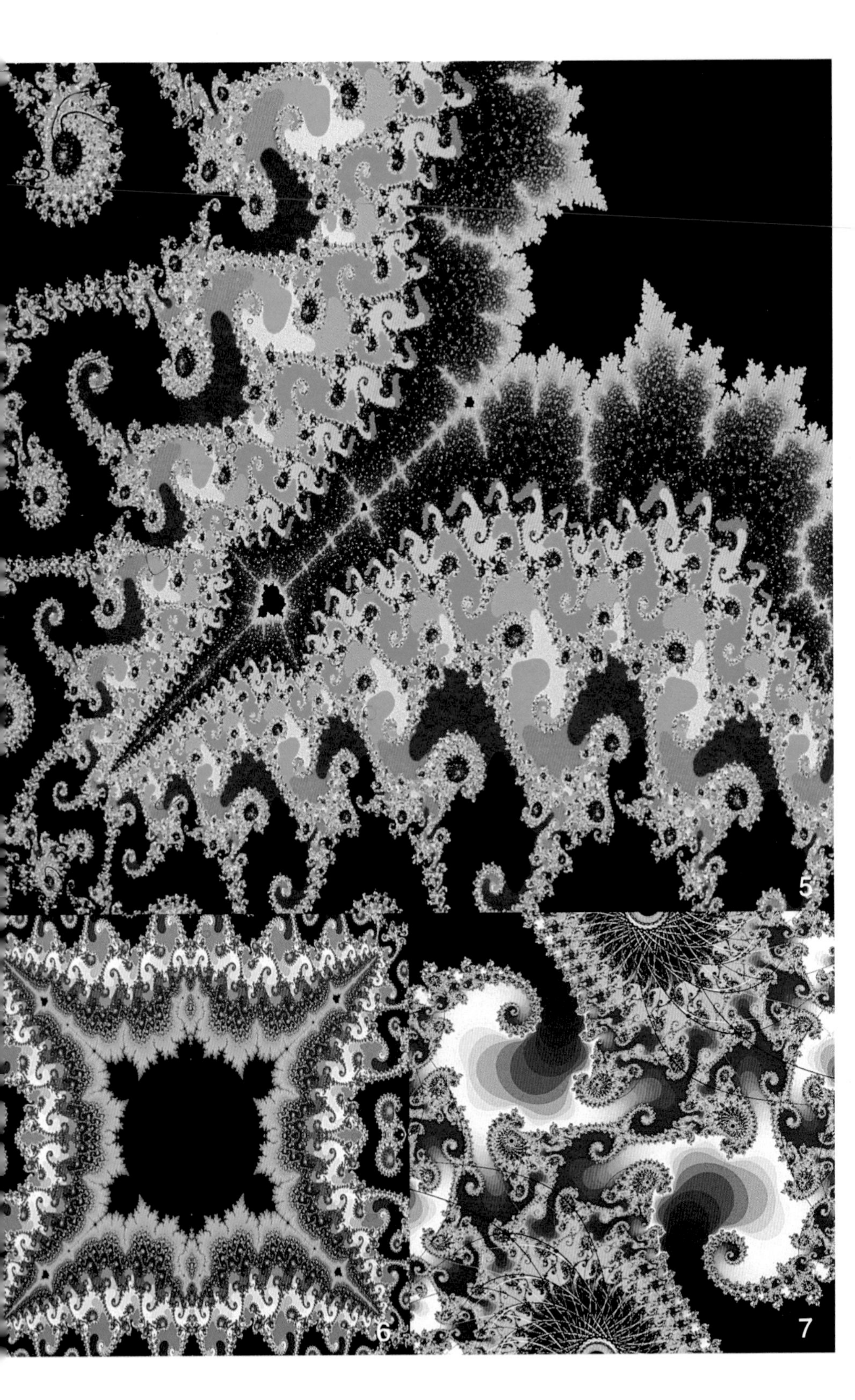
5
6
7

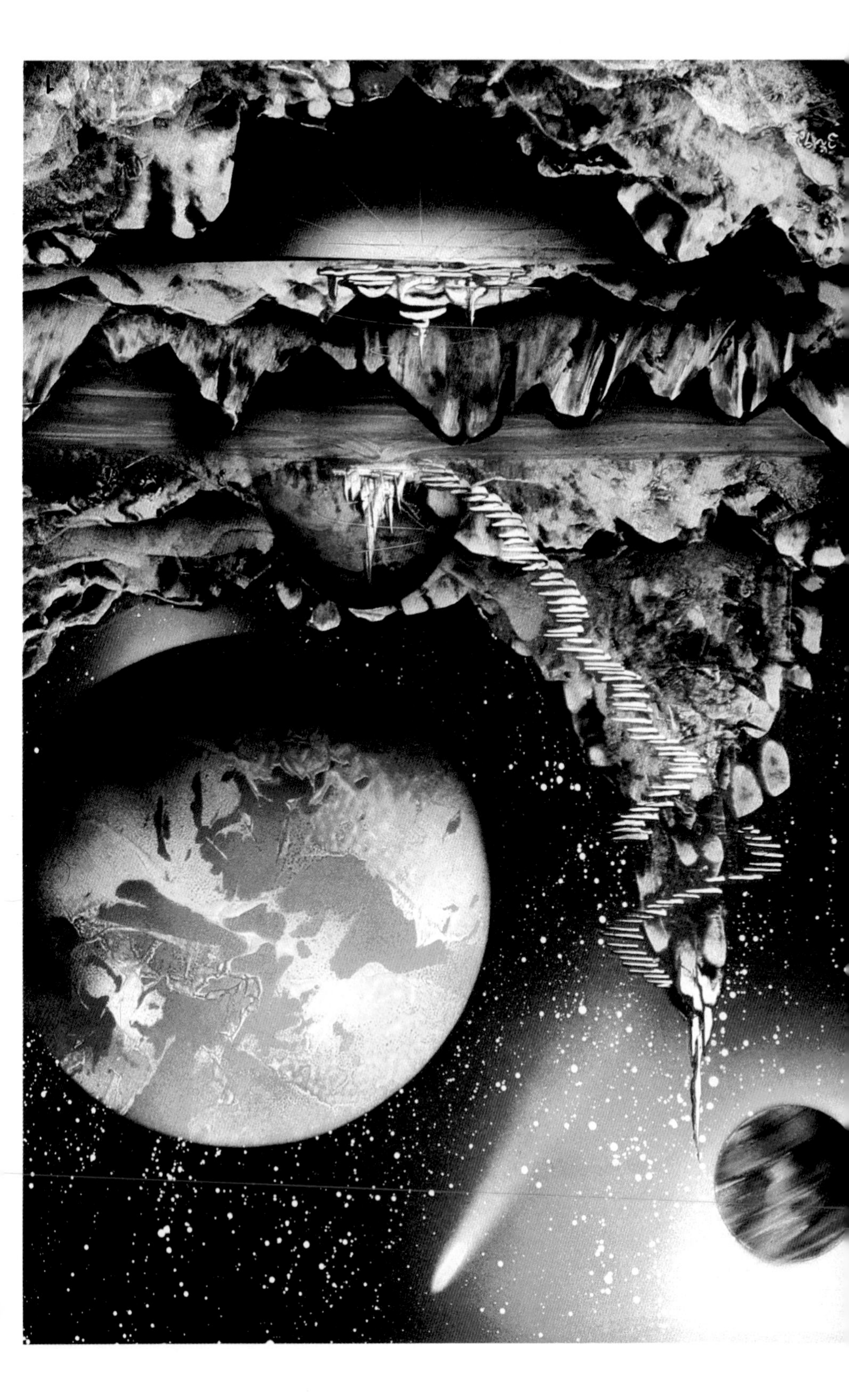

"Yeah, except for two things. The patterns are beautiful. Hard to believe they're just rubbish. Of course, we don't know what *they* consider beautiful."

"And?"

"Remember, they're limbless. No arms. They couldn't 'sling' these patterns over any walls. They have the Prohaptors to help them move things around."

"Uggh, Prohaptors. Don't remind me."

After another half hour of travel, we reached the towers and began to walk along a central plaza. The paving of the plaza was silver, and the towers where shiny crimson. Scintillating dust particles floated above us.

"I wonder where this dust suddenly came from?" I said. The dust rose so high into the red sky that the last glitter of the tiny particles was lost in the glacial whiteness of the air pocket's ceiling.

"Interesting houses," Kalinda said.

"Yeah, but not all that pretty. I expected more from a mathematical civilization." Some of the nearby houses had ice roofs. Others had thatched roofs. All the structures were squat with very small windows – rigid, unimaginative, ugly. A small gold Mandelbrot-set mezuzah was set into the doors of several of the houses.

"This place looks medieval," I said.

"Do you think the Latööcarfians live here?"

"Don't know."

As we walked, we came to a building much more lofty, majestic, and grandiose than the previous huts. The building appeared to be constructed of clear, flawless ice arranged in a nested triangular object known to mathematicians as a Sierpinski gasket. Having minored in fractal geometry in graduate school, I was particularly interested in the intricate conglomeration of triangular windows.

"It must be easy to get lost in there," Kalinda said.

"I think that this fractal building must be the Royal Palace of Yars Kothek. He's the ruler of what they call the Third Empire, and is supposed to be the smartest of the bunch." I'd heard lots of rumors about the Latööcarfian civilization, and I wondered how much of what I knew was based on fact. "Their heads are made of aluminum gallium arsenide. Know about that substance?"

"Sounds familiar."

"It's a semiconductor material similar to those used in computer circuits. Helps them think faster than us organic brain types. I've heard that the Latööcarfians spend all day thinking about ornate mathematical designs."

"Designs like the ones we keep seeing on the stones?"

"Yep. And status in their society depends on the beauty and intricacy of their thought-patterns. Wouldn't it be great if we could persuade the King to come back with us to visit Earth? Our two cultures could learn so much from one another."

The walls that protected the Sierpinski gasket palace rose from thirty to forty feet in height, and were so thick in places that they contained spacious galleries of mathematical art. Near the walls was a large pillared porch under which we could see sleeping Latööcarfians. "God, don't they look weird with their bulbous heads?" I said to Kalinda. "I think those are the slaves. They have a rigid caste system."

"I like their art," Kalinda whispered. Directly in front of us, on a raised slab of polished white stone, was a doughnut-shaped object I recognized as an umbilic torus, and it was embossed with a delicate fractal pattern.

"Uh-oh, there's some Prohaptors." One Prohaptor looked at us and then returned to his work.

"They don't seem to care about us. Maybe they're friendly."

Kalinda and I wandered over to a few Prohaptors who were setting up a device with many platforms.

"I wonder what that is?" Kalinda whispered to me.

"Looks like a harmonograph." In versions I had previously seen, the device required two pendulums arranged so that one moved a pen and the other moved a table to which paper was attached. The combined effect of the two pendulums produced a complicated motion which steadily decayed due to friction. Each path, on each revolution of the pen, was a short distance away from the path on the previous revolution. The whole movement tended eventually to a point. The patterns it traced out were beautiful.

"Nina looks like she's being hypnotized," Kalinda said. From her position on my shoulder, Nina followed the back and forth motion of the harmonigraph with her tiny black eyes.

"This is a lot more complicated than harmonigraphs I've seen on Earth," I told Kalinda. The Ganymedean version had a pen which oscillated on a platform which oscillated on another platform which oscillated on another platform, and so on for ten different platforms. The curves this produced were so complex and irregular that I was not sure if I'd call them beautiful or ugly.

"Let's see if we can find the King," Kalinda said.

"OK, first we've got to find the door to this place." We began to walk around the Sierpinski palace in search of an entry-way.

"How about that?" Kalinda pointed to a large triangular aperture. Colored light poured through the ice door which sealed the opening. Light also filtered out through some of the thin ice windows of the palace. It colored the ground in red and green hues and imparted a rainbow glow to the wisps of mists surrounding the castle.

I gazed at Kalinda. "You look strange in the colored light. Strange – but beautiful." Rainbow light started to come from nodules in the castle walls.

"You look pretty weird yourself."

On the huge icy triangular door was a large silver knocker. "Let's give it a try," I said. Twice I knocked, and finally the doors somehow opened by themselves. We hesitantly stepped inside and looked around. Engraved on the inner palace walls were fantastic mathematical patterns I recognized as chaotic attractors.

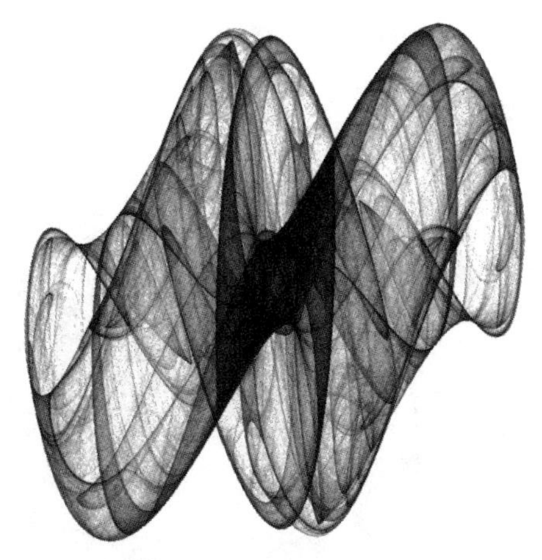

"The patterns look like the ones we just saw on the rocks," Kalinda said. "Except these are bigger."

I went closer to examine the large and exotic mathematical frescos which resembled swirls of smoke. "Look to the back of the hallway," I whispered. Upon an emerald studded, trigonal bi-planer stone throne, in unearthly splendor, sat Yars Kothek, ruler of the Third Empire. A million tiny lights twinkled on the surface of his bulbous forehead. The lights seemed to trace out his chaotic, beautiful mathematical designs as he thought of them. They reminded me of fireflies flying at night.

"How do you think his head is able to make light?" Kalinda whispered to me.

"Gallium arsenide emits light when exposed to an electric current. Maybe that explains it." All the recent walking we did had relaxed me, but now I felt a tightness in my stomach. It was the kind of tension anticipation brings, and I welcomed it. "Looks like that Latööcarfians of lesser status are lined up on both sides of the King. Their chaotic head patterns are beautiful, aren't they? But they don't have the symmetry of the King's pattern."

If mathematics were an art contest, than these chaotic attractor patterns would surely win the prize. The sheer complexity and variety of the dusty patterns produced by simple computations seemed more beautiful than other fractal forms I had seen.

"Let's take a look around the palace before we go any closer to the Latööcarfian King," I said.

"OK. Let's hope they don't mind us snooping around."

"We'll do it quickly. He can't move anyway." I paused and did a 360 degree turn to view the entire palace. Not a single creature moved. "Hey, this place gives me the creeps." As we looked around I heard a muffled drum beat that seemed to come from far in front of us. But then I realized it was the beating of my own heart.

Around the King's central receiving room we could see a labyrinth of chambers. As best as I could tell they were workshops, storerooms, administrative offices, servants' quarters, bathrooms, and theaters. Some of the rooms were adorned with minimal surface sculptures, möbius strip vases, and other mathemat-

Figure 34.1. *Inside the King's fractal palace.* (Drawing by Paul Hartal.)

ical statues. We saw some huge stone Klein bottles adorned with pretty triglyphs and half rosettes. As we got closer to the King we saw his throne covered with spirals and frets painted to simulate various Julia sets. The ceiling was adorned with Pentelic tiles forming a picture of drunken Prohaptors of all sexes swinging on Hilbert and Peano curves.

"Look over there," Kalinda pointed.

"Wild," I said suppressing a laugh. I saw statues of several buxom Prohaptor women with shapely arms standing amidst fields of flowers and parabolas. We also saw a small chapel-like area with a Mandelbrot set shape at the center of the altar. Perhaps their religion revolved around this complex mathematical object.

"Let's go into the chapel," I said. Kalinda and I held hands as we ascended stairs into an entry-way of material reminding me of Italian marble and smoked glass. As we finally neared the alter, the chapel looked more like a solarium with a tile floor and a trompe l'oeil ceiling than a holy place.

"How would you like to go to this church?" Kalinda said. Perky ceiling fans were in abundance.

"I wonder if those fans are supposed to cool the congregants' semiconductor heads as they think their deep mathematical thoughts."

Kalinda held up one of the prayer books. "This is interesting. The writing in this book looks vaguely like Hebrew. Mathematical shapes are on some pages."

וַיַּעַשׂ אֶת־הַיָּם מוּצָק עֶשֶׂר
בָּאַמָּה מִשְּׂפָתוֹ עַד־שְׂפָתוֹ עָגֹל ׀ סָבִיב וְחָמֵשׁ בָּאַמָּה
קוֹמָתוֹ וְקָוֹה שְׁלֹשִׁים בָּאַמָּה יָסֹב אֹתוֹ סָבִיב׃

"If you think that's interesting, take a look at this!" I motioned Kalinda to follow. We went into a side room. Standing before us was a latrine covered with ornately inscribed equations relating to chaos theory and dynamical systems theory.

"Only a wealthy, stable civilization with a lot of leisure time would create something like this!" I said.

"Let's get back to the King." We left the chapel and slowly walked to Yars Kothek. I looked into the Latööcarfian's flaming magenta eyes. His silver beard trailed along the polished calcite floor. Oh those eyes. They glowed with an ancientness which transgressed the sanctity of the Silurian Epoch. My eyes were transfixed upon his all too mesmerizing glance. On his beard was a precious item for the Latööcarfians – a violet jewel, shaped like a miniature Mandelbrot set. It hung from a chain of shimmering hairs. In fact, as I got closer I saw that his beard was not a beard at all – just flowing chains of silver hairs which he wore around his neck.

"What are the beautiful chains made of?" Kalinda whispered to me.

"The hairs are from a silver-haired Zooz." I had read that the Zooz population survived the demands of the luxury trade because the Prohaptors released the Zooz, after plucking their hairs, to produce more hair and more Zooz.

"Look at his eyes," Kalinda said quietly. "I – I think we should leave *now*."

His magenta eyes possessed a strangely disturbing power. They seemed to be looking into her rather than at her, as if he were carefully examining her innermost secrets. His head had many throbbing blood vessels containing electrorheological fluids. The fluids changed from liquid to solid and back again in the presence of an electric field.

Yars Kothek came closer – not by his own efforts, for he had no legs – but rather he was carried by a voluptuous hermaphroditic Prohaptor dressed in a green metallic gown. I noticed that the Prohaptor also carried a large pear-shaped chunk of gallium arsenide in a pouch at her side. Judging from an opening on the side of the King's head which matched the shape of the piece in the Prohaptor's pouch, I guessed that this might be an auxiliary brain which could supplement the King's brain power when he needed it.

The Prohaptor gazed at us for a few seconds, and then walked out of the palace. She said only two words, "Follow us." Neither the Prohaptor nor the King commented on little Nina perched on my shoulder.

"Now what?" Kalinda said. "Where in hell does she want to take us?"

"Let's follow. If the King wanted to hurt us, he could probably have had us killed by now."

As we walked through the central plaza, I saw some young Prohaptor children sucking on things resembling lemon lollipops. "Somehow this scene reminds me of earth," I said. As we got closer, we could see that on the top of the lollipop stick was a triangular Koch snowflake curve.

"Koch curve candy?" Kalinda said with a laugh. "This place seems a bit preoccupied with mathematical shapes, don't you think?"

I nodded. "The next thing we'll find is that they have fractally shaped sexual organs."

"Don't think so. It wouldn't be evolutionarily useful."

"I hear something. Music? It's coming from down the street."

"Yep, look. A group of Prohaptor musicians."

"Evidently not all Prohaptors are warriors," I said looking at the motley assortment of Prohaptor troubadours. Their strange and magnificent musical instruments splashed mellifluous sounds at us, as if from a fountain. One stringed instrument resembled a butterfly-shaped Lorenz attractor. This wiry shape was held in one hand and plucked like a harp by the other. I watched closely as the Lorenz player gazed at his instrument with colored lights and keyboard-like buttons. He wore a headset and a microphone. Many of the drums had fractal edges requiring the musicians to strike the drum at precise locations or else the sound would be quickly damped by the intricate edges.

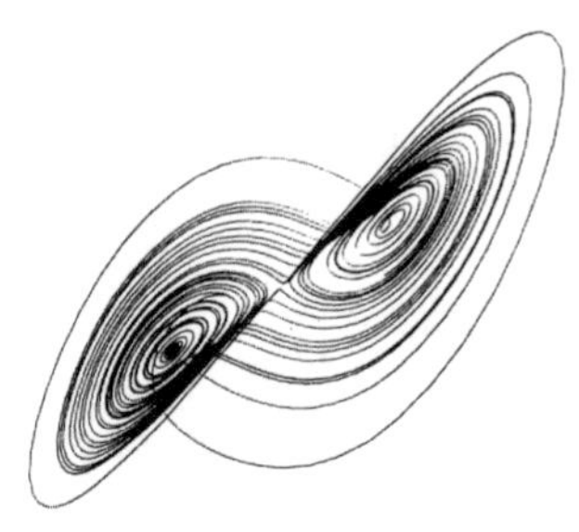

A small female Prohaptor beat upon what looked like a giant antihistamine capsule as perspiration poured from her brow. Kalinda looked like she was captivated by the unusual sonic experience. Even Nina the Navanax swayed with the sounds. The chromatophores in her flesh changed her color to umber.

"Think they could do *Melancholy Baby*?" Kalinda asked. I smiled. The music sounded more like Bulgarian folk music, played backwards.

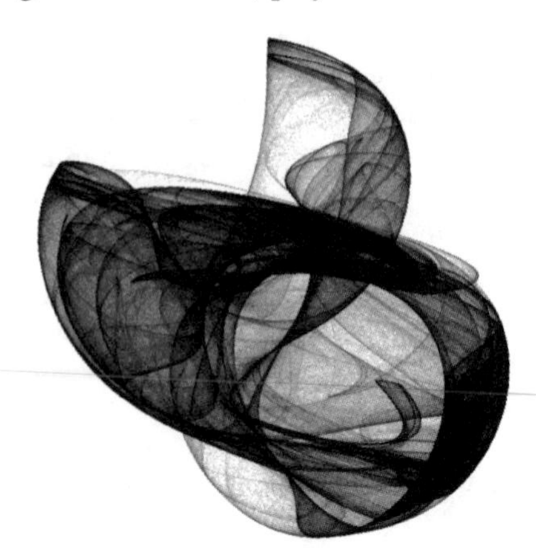

Chapter 35

The Glass Girls of Ganymede

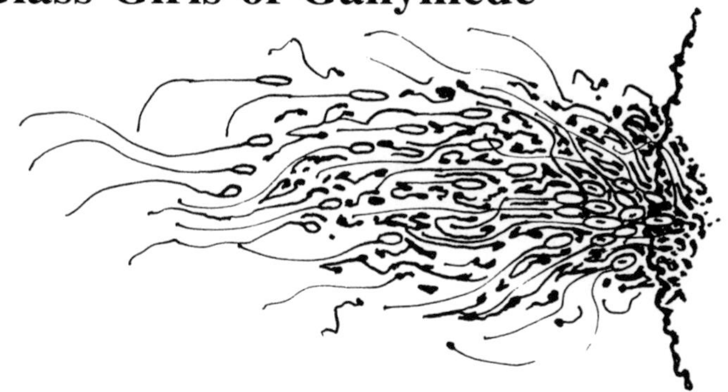

"Please follow us," the King's Prohaptor said to Kalinda and me. The Prohaptor stared at us with an eerie, out-of-focus gaze as she angrily pulled on her pendulous snout. We turned away from the fractal-playing musicians and walked onto a field of ice.

Suddenly a white hairy animal with two large ears and seven maroon eyes ran close to Kalinda. It was about five-feet tall and stank of urine. A semisolid fluid pooled beneath each of its eyes. It smiled at us and blinked rapidly as if surprised.

"Watch out," I cried.

"It doesn't look dangerous," Kalinda said as she ran across the ice field to follow it.

"Get back here!" the Prohaptor said. The muscles on her arm rippled with anger.

Kalinda was just in time to see the white hairy animal pop down a large hole under a crystal tree. In another moment down went Kalinda after it. Did she ever

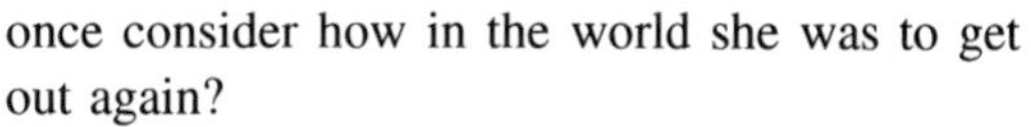

once consider how in the world she was to get out again?

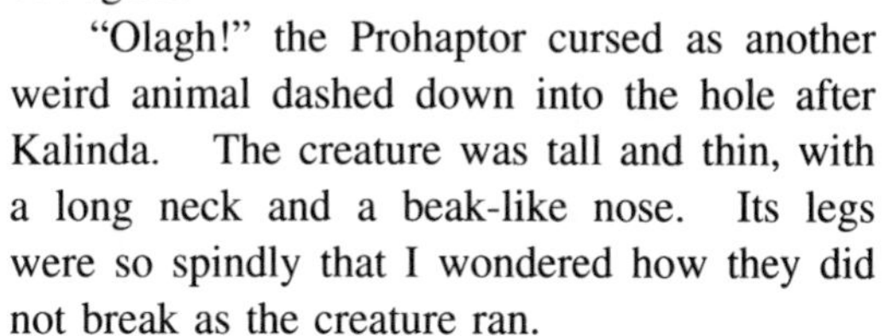

"Olagh!" the Prohaptor cursed as another weird animal dashed down into the hole after Kalinda. The creature was tall and thin, with a long neck and a beak-like nose. Its legs were so spindly that I wondered how they did not break as the creature ran.

"Kalinda," I called to her, but there was no response. Having no choice but to follow her, I went into the hole after her. The hole went straight on like a tunnel for some way, and then dipped suddenly down, so we didn't have a moment to think about stopping ourselves before we found ourselves falling down into what seemed to be a very deep well. Down. Down. Down. Would we ever stop falling? We must have been a mile beneath the subterranean air chamber when finally we slid down upon a heap of sticks and dry crystals. Evidently the King and the Prohaptor did not follow us.

"Kalinda, are you OK?"

"Yeah."

"What got into you? It's not like you to run after –"

"I don't know what got into me either, but I just *had* to. And I don't see the creatures anywhere."

"Looks as if we're in a smaller subterranean air chamber beneath the large one inhabited by the Latööcarfians."

"How are we going to get out of here?"

"Good question."

We stood near a clear lake. Around us, lining the walls of the cavern, were about twenty adolescent women. They looked somewhat human but had flat chests, long flanging ears, and hands with only two large fingers. Protruding from their foreheads was a small tusk.

"They're not moving," I whispered. Kalinda went right up to one, and touched her.

"You sure are brave today," I said.

"Her skin is like glass – hard as a rock and transparent." We looked through the transparent skin into the living guts of the girl.

"Looks like eggs inside of her," I said.

"Right, and here's her liver and stomach." Kalinda suddenly stepped back. "Look at that." In one cellophane-like organ buzzed a bunch of brown wasp creatures with multiple stingers on their abdomens. From their tiny mouths protruded long snaking tendrils. "Uh-oh." Kalinda pointed to a dark corner. "Some of the glass girls are moving." A few of the glass creatures were engaged in a mating embrace known as amplexus.

"Kalinda, I heard fables about these glass girls. They were once called *Reobatrachus*, and the way they make babies is supposed to be really weird."

"But they're humanoid – how weird can it get?"

"After the male has fertilized their eggs, the female swallows them and broods them in her stomach."

"How come stomach acid doesn't destroy the eggs?"

"From what I've read, which may be just speculation, the nurturing females stop feeding during the breeding period. The egg capsules secrete a prostoglandin which stops the mom's stomach from secreting hydrochloric acid. The stomach is transformed from a digestive organ into a protective gestational sac! The mother goes without food for a month. During birth the glass girl's esophagus dilates and the young glass creatures are shot from her mouth."

"They seem to be minding their own business now. Let's take a look at the lake."

The lake water was calm but with enough of a ripple to lead me to believe there was life beneath its black surface.

Kalinda stepped closer to the lake. "Maybe in this subterranean sea there are even more unimaginable creatures that'll change our ideas about life and evolution."

"Right, their bodies could contain chemicals that cure cancer," I said.

"Or chemicals more intoxicating than LSD, or more deadly than rattlesnake venom."

"Wow, there's an interesting specimen in the lake." The creature looked like a spiral soda bottle, except that its digestive system hung down from a pouch that trailed along, pulsing. Its eyes were translucent globes filled with a luminescing milky substance.

"Garth, I think the glass women are also interested in mathematics."

"What makes you say that?"

"Look over there. On the wall. Models of various conic sections." Seated on various platforms were glass models of cones, cylinders, ellipsoids and related conic shapes – but nothing more advanced than that. Maybe glass people studied mathematics when they weren't engaged in amplexus.

"Kalinda, the glass women don't seem to be too interested in us. Before we try to figure a way out of here, let's see if there is something to eat in this lake." I dug into my backpack and placed some Cyalume chemical lights into traps and lowered them into the water on a long polyethylene rope. I thought the light would attract animals in a world where bioluminecent bacteria were everywhere. Some animals might be drawn to the strangely colored lights simply out of curiosity.

Meanwhile, Kalinda went closer to the glass women to observe their transparent bodies. As I watched Kalinda, I saw the eyes of one of the girls tracking Kalinda's motions, and then the glass girl smiled.

"Hello," Kalinda said, as she waved her hand in a greeting. Unfortunately, Kalinda slipped on the ice and toppled over the smiling glass girl. As the girl went down, I saw a strange, nervous expression cross her face. Kalinda tried to grab her, but couldn't stop her fall. Then the girl's organs, eggs, and wasps tumbled out and shattered on the cold Ganymedean ground. Her gestational sac pulsed for a few seconds and then was still.

Her esophagus danced.

The wasps went crazy, buzzing as they flew at Kalinda, their multiple stingers poised and ready to strike. From a distant part of the cave came a number of mobile male glass creatures with deep voices. It seemed they were ready to aggressively defend their girl's eggs.

I ran to Kalinda and helped her up. "Those glass men look pretty damn dangerous."

"What are you talking about?"

"Look at their hands."

"Holy mackerel."

The glass men wielded needlelike spines at the base of their glass thumbs – probably to slash the skin of intruders. On their index fingers were crescent-shaped hooks made of bone. The males were also equipped with long, pointed tusks that protruded from their lower jaw. Their bodies were transparent. The closest glass man jumped and snapped at us.

"We're sorry. It was an accident!" I screamed at the glass men in several Ganymedean languages.

"You're a dead man," one of the men said to me.

Kalinda would soon be stung by the wasps, and I would soon be crushed by the glass men. In high stress situations like this, the adrenaline kicks in. I was completely surrounded. There was no avenue of escape. I saw the intestinal tract of the nearest glass man undergo peristaltic contractions. I reached into my backpack for my sword. When I closed my hand around the fractal sword, it completed me and made me whole. From behind me, a glass man struck me with his hand. My head erupted in blinding white pain as I felt one of his needlelike spines tear into me.

"Don't you realize it was an accident?" I jumped to the side, turned, and launched a front snap-kick to his groin. He went down. My foot slammed into another glass man's stomach, choking off his breath. He fell to the ground and shattered.

Kalinda jumped into the lake to keep the wasps away. I hoped there were no dangerous creatures beneath the black liquid.

Another glass man attacked. A shoulder throw sent him crashing into an icicle near the wall of the cavern. My fractal sword decapitated another. I didn't want to die under this strange cold world, so I tried something daring. Rather than run away, I rushed directly at the remaining standing glass men and executed a few tae kwon do maneuvers. They all ran away into the darkness.

"Garth, help, the wasps," Kalinda shouted. Wiping the blood away from my head, I quickly withdrew an aerosol bug repellant from my back pack and started spraying madly at the wasps. It seemed a dumb thing to try, but it was the only thing I could think of in the heat of the moment. In seconds, however, the wasps began to lose interest in Kalinda and then plummeted into the black lake. Kalinda swam quickly away from the half-dead, floating wasps.

"Looks like we're OK for a while," I said a little too soon. Jellyfish creatures with purple bladders floated to the surface of the black pool. They looked a little bit like Portuguese men-o'-war. On earth my subspecialty was teuthology, the study of squids, so I was naturally drawn to the strange looking creatures. Seconds later a vast purply mass, perhaps their mother, joined the jellyfish. It seemed almost plant-like, and it trailed a long stem that may have anchored it to the bottom of the pool. Its myriad fractal arms curled and twisted like a nest of boa constrictors.

"Quick, swim to me," I cried to Kalinda. The sheer physical smoothness of the jellyfish was alien, intimidating. Soon something resembling a green and red möbius ribbon with blue eyes and a pompadour of greasy hair came out of the depths and wriggled closer to her.

"I'm swimming as fast as I can," Kalinda gasped. She came closer to me. Closer. Her arms were out, reaching for me. Her eyes pleaded for help, and she was moaning. Several of the spiral soda-bottle creatures also emerged and drifted towards her. The translucent globes of their eyes grew slightly in size.

"Garth!" Kalinda's fear became panic as her screams escalated in volume. Just as I grabbed hold of Kalinda's two arms the mother jelly fish grabbed hold of her left leg. Kalinda kicked. I pulled. The jellyfish pulled. Kalinda screamed. Finally the jellyfish let go of Kalinda, leaving a trail of abrasions along her ankle. She fell into my arms.

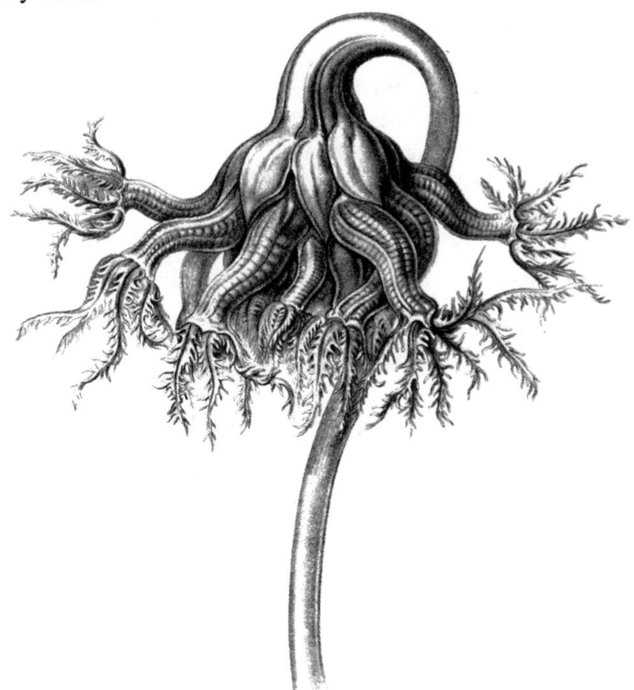

"Let's move as far away from the pool as we can," I said.

Kalinda panted. "That was close."

We spent the remainder of the day resting amidst various organs which had tumbled to the ground from bodies of the glass people. Occasionally an organ still pulsed.

"What a gross scene," Kalinda said.

"You say that after you spent most of your life in an intestine?"

"That was home. This is different." She kicked at the lobe of a liver. The lobe vibrated a little as if it still had a life of its own.

"I'm curious about what they eat," I said. I opened up one of the stomachs which was lying nearby. "Look here, Kalinda." I held open the splayed stomach as its contents dropped to the floor. The glass girls practiced entomophagy – they ate insects.

"Not very appetizing," she said.

"Actually, outside of Europe and North America, most people on Earth eat some insects."

"Africa?"

"Right. In parts of Africa, more than 60 percent of dietary protein comes from insects. Grubs and caterpillars have a lot of unsaturated fats."

"Doesn't appeal to me."

"Did I ever tell you what I ate at the banquet hosted by the New York Entomological Society?"

"Do I want to hear this?"

"Sure, you'll find it interesting: chocolate cricket torte, mealworm ganoush, sauteed Thai water bugs, and wax worm fritters with plum sauce." I didn't tell Kalinda about the time I went to a movie theater in Columbia where roasted ants were eaten like popcorn. Nor did I tell her that honeypot ants, with their transparent abdomens distended with peach nectar, were delightful sweets.

"Garth, something's moving in the lake." I looked and saw the purple jellyfish massing at the edge of the lake nearest to us.

"Let's get out of here," Kalinda said.

"Not a bad idea."

"You have a few spikes and ropes in your pack. Think we can use them to climb out of the hole?"

"I don't know. Seems awfully steep and slippery."

"Any better ideas?"

"Nope. Let's give it a try." I withdrew a metal spike from my pack and banged it into the ice with a hammer. I gave it a tug. "Seems to be holding."

We began the long journey back up through the tunnel connecting the glass girls' chamber to the Latööcarfian one. When we finally arrived at the top of the tunnel, the King and the Prohaptor were still there waiting.

"It's about time," the Prohaptor screamed at us, her stinking spittle flying in our faces.

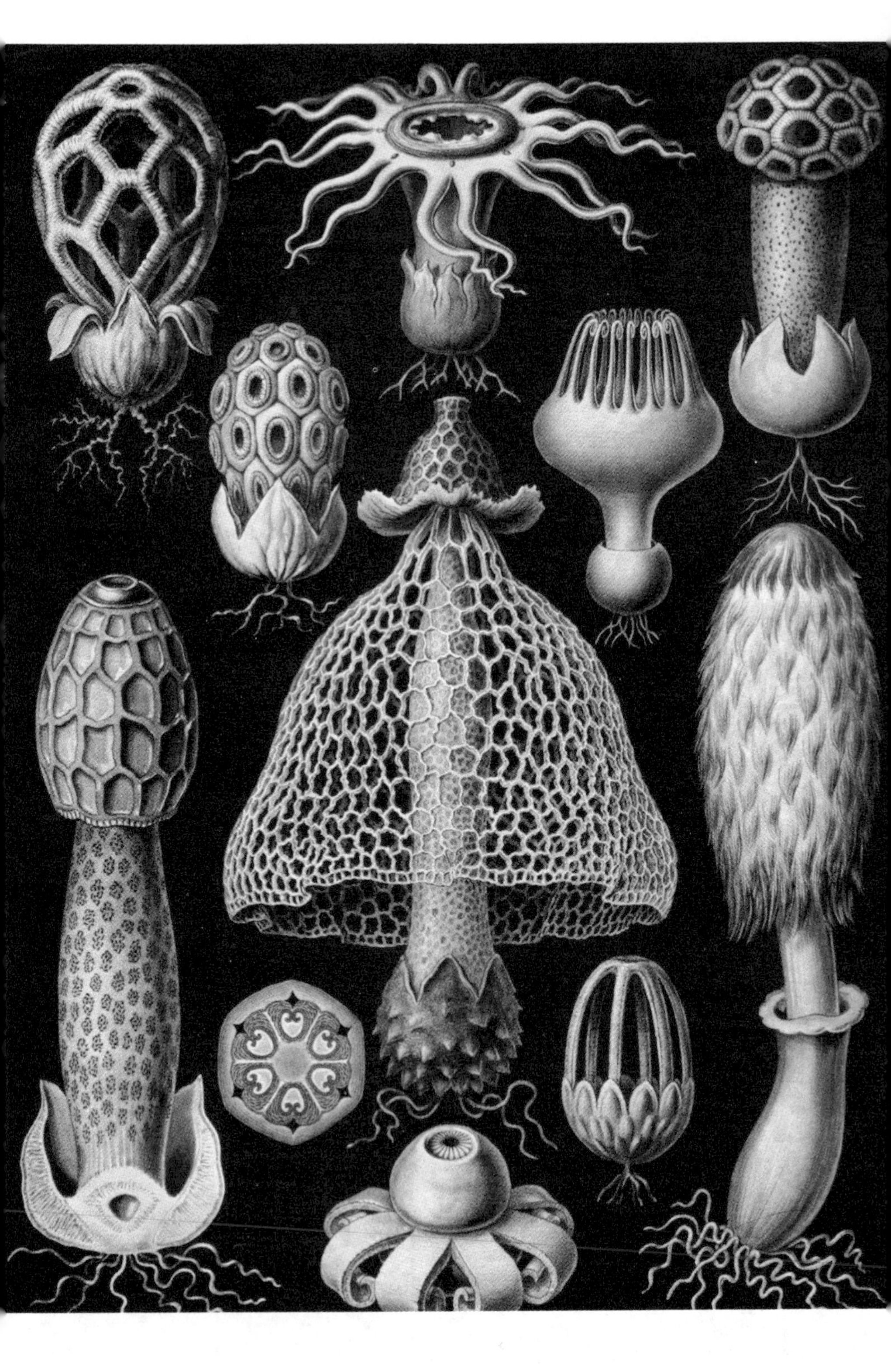

Chapter 36

Death-Fungi and Zinc Ants

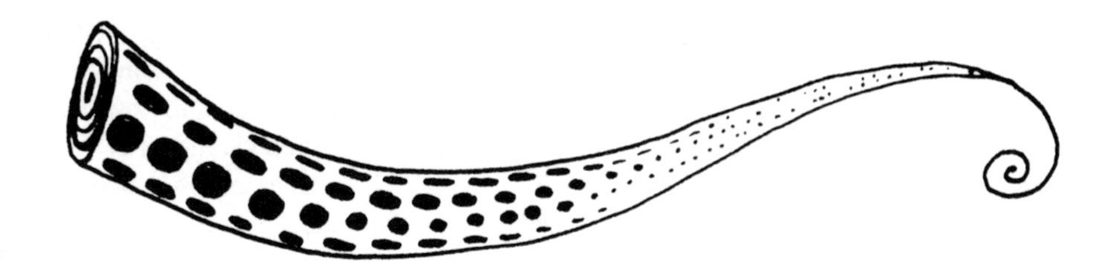

"The forest leans into the man's sleep. It cannot dream for itself."
Rosanna Warren, *The Field*

"'I was thinking,' Alice said politely, 'which is the best way out of this woods: it's getting so dark....' But the fat little men only looked at each other and grinned."
Lewis Carroll

Kalinda and I followed the Prohaptor-carrier and Yars Kotheck into the darkest heart of Ganymede, where little light penetrated the forest mists. From what we could see, the blue of the sky was fast turning to beige. The dwindling light in the forest tinted the ground the color of salmon. Occasionally we heard strange animal cries and the terribly harsh death-whine of Ganymedean insects. This was a gloomy place, except for the myriad luminous crystal flower blossoms on the forest floor.

"The flowers smell like roses," I said.

Suddenly the flowers moved! Then they turned black, as if someone had turned off a light switch inside them.

Kalinda trembled and paused for a moment, hesitant about saying her next thought. I said only one word, "Spooky." A wave of grayness passed over me, a kind of dark premonition.

To our left I saw gas oozing from a mist-covered swamp, and I heard a tinkling sound as tiny crystals began to rain down upon us.

"C'mon," said the Prohaptor. "Keep following me. Don't pay attention to the crystal shower."

"Kalinda look at that plant." I stooped down to pick up a piece of Ganymedean death-fungi. "Supposedly this strange fungus makes one sleepy when placed by the ear." I ran my finger along its flaccid sac and nerve-complex. The fungus wriggled in response.

With impatience in her voice, and without any explanation, the Prohaptor said, "Drop it." I hesitated, but then let go of the globular brown shape with its weird giant fruits.

"Wait," said Kalinda. "I want to see what's inside it." She lifted the fungus from the ground and split it open to expose a rotting pulp teeming with white specks. A complicated network of tunnels permeated the fungus. She squinted at the specks in the fungus. "There something moving inside." I came over and saw that there was an army of small arthropods with poodle-like heads. One of the poodles looked up at Kalinda and made a tiny barking sound.

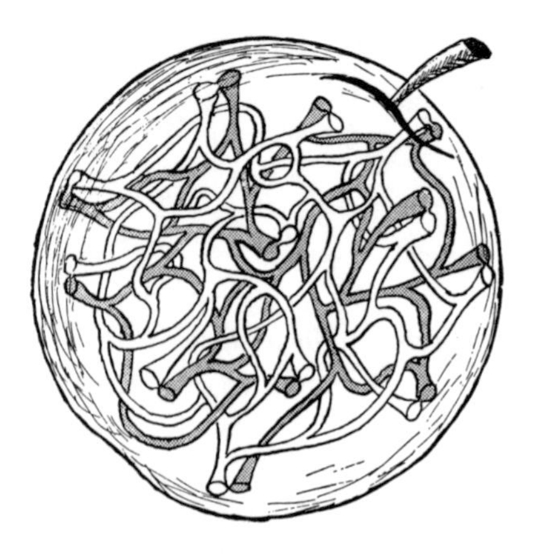

"Garth, look, there's two species of mini-poodles. One's white. The other's black."

I looked closer. "They don't seem to like one another because their channels inside the fruit don't join."

"Right." The double network of channels tightly intertwined, virtually filling up the entire interior of the fungus.

"The only way to get from one half of the labyrinth to the other seems to be at the surface of the fungus," Kalinda said. She pointed to the openings where the two tunnel types met the surface.

"Weird." I wish we had time to study this and figure out all the topological properties of the tunnel networks. Mathematically speaking, it was as if the space inside the fungus were formed by the overlapping of two independent spaces connected only at their common surface. "Why would they make such a network?" Kalinda quickly handed the fungus back to me. A few of the mini-poodles fell to the ground and scurried beneath the soil.

"Please place the fungus back where you found it," the Prohaptor said with an aggressive machine-gun voice. I carefully returned the fungus to the ground so as not to further disturb any more mini-poodles.

"There's all kinds of fungi here," I said with an astonished voice. The forest floor was covered with fungi consisting of elongated bodies and hoods resembling umbrella lattices.

We continued to walk. "Look, one of those bats," Kalinda pointed upward. "Is he chewing on one of the umbrella fungi?" A neon bat landed on a branch of a tree. The branch bobbed, raining crystals on my head. We were quite for a few minutes. White flowers continued to turn black as we passed.

In a bright clearing in the forest we came to a small conical mound of translucent, amber crystals. The Prohaptor, the King, Kalinda, and I took a closer look.

"The crystals are moving," I said.

"No, wait, I don't think so," Kalinda said. "They're not moving but being pushed by tiny Ganymedean ants."

I brought out my magnifying glass to take a closer look. Except for their heads, which resembled hornets' heads, the ants seemed to be nothing more than silicon-like chips with moving legs. On their backs were solar cells which may have controlled their microprocessor brains.

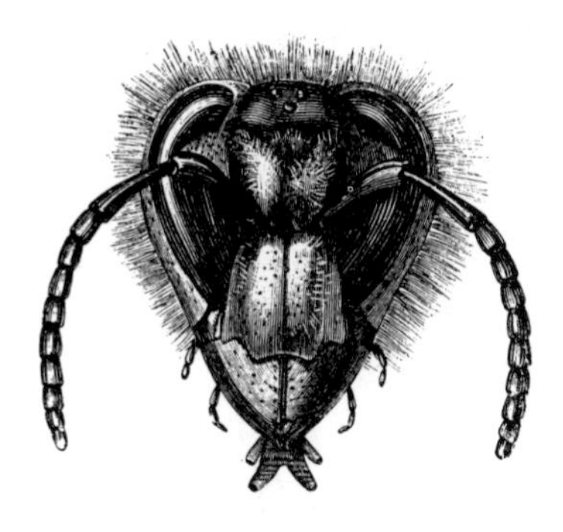

"What are they made of?" I asked the Prohaptor.

"Don't you two ever shut up?" she said.

"We're scientists. Natural curiosity."

The Prohaptor paused but then threw up her hands in exasperation. In doing so she nearly dropped the King. "OK, if it will keep you quiet, I'll tell you. The legs are composed of zinc oxide," she said brusquely. "It's a piezoelectric material which expands and contracts when exposed to a voltage. Like Latööcarfian mouths, these mechanical legs extend and curl in response to tiny electrical currents supplied by the microprocessor brains."

"Thanks," I said. "We only want to learn about your world. We're peaceful."

I watched as the six-legged insect walked as a terrestrial ant does, always lifting the middle leg on one side together with the front and back legs of the other.

The King finally opened his mouth. He seemed amused by our curiosity. "We distribute armies of Goobers [Ganymedean ants] in the crystal forests to hunt and kill crop-eating insects. They also serve another purpose."

The King's piezoelectric mouth grinned as the Prohaptor popped a handful of the tasty morsels into the King's mouth. I heard a crunching sound, as the zinc oxide ants were devoured alive. The ants made a horrible whining sound. A few tiny Goober legs dangled from the King's elliptical mouth.

The little legs twitched. I changed the subject.

"The 'Goobers' have no eyes. How do they find other animals?" I said.

"This is your last question," the Prohaptor replied bitingly. "Look, each ant contains a tiny spiral microantenna made of gold. It picks up infrared radiation. Because the spiral is as small as a grain of sand, it receives small wavelengths. The ants use them to detect heat from animals, and to see in the dark. We let them crawl in our bodies to detect changes in body temperature indicating the presence of a tumor."

I imagined an army of ants crawling in the King's digestive and circulatory system looking for tumors, polyps, and growths. I wondered if the ants could survive in the large veins and arteries containing electrorheological fluid. What if an ant got stuck and couldn't get out?

We walked another few miles in silence.

"Can we rest?" Kalinda finally asked.

All of a sudden we saw a motion in the woods near where Kalinda stood. From between the labyrinthine tangle of aerial tree roots I saw a large animal with three-fingered limbs. Its polyhedral eyes seemed to float in front of its silver-blond hair.

"My God," Kalinda said, "what –"

"Leave it alone," the Prohapter said.

Suddenly the silver haired "Zooz" turned and emerged from behind the tree. Kalinda was about to be trampled into human hamburger.

"Garth," she screamed.

Nina the Navanax jumped from my shoulder into my hand. With my free hand I withdrew the fractal sword. Damn, I was too far away. I threw it at the beast but missed. The sword went sailing into the forest underbrush and was lost.

The Zooz touched Kalinda with its long arm, and simultaneously placed its massive mouth and tendrilous tongue on her belly. Its 6-foot long tail also curled around the Prohaptor's hip.

"Get this friggin' thing off me," Kalinda screamed.

Just as I was about to smash the Zooz's head with my foot, several other Prohaptors ran out of the forest and thrust spears into the Zooz's belly.

"The tips of the spears have poison," the Prohaptor said with a grin.

"It doesn't seem to work very fast!" I said. The Zooz's tongue shot out at one of the Prohaptors with the speed of a chameleon's tongue, enveloped him, and turned him into a nearly-liquid mass of chewed flesh.

Maybe I could leap onto the Zooz's back. With a deep breath, I jumped, but I didn't land quite right. The pain in my testicles streaked up to my stomach.

"Get away from her," I screamed, kicking the beast. The Zooz trembled and collapsed as I jumped off. It died an excruciatingly slow death from the poison.

"Look – Goobers," Kalinda said with an indrawn gasp. She pointed to the ground near the dead Zooz. Hundreds, maybe thousands, of zinc oxide ants crawled on their tiny piezoelectric legs from the underbrush. Soon the Zooz's body was covered by an undulating mass of gregarious Goobers. They chomped. They pulled. They tore. With lightning efficiency, they dismembered the corpse and carried away the morsels of flesh back into the forest.

The Prohaptor and King allowed us to watch the Goobers until all that remained was a silver skeleton and a pelt of shiny silver Zooz hair. The Prohaptor folded the pelt of hair, and took it with her.

"What do you think the fur's worth?" I said to Kalinda.

"On Ganymede, or back on Earth?"

"Here."

"Hard to say. We don't know anything about their monetary system. Obviously they value the silver hairs."

"Wish we could bring a pelt back to Earth," I said.

"Don't ask her. She doesn't seem to be in a good mood."

"Please follow us," said the Prohaptor. Her voice sounded like a dead girl's voice.

Chapter 37

Starfish Soup

"I am reminded of a French poet who, when asked why he took walks accompanied by a lobster with a blue ribbon around its neck, replied, 'Because it does not bark, and because it knows the secret of the sea.'" Anonymous

Soon we came to a sandy beach. Nearby a device resembling a transistor radio played a strange song. Several beach towels, flaccid with moisture, hung on the jetties. In the distance a dog-like animal barked. A Prohaptor fisherman saw us and raised his hand – apparently as a greeting to our party. In a land far from the realm of most Latööcarfians, far from their teaming wharves and bright bioluminescent cities, a phosphorescent sea ebbed onto an orange beach.

"Skulls," I said and pointed to the ground. The ancient beach was dotted with mounds of broken bones, skull fragments, and mandibles.

"What are those?" Kalinda said. She kicked at a large object.

"The exoskeletons of some long-dead Heñtriacontañes."

On the beach I saw a wormlike animal feeding upon some slime. It was quite different from the pipe-world worm. The entire body, yellowish in color, was covered by a cuticle which in many places was prolonged into spines. The body was divided into segments.

"Kalinda, take a look at this," I said. The creature's mouth, at the anterior tip of the first segment, was guarded by a circle of spines. When resting, the head withdrew into the front end of the body.

The Prohaptor picked the spiny worm up and said, "Alternate thrusting forward and withdrawing of its piezoelectric head produces a squirming movement which is its chief means of locomotion. The creature is called a *Kinorhyncha*."

Wherever we disturbed the ice and sand, the burrowers were liberated. They reminded me of marine worms on Earth who fixed themselves to human skin and secreted a poison that caused pustulant sores to form.

"May I take this specimen?" I asked. Without waiting for an answer, I brought out a specimen jar. With forceps in hand, I placed a Kinorhyncha worm in the jar for transport back to earth.

Figure 37.1. *Ganymedean fractal starfish.* (Adult pictured at left, juvenile at right.)

"You oaf," the Prohaptor said as she removed the specimen jar from my pack, opened it, and let the creature emerge. It promptly burrowed into the orange sand.

"Hey, I'm sorry," I said. "I didn't know the worm meant so much to you." The Prohaptor slapped me in the face. My fractal sword was lost in the woods, but I automatically reached for my sword from Earth. Unfortunately, it was still in my knapsack. The Prohaptor snarled.

It was nighttime on Ganymede, and the sky deepened from purple to sable. I looked up. I thought I could perceive the incredible lamp of infinitely distant stars through the icy ceiling of the air pocket. The ice must have been unbelievably clear in this section. The more recognizable constellations were unmistakable. Orion's square shoulders and feet, the beautiful zigzagging Cassiopeia, and the enigmatic Pleiades all reminded me of life back on Earth where I studied the constellations long ago. I even saw Aldebaran, the red star in the constellation Taurus. I remembered the time I first saw Aldebaran as a boy growing up in Ajaccio, a town in Corsica, the birthplace of Napoleon. We moved to the U.S. when I was seven.

After some time, the old fisherman began to gather from the beach the Ganymedean version of starfishes. After meticulously removing their spines and madraeporites, he threw their naked bodies into an extremely small black caldron of steaming water. He sniffed at the ever-swirling mixture of starfish soup.

"Aye. T'is good!" the Prohaptor remarked as he smelled the liquid. To me it had the odor of urine and lime. He also had a dish of a chalky-looking tapioca substance.

Yar Kotheck beckoned for us to eat the pee-stinking bouillabaisse.

"Do we dare eat that stuff?" Kalinda said.

"I'm starved. I'll try a little bit." I felt queasy about the smelly food, but would just have to get used to it.

"Uggh," said Kalinda as she sipped at the brew.

"It's not all that bad is it?" I said. We sipped at the soup for a few minutes. It acted like a triple espresso, making us instantly wired and more anxious. I turned my attention to the Prohaptor and King.

"May I take the Ganymedean starfish to my home world?"

The Prohaptor stiffened and grunted. Neither the fisherman nor Yars Kotheck responded. I had heard that Ganymedeans disliked alien intrusion in their well-ordered society. They were also protective of their wildlife, having once had the ecology of their oceans destroyed by invaders from Callisto. Perhaps they were so angry with our presence that they had lured us to this remote place in order to conveniently dispose of us after assessing our behavior.

"Garth, the fisherman is crying." I looked over and saw silent ellipsoid tears ebbing in the membranes protecting the old fisherman's eyes.

"I don't like this," Kalinda said. "Maybe he knows something's going to happen to us."

"Why would he care what happens to us?"

The fisherman quickly gathered up his kettle and utensils and scurried away into the hell-black night. We watched him walk away, and in the distance we saw the northern edge of the air pocket, a glimmering aurora pulsing in glacial whiteness.

The fisherman walked into the ocean. I thought there was something very queer about this ocean, as every now and then I saw strange ripping movements upon its surface. I imagined that glow worms, beneath the mud under the old fisherman's feet, sensed his presence with the mottled discs of their heat receptors, and scurried away.

"Where is the Prohaptor fisherman going?" I asked the Prohaptor carrier and Yars Kotheck. Neither answered. The King's eyes revealed nothing whatsoever of his own feelings. I tried again in the Prohaptor's grandiose linguistic style, while smiling my most ingratiating smile.

"Where goest the Prohaptor fisherman?" Still no answer. The Prohaptor's eyes remained unreadable beneath hairy eyebrows that were knit together in a perpetual frown. I withdrew my French cup hilt rapier from my backpack and placed it in its scabbard, ready for action. Kalinda had her two daggers ready. A faint, mechanical smile was on her face. I looked into the Prohaptor's red eyes, and tried a final time to engage her in conversation.

I gritted my teeth and said, "Where goest the Prohaptor fisherman, you baboon?" The Prohaptor did not say a word, but merely looked at me with contempt. In the sky were a myriad of sounds.... voices, murmurs, whispers. I felt as if I had a chased a white rabbit down a hole into a weird and considerably unfriendly Wonderland.

A few neon bats ever-so-slowly circled overhead, as if suspended in space. Soon a few more bats joined them. As the bats came closer, I caught a glimpse of their sunken faces. Their crimson eyes seemed to radiate hatred and torment. As hard as it was to believe, they started tracing complex fractal patterns called Hilbert curves in the sky. Did these Hilbert bats require a special intelligence to fly in such patterns, or were their designs simply the result of animals moving together in groups, like geese flying in a "V" formation or ants and termites building fantastically complex tunnel systems?

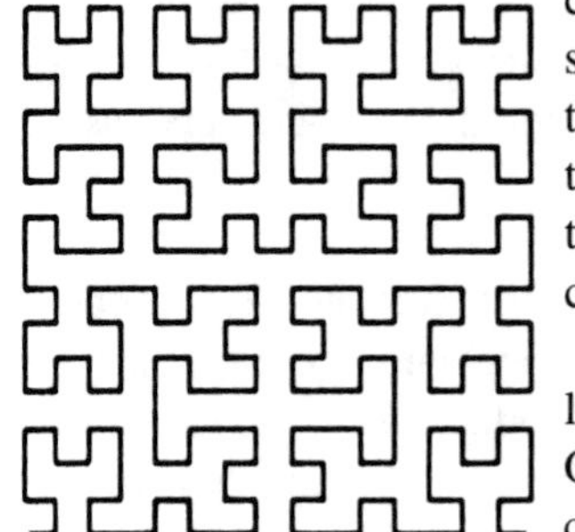

In the distance were dull pulses of thunder. I looked at Kalinda and saw goosebumps on her arms. Occasional preternaturally bright lightning in the sky cast stroboscopic patterns across the ground. This Ganymedean world was making me more nervous every second.

It was time to capture Yars Kotheck, the Prohaptor, and a few Ganymedean starfish for transport to my employer's intergalactic zoo – and get the hell out of here. My plan was to tranquilize them and then place them in some large nets I had brought with me, and hope they would survive the long journey back to earth. I thought I'd have no reservations about treating advanced lifeforms in this crude manner, because the scientific and financial rewards were so great. Yet I hesitated, and felt some remorse. I didn't know what effect my removal of the King would have on Latööcarfian society. I would be uneasy if my actions adversely affected many innocent individuals. On the other hand, both the Prohaptors and the Latööcarfians seemed to be a pretty mean bunch of creatures.

The Prohaptor's foot suddenly slammed into my stomach, choking off my breath. I fell to the ground, as colors exploded in my brain, and I felt the icy Ganymedean ground on my cheek.

"Garth!" Kalinda turned on the Prohaptor with a sudden flash of defensive spirit. The Prohaptor kicked her hand but she didn't drop the dagger. She slashed at the Prohaptor's eye, making her scream. A pool of blood, not yet congealed, bloomed from the Prohaptor's right eye.

"Aaah," the Prohaptor yelled again, her voice a high sonic stiletto. She reached for Kalinda, grabbed her arm, and broke it over her knee. I decided to dispense with any pleasantries and meet fire with fire. As the Prohaptor reached for Kalinda's other arm, I quickly kicked the Prohaptor's leg out from under her with a pile-driving blow of my right foot. The Prohaptor fell. I then knocked the King to the ground and slammed his forehead with my prosthetic arm. Why make little problems when you can create a holocaust?

The King cried, flashing into sudden fury. His face became red and blotchy with anger, and his mouth opened wide revealing an interior which glowed crimson from bioluminescent bacteria.

The Prohaptor's leg was only slightly bruised by my kick. She started laughing at me as if there were no way I could stop her. It was a deep laugh, like a hyena's tuned several octaves lower.

"Shut up," I said. "Take a look at this and shut your trap." I brought out the specimen jar which contained Ka and Da, the two Prohaptor brain parasites we had removed from the dead Prohaptor warrior who attacked us. I held the jar up to the Prohaptor. "What do you think of these two critters?" Her laughter stopped as though I turned a valve in her chest. She stared at the Ka and Da automatons as she recognized their origin. The enraged Prohaptor responded by running towards me.

I withdrew my sword from its scabbard, but I was slow. The carrier Prohaptor evaginated her digestive track through her oral cavity. The inside-out pouch came closer. I peered into her now colorless eyes, my image many time reflected as if I stood on the periphery of some gigantic crystal, alone in a field of darkness. In the Prohaptor's armpit, crushed like a squirming beetle upon dry concrete, was a parasitic organism. I thrust my sword at both the pouch and the parasite.

Hartal

Chapter 38

Attack of the Attractors

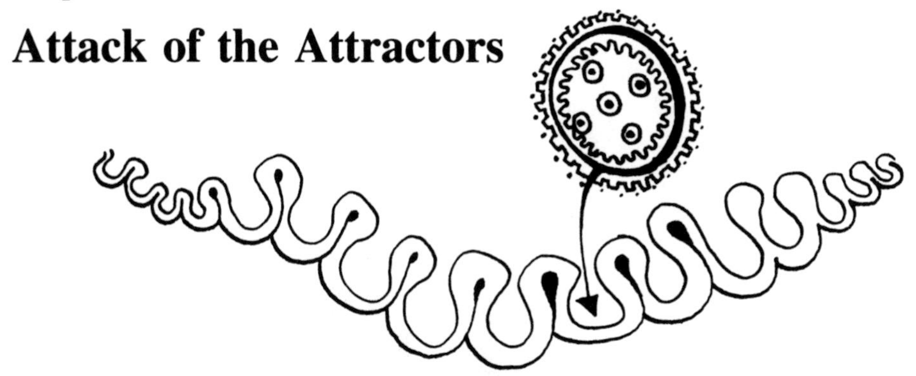

"Everything that has beauty has a body, and is a body; Everything that has being has being in the flesh: And dreams are only drawn from bodies that are."
D. H. Lawrence, *Bodiless God*

"Garth, the sky – something's coming!" said Kalinda. It was then that the chaotic attractors attacked me. They looked like glittering dust particles. I knew that the Latöökarfians dreamed of mathematical patterns, but did not realize that these dreams could take on a life of their own. Perhaps they were metallic dust particles which were somehow directed by the King via electromagnetic signals, or perhaps they were independent living organisms. Whatever their origin, the particulate matter in the skies formed patterns and swirls in the wisps and eddies of wind.

"Ha, you fools," the Prohaptor said to us. Broad grins flashed across her double snouts.

The particles came closer to Kalinda and me. We backed away. Some of the patterns were symmetrical, like pinwheels, while others formed patterns that defied description. The Prohaptor stood back, watching as she smiled. It was as if there were ten-thousand silver gnats glimmering like snowflakes of mercury. They hovered around my nostrils as if to mock me. I swatted them with my hands, but the patterns simply reformed and came even closer. They invaded my nasal canal, and I thought I smelled butterscotch.

"Garth, they're trying to get inside of me. My mouth. My ears!" There was a thick vein of indigo light, a flood of unimaginable brilliance – and my beautiful Kalinda writhed in the unholy blaze.

"Kalinda!" I screamed.

It was then that I exploded, my duodenum and small intestine shattering outward to angular infinity. Kalinda's eviscerated bowel-remnants quivered upon the Ganymedean ground and took on the aspects and coloration of pimento cheese. I shoved my fist into my mouth, overwhelmed with horror. Feeling both cold and numb, with the screams of Kalinda still echoing in my head, I looked up into the Prohaptor's eyes and heard a sound like a ripe melon bursting.

The Prohaptor was stepping on my left kidney.

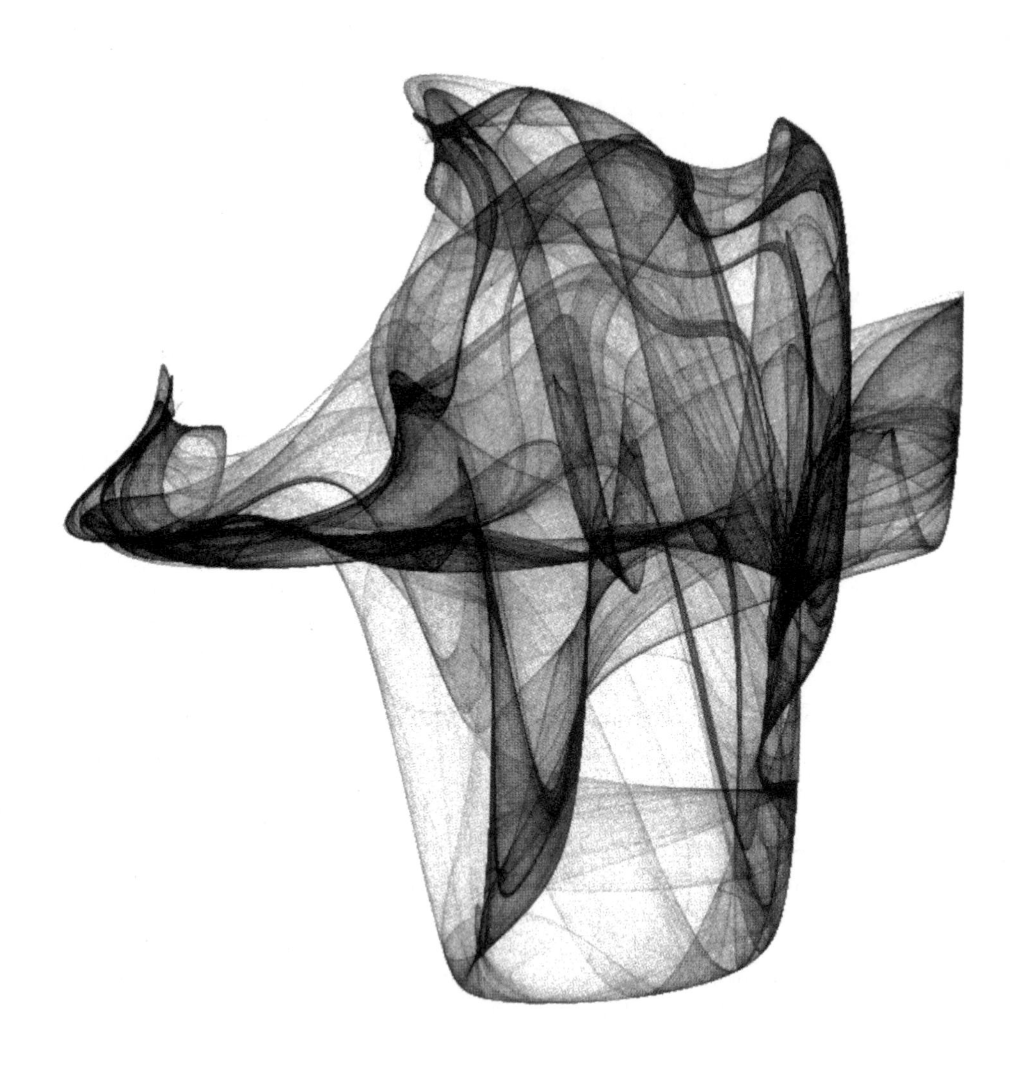

Chapter 39

Death and Worm Cages

"You have said that the land is a dream for you – and that you fear to be made mad. But madness is not the only danger in dreams. There is also the danger that something may be lost which can never be regained." S. R. Donaldson

Before arriving on Ganymede we had consumed several salubrious medications to help our bodies in the event of injury. When they had pumped blood into my veins, it was heated to 99 degrees and laced with phenyl tertiary butyl nitrone. They also had injected various cocktails of free radical scavengers and lazeroids. These tonics were the only reason we were still alive. We'd soon be dead, however, if we did not get medical attention. These chemicals would only keep our devastated bodies alive for a day or two.

I think that the Ganymedean leader realized my plan was to capture him, and he could not allow this. Aided by the tiny chaotic attractor creatures, he had attacked me and had won.

"Arrgh..." I moaned as I continued to feel pain in my legs even though they were no longer attached to my body. I had once read about "phantom limb phenomena" where people who had lost a leg still perceived the limb as though it were still there. I experienced this myself months after the nautiloid water beings consumed my own arm. I knew that phantom pain arose from excessive firing from spinal cord neurons that have lost their normal sensory input. In order to decrease my pain, I chanted a few mantras hoping to halt the transmission of impulses though my spinal cord, my brain's thalmus, and the somatosensory areas of my cortex, but to no avail.

I suddenly vowed to make Yars Kotheck reconsider murdering me, and if he could not be persuaded, at least take revenge on him. But what could I do? I could not move.

I was filled with such bitter sorrow, not for me, but for Kalinda. But beneath the pain was a vast, white-hot core of growing rage.

Then I had an idea. I was able to dislodge my prosthetic arm from my body and control it via its infrared signal link. I sent a command to the disembodied arm:

Go to the King.

The Prohaptor and the King did not notice my arm slowly crawling to them. When it was almost by their side, the Prohaptor turned in its direction.

"Yaaah," I screamed to distract them and attract their attention away from the crawling arm. When it finally reached the King, I made it hop up and grab his neck with a force so great that the King and Prohaptor could not remove it.

"Get it off of him!" the Prohaptor screamed.

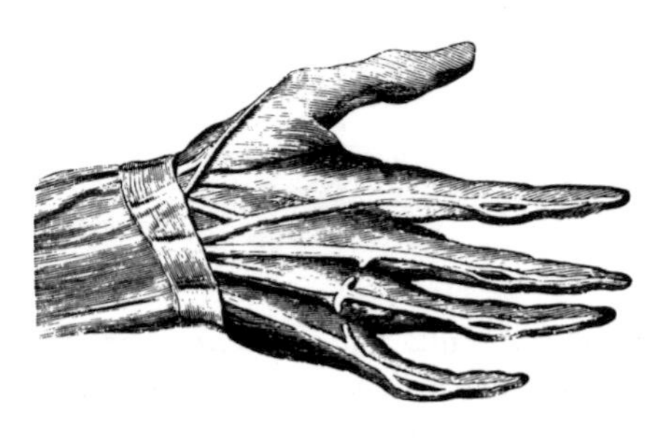

I realized that I could not persuade them to let us get off this moon. Therefore, I sent a signal to the arm.

Continually squeeze the King's neck with increasing pressure.

"Noooo," the Prohaptor yelled.

I heard the splintering sound of gallium arsenide. The Prohaptor screamed again, and the King made little choking noises. The Prohaptor continued to wrestle with the shaft of my remote-controlled arm, trying to remove or disable it, but it was too strong for her. Jumping up and down like a chicken without a head, the Prohaptor went wild. She hacked away at the synthetic flesh of my prosthetic arm with her long sword, yet by some miracle my hand could not be dislodged from the King's throat. Then there was a cracking sound, not in my arm, but in the King.

I looked down. The King's piezoelectric vocal cords, with some attached vasculature, lay on the ground and for a few seconds continued to scream, and at last were silent. Electrorheological fluid dripped from some of the torn arteries. The King's mouth and eyes opened and closed spasmodically and then stopped. His magenta eyes faded to ochre.

The Prohaptor ran to me and started slapping my face, back and forth, back and forth.... The pain was excruciating.

The Prohaptor suddenly stopped her beating. Why? I followed her gaze. She was staring at our Aysheaia worm from pipe-world which intently watched us. Perhaps the Prohaptor had never seen such a creature.

Was there anything else to do to save myself? I looked around and noticed that my artificial arm was close to the cracked specimen jar which held the fractal spider. I sent another mental signal to my arm:

Grab the spider and throw it at the Prohaptor's face.

My arm attempted to respond but was stuck. I cursed. Then I smiled. For some unknown reason, the Aysheaia worm slithered over to my arm and gave it a nudge to get it started. The arm finally responded and tossed the spider onto one of the Prohaptor's snouts. The Prohaptor screamed and the spider quickly scuttled down her throat. For the first time I saw fear in the Prohaptor's eyes. She screamed in the madness of extreme agony and clutched at her head. Perhaps the spider was travelling upward through her nasal cavities into her brain.

Her mouth dropped opened and she gagged, as if she wanted to scream but couldn't. Suddenly I saw the Prohaptor's Ka and Da run from her ears, down her body, and then into the surrounding underbrush.

"Come back here, you scoundrels," the Prohaptor yelled at her own Ka and Da brain symbionts. She took a few steps toward them, seemed to dance the watusi, and collapsed to the ground. She got up again, danced like a crazy woman, and fell down again. Apparently without Ka and Da she was losing control of her brain. The Prohaptor continued the repetitious gyrations until she fell down dead for a final time.

"Yes!" I screamed in triumph. Hopefully, if we could survive for a while, a rescue mission would be launched and both Kalinda and I would be medically resurrected. Nina stayed by my side, and the pipe-world worm stayed by Kalinda, occasionally fending off marauding Goobers who tried to crawl on Kalinda using their zinc oxide legs. Occasionally I heard the chomping of their tiny metallic jaws.

I thought we would make it, that we would survive, but was suddenly disappointed when I heard hoofbeats in the distance. A cloud of dust. I looked to my left and saw an army of Prohaptors coming toward us. Their fierce countenances burned into my soul like a hot poker boring through ice. The game was lost, the lights were fading. The feeling I had was overpowering: a clammy, heart-wrenching, chest-tightening helplessness. I saw Ka and Da jumping for joy.

"Garth," Kalinda whispered. Her voice trailed away like a dying flower. Every breath was punctuated with a respiratory wheezing.

"Don't talk. Save your energy." I looked at her face and saw terror born of the conviction that each breath could be her last. I tried to distance myself from the humiliation and the pain. The Prohaptors would soon be upon us.

One of the Prohaptors pulled out a long weapon resembling a rifle. He pointed it our way. To my shock, the rifle did not propel bullets but rather wasps – the same kind of wasps Kalinda and I found residing in the transparent organs of the glass girls of Ganymede. One of the wasps came closer to the Aysheaia worm. The worm swiveled its eye around and around in an attempt to visually track the wasp. Finally, with a dreadful buzzing, the wasp dived at the worm and stung it. An instant later the worm's bright eye turned an opalescent shade.

The worm did not die, but it did something strange. It began to thump on the ground with its long tail. *What's happening?* I thought.

Boom. Boom.

Boom. Boom.

Over and over again. It made a flat sound, as if someone were banging on a bongo drum. I heard the sound echo across the hills. Then the ground began to shake. The ice splintered.

It sounded like the tinkling of ice cubes. Like corks popping from champagne bottles, thousands of Aysheaia worms burst up through the ice. The ice around us shattered.

"Kalinda!" I cried. No longer did I hear the tinkling sound of ice, but rather of bells. The worms came close to one another, aligning themselves head to tail forming a living lumbricoid network that grew in size as the worms fastened themselves to create a complex weave. As the Aysheaia intertwined their bodies higher and higher off the ground, creating a cagelike structure around us, the sounds rose also, lifting in great waves. Church bells. Big bells. A clanging diapason of resonating worms. I moved my head to help me look in all directions.

The worms were forming a huge geodesic dome around us with their bodies, completely walling us off from the Prohaptors. As their bodies tightened, the sounds rose in pitch. At each vertex of the structure a worm pointed outward and hissed at the Prohaptors. Why were the worms helping? Was it because the wasp stung their friend – the worm I once fed and carried in my pocket? Did the worms hate the Prohaptors for some historical reason?

The Prohaptors stopped dead in their tracks. Their leader gasped. A few in front fell off their stallion-like beasts. The Prohaptors looked at our protective enclosure of resonating, rebarbative worms as if it were a nightmare beyond anything they had encountered, and they rode away. Those who had fallen off their riding beasts turned and ran. Ka and Da retreated with them as they stumbled and slid on the icy ground.

"Kalinda," I shouted. No answer. "Kalinda, I love you." I commanded my arm to crawl to her, to take her pulse using its bioelectric sensors. It gently grasped her neck... I waited for my arm to return a verdict... Breathing heavily I began to shiver.

She still lived.

I looked into her half-open eyes. The scintillating eyes of a doe. Like little shiny chips of quartz. They were the only reality in a shifting world.

Long has paled the sunny sky:
Echoes fade and memories die...
Still she haunts me, phantomwise.
Never seen by waking eyes.
– Lewis Carroll

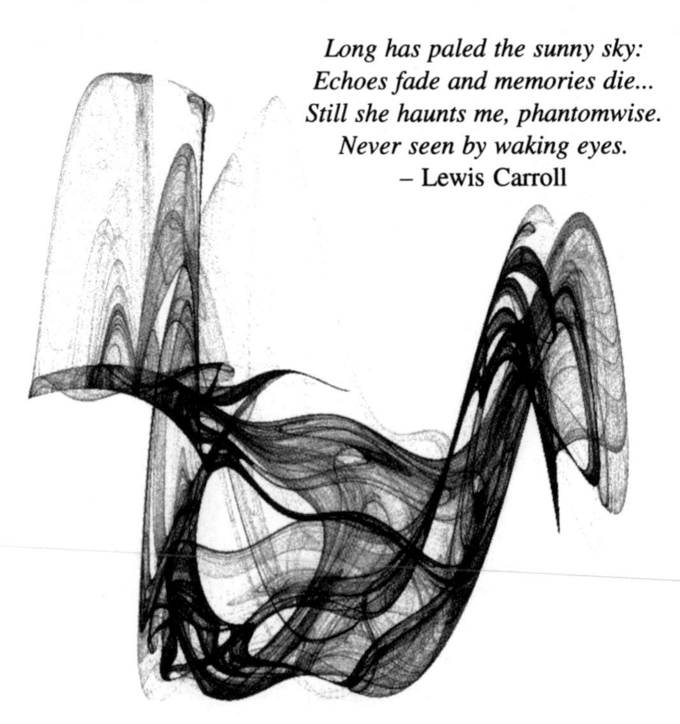

Part III

APPENDICES

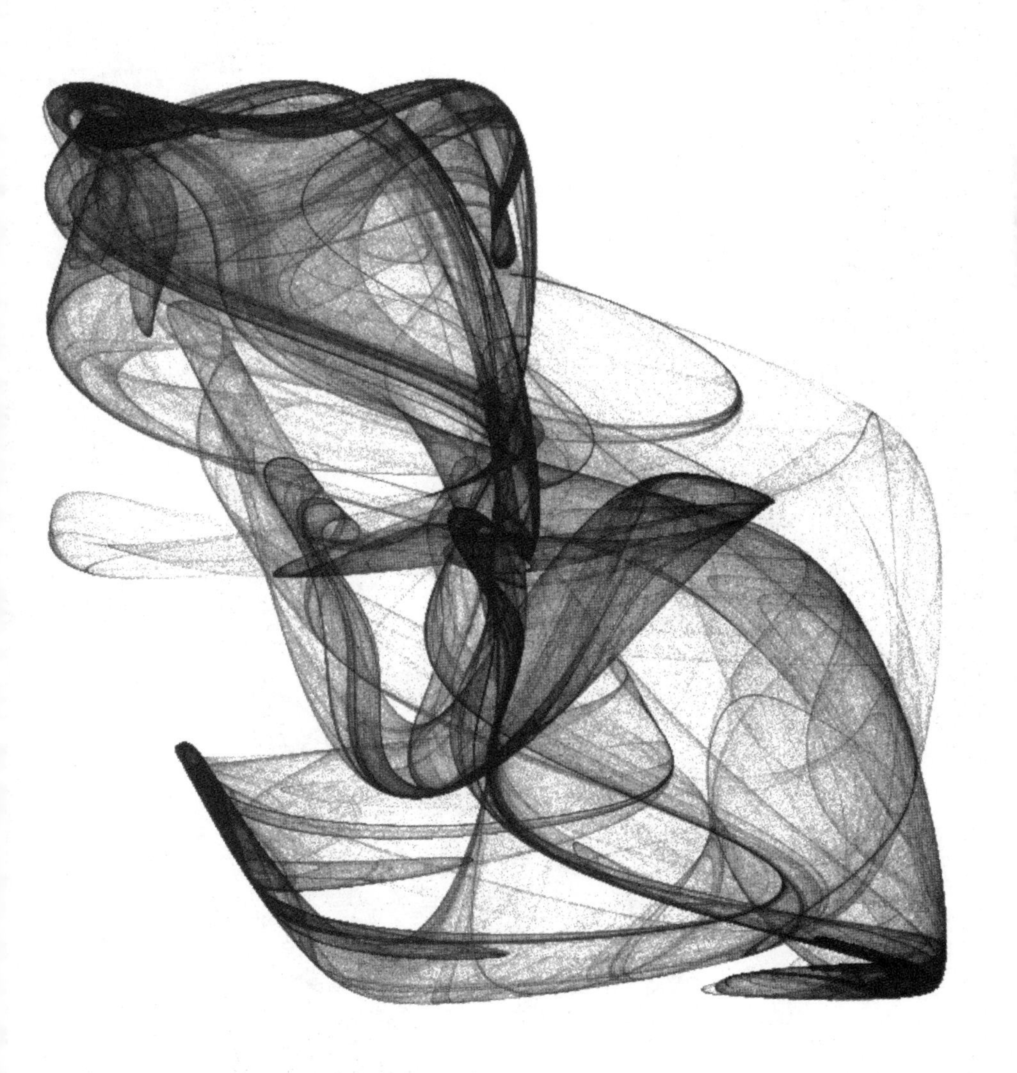

"Though the source be obscure,
still the stream flows on."
– Henri Poincare

Appendix A

Mutations of Equations

"The essence of mathematics is its freedom." Georg Cantor

Occasionally, due to reasons not entirely clear, a small percentage of Latöőcarfians exhibit mathematical patterns which are generated from equations slightly different from the standard Latöőcarfian formulas given below:

$$x_{t+1} = \sin(y_t b) + c\,\sin(x_t b) \qquad (A.1)$$

$$y_{t+1} = \sin(x_t a) + d\,\sin(y_t a) \qquad (A.2)$$

Variations include "Mutation α":

$$x_{t+1} = \sin(y_t b) + \sin^2(x_t b) + \sin^3(x_t b) \qquad (A.3)$$

$$y_{t+1} = \sin(x_t a) + \sin^2(y_t a) + \sin^3(y_t c) \qquad (A.4)$$

As with the standard Latöőcarfian formulas (see Chapter 3, "Modern History") these equations determine the position of points in intricate patterns as a function of time, represented by the subscript symbol t. Latöőcarfians who exhibit patterns based on the mutated equations are often viewed with scorn and derision, when it's possible to detect their transgression. There is great debate as to whether this change is an entirely voluntary act, or is a compulsion resulting from an inherited mutation. Some Latöőcarfian sociologists claim that family and environmental influences from birth are the cause of the aberration. To this day the debate over "nature vs. nurture" continues. Many of the mutants are relegated to the cotter and crofter class in Latöőcarfian society.

Here are some additional mutations. Mutation β is:

$$x_{t+1} = \sin(y_t b) + \sin^2(x_t b) \qquad (A.5)$$

$$y_{t+1} = \sin(x_t a) + \sin^2(y_t a) \tag{A.6}$$

Mutation Γ is:

$$x_{t+1} = |\sin(y_t b)| + \sin^2(x_t b) \tag{A.7}$$

$$y_{t+1} = |\sin(x_t a)| + \sin^2(y_t a) \tag{A.8}$$

A.1 Digressions

1. How do the symmetries of the patterns produced by the mutant Latööcarfians compare with those of the "normal" Latööcarfians?
2. Are the Lyapunov exponents of the mutants, on average, higher or lower than those for the normals. (See Chapter 14, "Interlude: Lyapunov Logomania.")
3. Do the mutants exhibit ghost siblings? (See 12.2, "Ghost Siblings.")

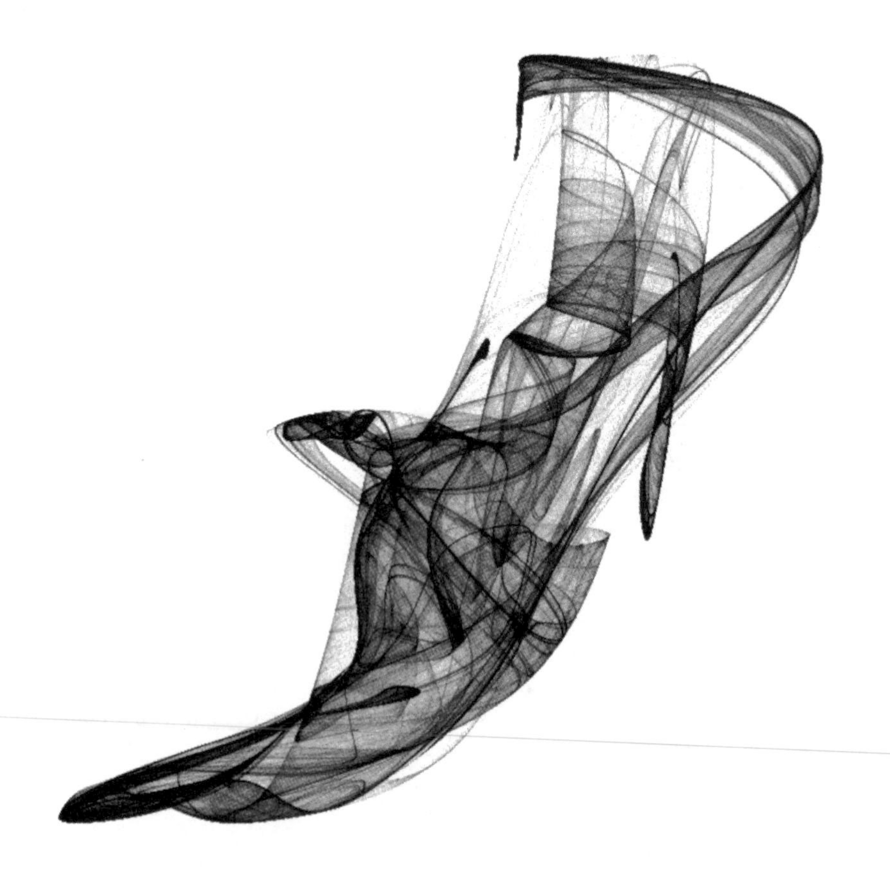

Appendix B

Color Plates and Frontispieces

"The eye, which is called the window of the soul, is the chief means whereby the understanding may most fully and abundantly appreciate the infinite works of nature."
Leonardo da Vinci

B.1 Color Illustrations

I produced the color computer graphics renditions in this book using custom C-language programs running on an IBM RISC System/6000.

Cover image: The computer graphic illustrating the cover of this book is a Julia set representation of $z \rightarrow z^2 + \mu$, with $\mu \sim (0.27334 + 0.00742i)$. ($z$ and μ are complex numbers.) These "mathematical feedback loops" have chaotic behavior with a sensitive dependence on initial conditions. As with other color plates in this book, the image was colored using a coloring program running on a IBM RISC System/6000. The program allows me to use a mouse to map data values in the image to colors. Colors generally indicate the rate of explosion for initial points in the complex plane. To create the fine-grained "stalks" I used a special convergence test described in my book *Computers and the Imagination.* (To produce the stalks, a dot is plotted whenever a trajectory wanders across a cross-shaped aperture centered at the origin.) For a wonderful collection of related Julia set images, see: Peitgen, H. and Richter, P. (1986) *The Beauty of Fractals* Springer: New York.

The color plates are described in the following:

1. Beginnings of the above-ground Latööcarfian civilization, painting by Alisa Franklin and Gerardo Amor (see Credits).
2. Julia set representation for $z \rightarrow z^2 + \mu$, where $\mu = (-0.09571678 + 0.6540306i)$. The picture boundaries are: $(-0.45258620 < z_{re} < -0.02586206)$ and $(0.27218390 < z_{im} < 0.64)$. These parameters were suggested by: Philip, A. G. Davis (1992) The evolution of a three-armed spiral in the Julia set, and higher order spirals. In *Spiral Symmetry*, Hargittai, I. and Pickover, C. (eds.) World Scientific: New Jersey. (World Scientific Publishing, Suite 1B, 1060 Main St, River Edge, New Jersey 07661. Phone: 800

227-7562, ISBN 981-02-0615-1). I iterated the mapping 2000 times and, as with other images, mapped data values to color using a custom color program running on an IBM RISC System/6000.

3. Same as cover description, but using different color map.
4. Lyapunov representation of the Latööcarfian chaotic mapping. (See section Chapter 14, "Interlude: Lyapunov Logomania.")
5. Mandelbrot set for $z \rightarrow z^2 + \mu$. (Motivated by experiments by Peitgen and Richter.)
6. Same as previous, but with artificial 4-fold symmetry. (Simply reflect the image through mirror planes.)
7. Julia set mapping for $z \rightarrow z^2 + \mu$.
8. Molecular model for spider web (Bucky-tube) strands, viewed from within the tube. (Coordinates supplied by Dr. Tom Jackman. See Chapter 23, "Fractal Spiders.")
9. Computer graphics representation of a 3-D spherical Lissajous curve. $x = r\sin(\theta t)\cos(\phi t)$, $z = r\cos(\theta t)$, $y = r\sin(\theta t)\sin(\phi t)$. Readers can try ratios of θ/ϕ such as 1/2 or 1/3. (See *Computers, Pattern, Chaos and Beauty*.)
10. Computer graphics representation of a 3-D spherical Lissajous curve. (See *Computers, Pattern, Chaos and Beauty*.)
11. Halley map. This is based on a numerical method for finding the roots of $z^7 - 1 = 0$. See *Computers, Pattern, Chaos and Beauty* for details.
12. 3-D computer graphic based on spherical Lissajous curves.
13. Julia set mapping for $z \rightarrow z^2 + \mu$. $\mu = (-0.39054 - 0.58679i)$. To create the fine-grained "stalks" I used a special convergence test described in my book *Computers and the Imagination*. The value of μ was suggested by: Peitgen, H. and Richter, P. (1986) *The Beauty of Fractals* Springer: New York.
14. Color composite of Jupiter's atmosphere. (Courtesy of Jet Propulsion Laboratory, California Institute of Technology.)
15. Asteroid exploration, by Beth Avary, Portola Valley, California. (See Credits.)
16. Mathematical art based on seashell growth formulas.
17. Mathematical art based on seashell growth formulas. A computer method called "ray-tracing" was used to created shadows and reflections.
18. Latööcarfian Pythagorean tree. On Earth, during the Second World War, A. E. Bosman (1891-1961) conveyed his wonder at the intricate geometrical shapes of nature by constructing line drawings of Pythagorean trees. (Using simple branching rules he drew them on the same drawing boards he used for designing submarines.) I produced this rendition using his basic ideas coupled with a random number generator to inject variation into the design. To produce a perfectly regular tree, start with a square and an isosceles right triangle from which two smaller squares sprout. Continue as diagrammed in the following:

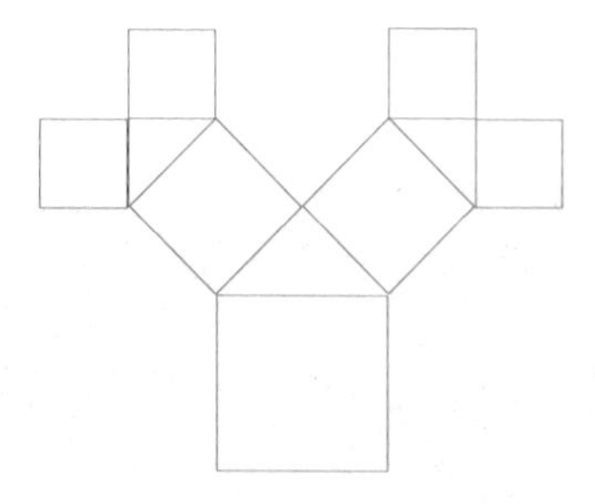

19. Beginnings of the above-ground Latööcarfian civilization, painting by Alisa Franklin and Gerardo Amor.

B.2 Some Figure Parameters and Descriptions

Various chaotic attractors are scattered throughout the book. The following are parameters you can use to create several of them, or closely related images. The Lyapunov exponent is also given. Mutations are explained in Appendix A, "Mutations of Equations." For interesting experiments, change the mutation while keeping *a, b, c* and *d* the same, and view the changes in the pattern.

Lyapunov	a	b	c	d	Mutation
The King					
0.486991	-0.966918	2.879879	0.765145	0.744728	Standard
Dreams of the Lords					
0.986274	-2.905148	-2.030427	1.440550	0.703070	Standard
0.833576	-2.951292	1.187750	0.517396	1.090625	Standard
0.510855	2.668752	1.225105	0.709998	0.637272	Standard
0.987196	1.380932	2.656301	1.157857	1.272576	Standard
0.839650	2.733940	1.369945	1.471923	0.869182	Standard
0.347174	1.008118	2.653920	0.599124	0.650700	Standard
Dreams of the Villeins and Serfs					
0.195531	-2.767266	-0.633839	1.352107	0.705481	Standard
0.462844	-0.299661	-2.315714	1.071856	1.404386	Standard
0.198499	-2.164647	-0.641713	1.277032	1.003342	Standard
Dreams of the Crofters and Cotters					
0.594922	2.039949	1.322977			Beta
0.253815	0.992187	-1.111942	-1.713095		Alpha
0.670506	0.805414	2.132054			Gamma
0.289456	1.966155	-1.005921	1.091876	1.466765	Standard
0.822533	-1.804285	-2.513108	0.957945		Alpha
1.163419	2.755364	2.791253			Beta

There are also many full-page, intricate spiral graphics scattered throughout the book. These are Julia sets for $z \rightarrow z^2 + c$, where c takes on values such as: (-.03249108, .79245342), (.4052405, .1467907), (-.74858089, .06360646), (.37953073, .20567030), (-.09571678, .65403036), and (-.74543, .11301). To enhance the quality of the images, I used a computational technique called the "Milnor-Thurston distance estimator method." An excellent reservoir of c values leading to beautiful Julia sets is Philip (1992) (see the reference for color plate 2).

Appendix C

Heñtriacontañe Games

Chapter 32, "The Imaginarium," discusses the Heñtriacontañes – a race of insectile philosopher-mathematicians on Ganymede. Status in their henotheistic society is based on the prowess with which an individual plays mathematical games and proves mathematical theorems.[25] The center of such activity is a building call the Imaginarium, which is shaped like a Mandelbrot set. As mentioned, there are various pleasurable rewards bestowed upon the Heñtriacontañe philosophers in proportion to the novelty of theorems they solve. This chapter describes a few additional, mathematically-based Heñtriacontañe games.[26]

C.1 Fractal Fantasy Game

The board for the *Fractal Fantasy Game* is a fractal nesting of interconnected rectangles (Figure C.1). So enthralled are the Heñtriacontañes with this game that the design has been cut into the roofing slabs of the Imaginarium building and in the floors of many Heñtriacontañe homes. The board for the *Fractal Fantasy Game* (FFG) is a fractal nesting of rectangles within rectangles interconnected with wiggly lines as shown in the diagram. There are always two rectangles within the rectangles which encompass them. The degree of nesting can be varied. Beginners play with only a few nested rectangles, while grand masters play with many recursively positioned rectangles. Tournaments last for days, with breaks only for eating and sleeping (and breeding, in the case of most Heñtriacontañes). The playing board illustrated in Figure C.1 is called a "degree 2" board, because it has

[25] One such game, the "Armacolite Parasite Game" is described in Chapter 32, "The Imaginarium."

[26] I welcome discoveries and observations on the games and puzzles in this section. To obtain the informal newsletter *Heñtriacontañe Puzzles*, which includes letters from readers and observations on the Heñtriacontañe games, write to me at the address given on the "About the Author" page. I'll publish a sampling of your comments in a future book.

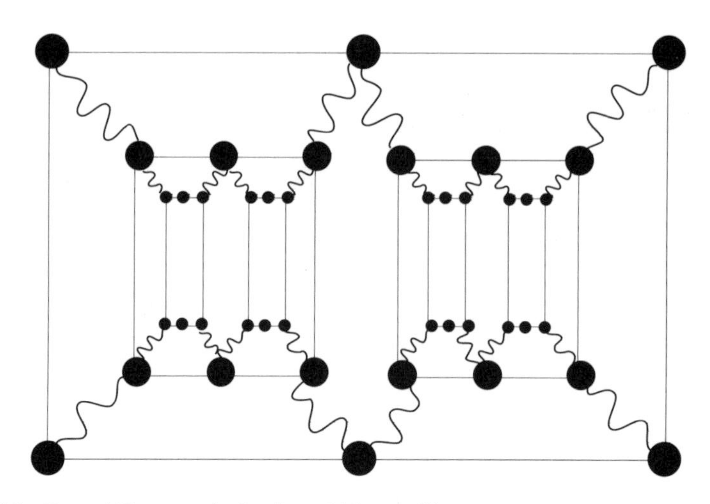

Figure C.1. *Fractal Fantasy playing board (degree 2).*

two different sizes of rectangles within the large bounding rectangle. Beginners usually start with a degree 1 board, and grand masters have been known to use a degree 20 board. One player uses white playing pieces (stones), the other uses black. Each player starts with a number of pieces equal to 1/2 the number of vertices (dots) on the board minus two. For the board here, each player gets 19 stones. With alternate moves, the players begin by placing a stone at points on the black dots which are empty. As they place stones, each player attempts to form a row of 3 stones along any one of the horizontal sides of any rectangle. This 3-in-a-row assembly of stones is called a *Googol*. When all the stones have been placed, players take turns moving a piece to a neighboring vacant space along one of the wiggly or straight connecting lines. When a player succeeds in forming a *Googol* (either during the alternate placement of pieces at the beginning of the game, or during alternate moves along lines to adjacent empty points) then the player captures any one of the opponent's pieces on the board and removes it from the board. (In some versions of the FFG, an opposing stone cannot be taken from an opposing *Googol*.) A player loses when he or she no longer has any pieces or cannot make a move.

Mathematicians and philosophers will no doubt spend many years pondering a range of questions, particularly for boards with higher nesting. Computer programmers will design programs allowing the board to be magnified in different areas permitting the convenient playing at different size scales. They'll all wish they had fractal consciousnesses allowing the contemplation of all levels of the game simultaneously.

C.1.1 Digressions

Many of the Heñtriacontañe philosophers have spent years of their lives pondering the following questions relating to the Fractal Fantasy Game. No Heñtriacontañe has succeeded in answering these questions for games with degree higher than 2. Various centers have been established and funded in order to answer the following research questions:

1. What is the maximum number of pieces which can be on the board without *any* forming a row?
2. Is there a best opening move?
3. If the large bounding square has a side one foot in length, and each successive generation of square has a length 1/6 of the previous, what is the total length of lines on the board?
4. If a Ganymedean fractal spider were to start anywhere on the board and walk to cover all the lines, what would be the shortest possible route on the board?
5. How many positions are possible after one move by each player?
6. How large would a degree-100 board have to be in order for the smallest squares to be seen? How many playing pieces would be used? What length of time would be required to play such a bizarre game?

C.1.2 Reader Comments

When the Fractal Fantasy Game was first published in *BYTE* magazine (1993, March, pg. 256) I received numerous mail from interested readers.

Martin Stone from Temple University suggests a distributed version of the game played over the Internet. He writes, "Imagine a multi-user recursive game server dedicated to the fostering of a greater intuitive understanding of recursive structures and permutations." David Kaplan from New York University points out that the game rules for the Fractal Fantasy Game are similar to a medieval game called Nine Man Morris played on a different playing board. Paul Miller notes that the Fractal Fantasy Game was discussed at the Boston Chapter of MENSA. He asks, "Can pieces of a Googol move out and back (thus forming and reforming the Googol)?" He suggests that the Googol pieces be allowed to move only if there is no other legal move. Alternatively, if a player moves a piece out of a Googol, he should not be allowed to move it back into the same place on the next turn. Michael Currin from the University of Natal (South Africa) suggests that the game be adapted to allow more than two players. Finally, Brian Osman, a 15-year-old from Massachusetts, writes:

> "I greatly enjoyed your description of the Fractal Fantasy Game in the March 1993 issue of *BYTE* magazine. However, I point out that some of what you said is almost impossible! I've calculated the number of rectangles and 'spots' for every size board, using the formula: $(2^{N+1}) - 1$, where N is the degree of

the board. From this, one can find the number of spots by simply multiplying by 6. Once you have this number, divide by two and subtract two to find the number of stones for each player. You have stated that grand masters have been known to use boards of degree 20. I've checked my calculations repeatedly, and this would require each player start with 6,291,451 stones! Assuming each opening move (only those to place your pieces) took 2 seconds, the players wouldn't be able to move until 291.2708797 days after they started the game. Am I missing something, or are your numbers as ludicrous as they seem to me? Please don't take offense at this. I still found your article very enjoyable."

C.2 Fractal Sponge Game and Hyperdimensional Zenza

Some Heñtriacontañe philosophers who master the Fractal Fantasy Game progress to the Fractal Sponge Game, a game with more difficult rules of play and greater potential for expanding their consciousness. Since most of the Heñtriacontañes are unable to master this game, there are only about twenty good players of the Fractal Sponge Game on Ganymede.

The game is played using pieces called *Prohaptors* on a board containing different-sized squares as shown in Figure C.2. (Prohaptors are fierce warriors on Ganymede.) Each square size is denoted by a different name:

1. The four largest squares on the playing board are called *Texcoco.* They form the channels through which the Prohaptors pass, and are not used (as the other squares are) as playing positions or stepping stones.
2. The nine intermediate sized squares are called *Texcoca* (filled in black in figure).
3. The 58 small squares are called *Texcaca.*

The topmost central square is marked *Ch'ut* which means both "entrance" and "exit." (The Heñtriacontañe philosophers do not distinguish between these two meanings.) The playing pieces, usually armacolite stones, represent Prohaptor warriors. Two players have four Prohaptors each. Each Prohaptor starts at the *Ch'ut* square and moves anticlockwise around the periphery of the porous sponge while stepping on the *Texcoca* and *Texcaca* in the channels formed by the four *Texcoco.* The goal of the game is to get one's Prohaptor pieces to return (or go beyond) the *Ch'ut* square after the anticlockwise travel.

The Heñtriacontañes use discs with a black and white side to determine the movement of playing pieces. Pieces are moved according to the orientation of four such discs. For convenience, you can use pennies. Throw four pennies and count as follows:

1. 1 head up - score 1
2. 2 heads up - score 2

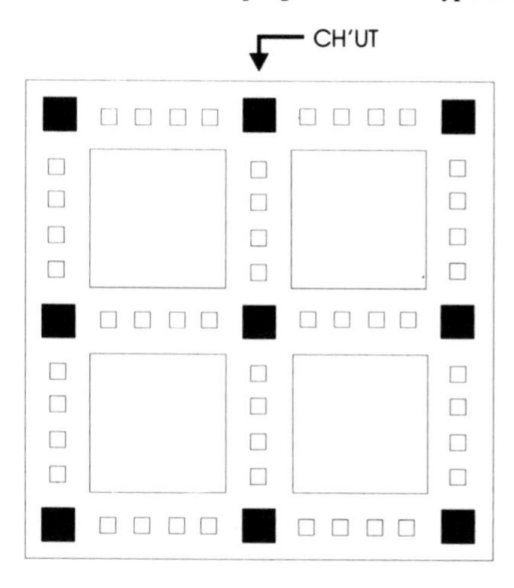

Figure C.2. *Fractal Sponge Game.* (In the authentic version, several adjacent boards are used, creating a fractal board in a game known as *Zenzo*, as described in the text.)

3. 3 heads up - score 3
4. 4 heads up - score 4
5. 0 heads up - score 5

If a score of 4 or 5 is thrown, then the player has an extra throw. The scores are added together and can be used for one Prohaptor or spit between two or more Prohaptors. For example, a player might throw 5, 4, 1. The possible moves are then:

1. One Prohaptor moves 10 squares.
2. Two Prohaptors move 9 and 1 squares, or 6 and 4 squares, or 5 and 5 squares.
3. Three Prohaptors move 5, 4, and 1 squares.

If a Prohaptor lands on any of the *Texcoca* squares at the bottom, left, and right of the playing board it must leave the outermost channel and travel to the exit square (*Ch'ut*) through a central route (which forms a cross), passing through the center *Texcoca*. Possible routes are then:

1. *Ch'ut*, left *Texcoca*, central *Texcoca*, *Ch'ut*. (Using this route, a Prohaptor piece passes through 19 squares, not including the beginning and ending *Ch'ut square*.)
2. *Ch'ut*, bottom *Texcoca*, central *Texcoca*, *Ch'ut*.
3. *Ch'ut*, right *Texcoca*, central *Texcoca*, *Ch'ut*.
4. *Ch'ut* to *Ch'ut* around the perimeter of the board.

If one Prohaptor lands on another of the player's own Prohaptors, the two are joined together and move as a single piece as a family unit. Similarly three or four Prohaptors can join together, and move as a single piece.

If a Prohaptor lands on a square occupied by an opponent, the opponent is sent back to the start if the piece landing is made up of at least as many Prohaptors as the occupying piece. A piece cannot land on a square if the opponent already occupies that point with a piece of higher order (a bigger family).

The game just described is the beginner's version of the Fractal Sponge Game. When a Heñtriacontañe child is 7 years old, he or she is taught the game and spends the childhood years attempting to master it. At the age of 13, a lavish ceremony is held by the parents of the child (somewhat like a Bar Mitzvah on Earth), signifying that the child is ready to progress to a more advanced, fractal version of the game called *Zenzo*. As suggested at the beginning of this section, there are only a handful of master Zenzo players on Ganymede. A few thousand Heñtriacontañes play the Zenzo game in elaborate contests within the Imaginarium walls, attempting to achieve master status at the game. In this multidimensional fractal version, more than one playing board is used, each interconnected and accessible through the *Texcoca* squares. The boards can be placed on the ground from left to right as follows:

1. Board 1 (placed at left) – same as in Figure C.2.
2. Board 2 – same as Board 1, except that the number of small *Texcaca* squares in each channel is reduced from four to three. The number of *Texcoco* and *Texcoca* remains the same and surround the *Texcaca* as in Board 1. To produce this board, simply erase one *Texcaca* between each *Texcoca.*
3. Board 3 – same as Board 2, except that the *Texcaca* in each channel are reduced from three to two.
4. Board 4 – same as Board 3, except that the *Texcaca* in each channel are reduced from two to one.
5. Board 5 – same as Board 4, except that the *Texcaca* in each channel is reduced from one to zero. Only the *Texcoca* and *Texcoco* remain.

The game is played in the same manner as previously described on Board 1. All Prohaptors start at the *Ch'ut* of Board 1. Whenever a player lands on a *Texcoca* square he then has the option of traveling through an interconnection to an adjacent board.

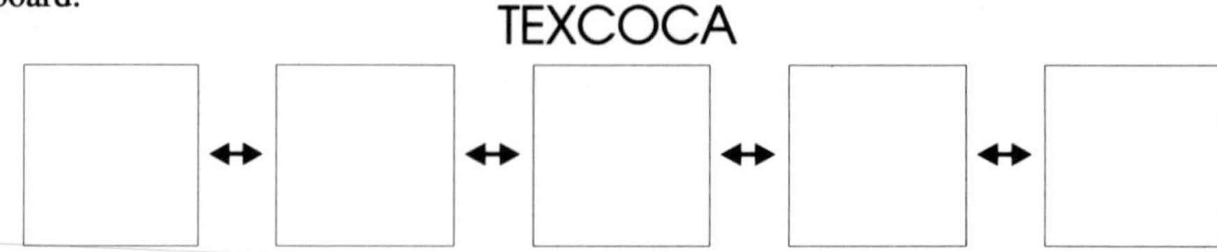

If a piece lands on a point occupied by an opponent, the opponent must start again and reenter the *Ch'ut* of Board 1. A player's goal is to get all of his pieces off of any of the playing boards via their *Ch'uts.*

C.2.1 Digressions

1. Does the player who goes first have an advantage?
2. What is the quickest possible game?
3. Assume no coalescing of multiple Prohaptors into family units, and assuming that no pieces are sent back, what are the total possible lengths of routes for the four Prohaptors together?
4. Investigate the game in higher dimensions. For example, use 10 boards, rather than 5. This requires that you start with a board with more *Texcaca*.
5. Change the rule in Zenza so that if a piece lands on a point occupied by an opponent, the opponent is sent back to the *Ch'ut* of the current board, rather than the *Ch'ut* on Board 1.

C.3 Fractal Triangle Game

Some of the Heñtriacontañe philosophers spend hours contemplating a number puzzle played on a board of nested triangles (Figure C.3). A single digit is placed at the vertices of each triangle so that the line segment connecting adjacent digits creates a two-digit number that is a multiple of 7. The digits in the two-digit number can be read in either order so long as one of the orders creates a multiple of 7. For example, a line connecting 5 and 3 is valid because 35 is a multiple of 7. As an example, place the digit "5" at the bottom left triangle in the following diagram:

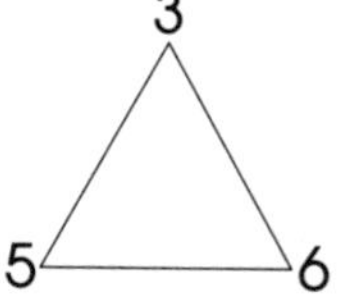

Next, fill in the remaining two vertices so that the two digit numbers are multiples of 7. In the example, we have 35, 63, and 56. Now add a smaller triangle creating three new vertices, and try to find digits that will satisfy the rules. In this example, the two digit numbers are: 56, 63, 35, 56, 63, and 35 for the outer triangle, and 63, 56, and 35 for the inner triangle.

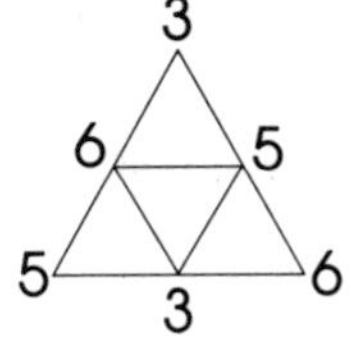

In this example, the next triangle can be a copy of the first large triangle:

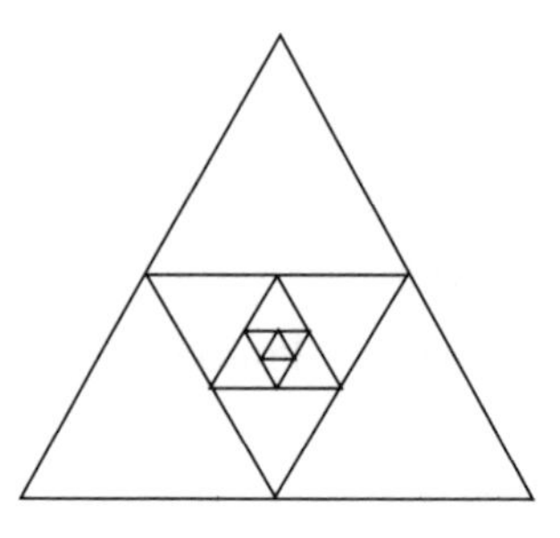

Figure C.3. *Playing board for the Fractal Triangle Game.*

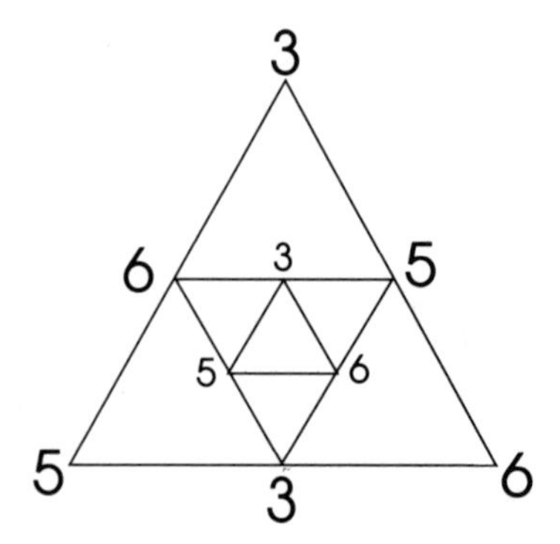

You can continue oscillating the values in this "cycle 1" solution which contains only one unique triangle, the 5,3,6 triangle. For decades Heñtriacontañe philosophers have been trying to find "cycle 3" and higher order solutions? Are these possible?

C.3.1 Digressions

1. Select a random number between 1 and 9. Place it in the lower left corner of the starting triangle. Can you make a cycle 2 fractal? Are there solutions for any starting number you choose?
2. What is the largest "cycle" solution that can be found?
3. The game can be played with two people. One player writes a digit at a vertex. The second player writes a number at another vertex. The first person who cannot place a digit because it would create a 2-digit number which is not a multiple of 7 is the looser.
4. Extend the game so that *either* multiples of 7 *or* 13 can be used to complete the triangle. How many nested triangles can you complete before violating the rules?

C.4 IQ-Block

An interesting example of cultural contamination occurred in the Heñtriacontañe land when a visitor from earth left behind a mathematical puzzle called *IQ-Block* (manufactured by Hercules, Item No. P991A, UK Registered No. 2013287, made in Hong Kong). The puzzle, schematically illustrated at left, consists of ten brightly colored polygonal pieces of plastic. The ten pieces fit together to form a square. Only one piece is shaped like a rectangle. The others are more complex. One is shaped like a "Z." The remainder are "L"-shaped. To play IQ-Block, first choose a shape you like, place it in the upper left corner, and do not change its position as you try to place the other nine blocks into the remaining space on the square playing board. The manufacturer boasts, "There are more than sixty different kinds of arrangements" of pieces that will fill in the square playing board. They also state, "It is an incredible game. Join us in challenging your I.Q."

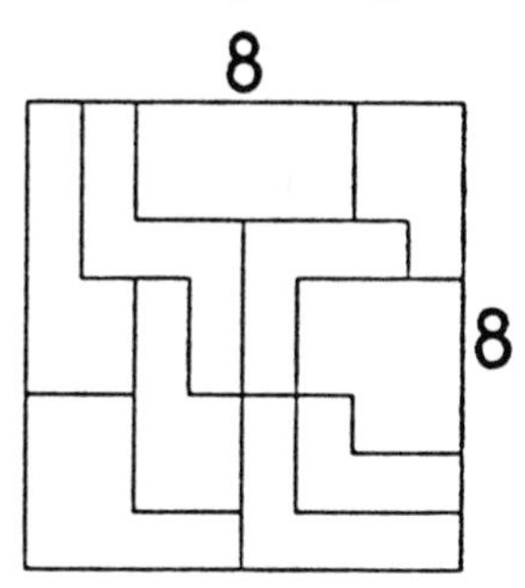

After the Heñtriacontañes found IQ-Block protruding from the ice in an area near the Imaginarium, they quickly translated the playing manual and passed out copies to all their citizens. They began to study the game and hold tournaments. During these tournaments Heñtriacontañe masters attempted to form as many possible different arrangements of pieces within the bounding square before a ten-minute alarm rang.

C.4.1 Digressions

1. The manufacturer of "IQ-Block" boasts that there are over sixty different ways of placing the pieces together to form a square. Is this correct? Just how many different arrangements are there? Some Heñtriacontañe philosophers argued that there are only ten different unique arrangements while members of another school of philosophy asserted that there are over one thousand ways of solving the puzzle! Who is closer to the truth?
2. On one particularly frigid evening, in a fit of frustration, a Heñtriacontañe master swallowed a polygonal playing piece to prevent the opponent from finding any solutions. Their multifaceted eyes locked in open warfare. Suddenly a blush of pleasure rose to his opponent's cheeks as she then created a square of slightly smaller dimensions. Can you create a square after removing a piece and using all the remaining pieces?
3. After giving this puzzle to five friends, I found that none could create a square by arranging the ten pieces. So, even if there are "sixty different ways" of solving this, readers should not despair if the task seems too difficult. Try IQ-Block on your own friends and colleagues, children and students, to see if

any can even find a *single* way of arranging the pieces to form a square. I look forward to hearing from readers regarding this intriguing puzzle.

Note: When I presented this puzzle to Joseph Madachy, editor of *Journal of Recreational Mathematics*, he remarked, "I say you cannot create a square after removing a single piece and using all the remaining pieces. The area of the complete Block is 64. The areas of the 10 pieces are 8, 5, 7, 7, 8, 6, 5, 4, 6, 8. Removal of one of these pieces is simply insufficient to produce the next smallest square (7x7=49 area). I haven't tried it, but it might be possible if two pieces are removed."

C.5 Heñtriacontañe Number Sequence

On one fine frigid day, a great Heñtriacontañe philosopher gave his Heñtriacontañe followers a number puzzle. "What is the significance of the following sequence of digits?" he asked them.

```
425260376469080434957
```

After rattling off the sequence, he suffered a heart attack and died. The evening air was as astringent as alcohol, as the Heñtriacontañes continued to study the sequence while the dead philosopher's body grew cold on the snowy ground. Even after a century of study, no Heñtriacontañe (or Earthling) could fathom the mystery of this sequence. Can you?

C.6 Apocalyptic Powers

One Heñtriacontañe philosopher presented the following number to his disciples and asked them what was special about it:

```
182687704666362864775460604089535377456991567872
```

After much discussion, one young Heñtriacontañe spoke: "It is the first power of two that exhibits three consecutive 6's." The audience of Heñtriacontañes applauded with delight. In fact the "...666..." number is equal to 2^{157}. I've called numbers of the form 2^i, which contain the digits 666, "apocalyptic powers" because of the prominent role 666 plays in the last book of the New Testament. In this book called The Revelation (or Apocalypse) of John, 666 is designated as the Number of the Beast, the Antichrist. Are there any other apocalyptic powers for higher values of i, or is this the only one? Heñtriacontañes have enlisted the help of the Latööcarfians in the computational search for "double apocalyptic powers" which contain six 6's in a row, but to date none have been found. I will publish your discoveries on apocalyptic powers in a future book.

Appendix D

The 100 Strangest Mathematical Titles Ever Published

"I once asked Gregory Chudnovsky if a certain impression I had of mathematicians was true, that they spend immoderate amounts of time declaring each other's works trivial. 'It is true,' he admitted."

Richard Preston, 1992, *The New Yorker*

The following is a list of one hundred serious mathematical papers with strange, indecipherable, and/or amusing titles. Candidates for this list were nominated by students, educators, and researchers around the world. To be eligible for the list, all papers must have been published in scientific journals (preferably mathematics or physics journals) rather than popular magazines. Some book titles are also listed.

D.1 The Top Ten

The award for all-time strangest title goes to Dr. A. Granville for:

Granville, A. (1992) Zaphod Beeblebrox's brain and the fifty-ninth row of Pascal's Triangle. *American Mathematical Monthly.* April, 99(4): 318-331.

Second place prize goes to Dr. Forest W. Simmons of Portland Community College for

Simmons, F. (1980) When homogeneous continua are Hausdorff circles (or yes, we Hausdorff bananas). In *Continua Decompositions Manifolds (Proceedings of Texas Topology Symposium 1980).* University of Texas Press. (Not too surprisingly, the illustrations are reminiscent of bananas!)

Third Place Prize goes to the romantic S. Strogatz for:

Strogatz, S. (1988) Love affairs and differential equations. *Mathematics Magazine.* 61(1): 35. (This is an analysis of the time-evolution of the love affair between Romeo and Juliet).

Fourth Place Prize goes to A. Berezin:

Berezin, A. (1987) Super super large numbers. *J. Recr. Math.* 19(2): 142-143. (This paper discusses the mathematical and philosophical implications of the "superfactorial" function defined by the symbol \$, where

$$N\$ = N!^{N!^{N!}} \cdots \qquad (D.1)$$

The term $N!$ is repeated $N!$ times).

Fifth Place Prize goes to A. Mackay:

Mackay, A. (1990) A time quasi-crystal. *Modern Physics Letters B.* 4(15): 989-991.

Sixth Place Prizes goes to J. Tennenbaum:

Tennenbaum, J. (1990) The metaphysics of complex numbers. *21st Century Science.* Spring 3(2): 60.

Seventh place prize goes to T. Morley for:

Morley, T. (1985) A simple proof that the world is three-dimensional. *SIAM Review.* 27: 69-71. (The article starts, "The title is, of course, a fraud. We prove nothing of the sort. Instead we show that radially symmetric wave propagation is possible only in dimensions one and three.")

Eighth place prize goes to the encyclopedic Prof. Akhlesh Lakhtakia, from the Department of Engineering Science and Mechanics at Pennsylvania State University, for:

Lakhtakia, A. (1990) Fractals and The Cat in the Hat. *Journal of Recreational Math.* 23(3): 161-164. (Reprints available from: Prof. A. Lakhtakia, Dept. of Engineering Science, Pennsylvania State University, University Park, PA 16802)

Honorable mention goes to the following titles:

1. R.C. Lyness (1941) Al Capone and the Death Ray. *Mathematical Gazette.* 25: 283-287.
2. Englebretsen, G. (1975) Sommers' proof that something exists. *Notre Dame J Formal Logic* 16: 298-300. (The review (MR 51 #7803) by K. Inoue says "The author points out that F. Sommers's proof that something exists is invalid...")
3. Hale, R. (1978) Logic for morons. *Mind.* 87: 111-115.
4. Braden, B. (1985) Design of an oscillating sprinkler. *Mathematics Magazine.* 58: 29-33.
5. Pickover, C. (1993) Apocalypse numbers. *Math Spectrum,* 26(1): 10-11.
6. Bing, R. (1954) Locally tame sets are tame. *Annals of Math..* 59: 145-158.
7. Pickover, C. (1989) Chaotic fragmentation in Halley's Paradise. *Physica Scripta.* 39: 193-195.
8. Taylor, C. (1990) Condoms and cosmology: the 'fractal' person and sexual risk in Rwanda. *Social Science and Medicine.* 31(9): 1023-8. (This entry

would have been higher up on the list had it been published in a mathematics journal.)

9. Hoenselaers, C, and Skea, J. (1989) Generating solutions of Einstein's field equations by typing mistakes. *Gen. Rel. Grav.* 21: 17-20. (The authors made some typing mistakes entering the problem into a computer, and came out with new solutions to the equations.)

D.2 Runners-Up List

1. Coxeter, H. S. M. (1983) My graph. *Proc. London Math. Soc.* 46(3): 117-136. (The title refers to a graph which ironically came to be called the Coxeter graph even though prior work had been done on the graph by other researchers.)
2. Jazaeri, A. and I. Satija (1992) Double devil's staircase in circle maps. *Physical Review A*, July 15, in press.
3. Evans, D. (1980) On O_n. *Publ. Res. Inst. Math. Sci.* 16: 915-927.
4. Nottale, L. (1991) The fractal structure of the quantum space-time. In Heck, A. and Perdang, J. *Applying Fractals in Astronomy*. Springer: NY.
5. H. Araki, A. L. Carey, and D. E. Evans, (1984) On O_{n+1}. *J. Operator Theory.* 12: 247-264.
6. Wilf, H. (1982) What is an answer? *American Math Monthly.* May 89(5): 289.
7. Landini, G. (1991) A fractal model for periodontal breakdown in periodontal disease. *J. Periodont. Res.* 26: 176-179.
8. Freedman, A. (1977) Surgery on codimension 2 submanifolds. *Memoirs of the American Mathematical Society, No. 191.* AMS: Providence.
9. Brody, A. (1987) About irremediable multicollinearity. *Szigma* (Hungary). 20(4): 287-298. (Winner: best title of an international paper.)
10. Johnstone, P. (1981) Scott is not always sober. In *Continuous Lattices.* Springer Lecture Notes in Mathematics 871, 282-283. (The "Scott" in the title refers to Scott toplogies named after Dana Scott, and "sober" denotes a topological property.)
11. Lieb, E. (1967) Residual entropy of square ice. *Phys. Rev.* 162: 162-172.
12. Moss, J. (1972) Some B. Russell's sprouts (1903-1908). In *Conference in Mathematical Logic, Lecture Notes in Mathematics 255.* Springer: NY. pp 211-250.
13. Laver, R. (1978) Making the supercompactness of κ indestructible under κ directed closed forcing. *Israel Journal of Mathematics.* 29(4): 385-388.
14. Cutting, J., and Garvin, J. (1987) Fractal curves and complexity. *Perception and Psychophysics*. 42: 365-70.
15. Halmos, P. (1944) In general a measure preserving transformation is mixing. *Annals of Math.* 45: 786-792. Also: Rokhlin, V. (1948) In general, a measure-preserving transformation is not mixing. *Dokl. Akad. Nauk.* 60: 349-351.
16. Wall, W. (1990) On the notion 'derivational constraint of grammar,' or: the Turing machine doesn't stop here anymore (if it ever will). In *Studies Out in Left Field: Defamatory Essays Dedicated to James McCawley on the Occasion of His Thirty-Third or Thirty-Fourth Birthday.* (Mathematical linguistics.)

17. Chow, Y., Robbins, H., and Siegmund, D. (1971) *Great expectations: the theory of optimal stopping.* Houghton-Mifflin: Boston. (Note: in the theory of optimal stopping, one tries to make the expectation of a random variable as large as possible.)
18. Kac, M. (1966) Can one hear the shape of a drum? *American Math. Monthly.* April 73: 1-23.
19. Gordon, C, Webb, D., and S. Wolpert (1992) One cannot hear the shape of a drum. *Bulletin of the American Mathematical Society (New Series).* July 27(1): 134-138.
20. Mendes-France, M. (1988) Nevertheless. *Mathematical Intelligencer.* 10(4).
21. Gott, J. (1967) Pseudopolyhedrons. *American Math Monthly.* May.
22. Pedersen, J. (1975) Collapsoids. *Mathematical Gazette.* No. 408.
23. Thatcher, D. (1990) The length of a roll of toilet paper. In *Mathematical Modelling.* Oxford University Press: NY. (Teachers and students of mathematical modeling will find this book a rich source of examples ranging from insulating houses to basketball, and from modelling epidemics to studying the generation of windmill power.)
24. Conway, J. (1987) The weird and wonderful chemistry of audioactive decay. In *Open Problems in Communications and Computation.* T.M. Cover and B. Gopinath, eds. Springer Verlag: NY. (Conway studies the sequence: 3, 13, 1113, 3113, 132113, 1113122113, ...)
25. Mermin, N, Rokhsar, N. and Wright, D. (1987) Beware of 46-fold symmetry: The classification of two-dimensional quasicrystallographic lattices. *Physical Review Letters.* 58(20): 2099-2101. (An application of number theory to quasicrystals.)
26. Knuth, D.E. (1984) The toilet paper problem. *American Math Monthly.* October, 91(8): 465-470.
27. Feldman, W. (1931) *Rabbinical Mathematics and Astronomy.* 2nd American edition. Hermon Press, New York.
28. Milnor, J.(1956) On manifolds homeomorphic to the 7-sphere. *Ann. of Math..* 399-405
29. Munkres, J. (1960) Obstructions to the smoothing of piecewise-differentiable homeomorphisms. *Ann. of Math.* 72: 521-554.
30. Reznik, B. (1983) Continued fractions and an annelidic PDE. *Math Intelligencer.* 5(4): 61-63. (The term "annelidic" means earthworm-like. The paper starts, "If you cut an earthworm (annelid) in two, each half will regenerate its missing part and become a new earthworm.")
31. Pickover, C. (1990) Juggler geometry and earthworm algebra. *Algorithm.* Nov. 1(7): 11-13.
32. Pickover, C., Khorasani, E. (1991) Visualization of the Gleichniszahlen-Reihe, an Unusual Number Theory Sequence, *Math. Spectrum*, 23(4): 113-115.
33. Pickover, C. (1990) Results of the very-large-number contest. *J. Recr. Math.* 22(4): 249-252.
34. Kustin, A. and B. Ulrich (1992) If the socle fits. *Journal of Algebra.* April 147(1): 63-80.
35. Pickover, C. (1990) On the aesthetics Sierpinski gaskets formed from large Pascal's triangles, *Leonardo.* 23(4), 411-417.
36. Pickover, C. (1992) Pentagonal Chaos. In *Five-Fold Symmetry* I. Hargittai, ed. World Scientific: Singapore.

37. Pickover, C. (1989) A note on computer experiments with chaotic shattering of level sets. *Computers in Physics*. Nov/Dec 3(6): 69-73.

38. Pickover, C. (1988) Aesthetics and iterative approximation. *Computer Language*. November, 5(11): 53-57.

39. Pickover, C. (1987) Blooming Integers: An elegantly simple algorithm generates complex patterns (Mathematics and Beauty III) *Computer Graphics World*. (March), 10(3): 54-57.

40. Baum, P. (1993) Chern classes and singularities of complex foliations. *Proc. Symp. Pure Math. 27, Differential Geometry, Amer. Math. Soc.*, to appear.

41. Bott, R. (1962) A note on the KO-theory of sphere bundles. *Bull. Amer. Math.* Soc. 68: 395-400.

42. Bott R., and Chern. S. (1965) Hermitian vector bundles and the equidistribution of the zeroes of their holomorphic sections. *Acta Math.* 114: 71-112.

43. Bott R., and Milnor, J. (1958) On the parallelizability of spheres. *Bull. Amer. Math. Soc.* 64: 87-89.

44. Chapman, T. (1973) Compact Hilbert cube manifolds and the invariance of Whitehead torsion. *Bull. Amer. Math. Soc.* 79: 52-56.

45. Chern, S. (1948) On the multiplication in the characteristic ring of a sphere bundle. *Ann. of Math.* 49: 362-372.

46. Haeflinger, A. and C. Wall (1965) Piecewise linear bundles in the stable range. *Topology*. 4: 109-214.

47. Hirsch, M. (1963) Obstruction theories for smoothing manifolds and maps. *Bull. Amer. Math. Soc.* 69: 352-356.

48. Hirzebruch, F. (1953) On Steenrod's reduced powers, the index of inertia and the Todd genus. *Proc. Nat. Acad. Sci. U.S.A.* 39: 951-956.

49. Peterson, F. (1970) Twisted cohomology operations and exotic characteristic classes. *Advances in Math.* 4: 81-90.

50. Szczarba, R. (1964) On tangent bundles of fibre spaces and quotient spaces. *Amer. J. Math.* 86: 685-697.

51. Thomas, E. (1962) On the cohomology groups of the classifying space for the stable spinor group. *Bol. Soc. Mat.* Mex: 57-69.

52. Thurston, W. (1972) Non-cobordant foliations of S^3. *Bull. Amer. Math. Soc.* 78: 511-514.

53. Wood, J. (1971) Bundles with a totally disconnected structural group. *Comm. Math. Helv.* 46: 257-273.

54. Propp, J. (1983) Nim for three: an overview and an offer of alcohol. *Eureka*. 43: 41-46.

55. Propp, J. (1989) What are the laws of greed? *American Mathematical Monthly* 96: 334-336.

56. Wu, W.T. (1948) On the product of sphere bundles and the duality theorem modulo two. *Ann. of Math.* 49: 641-653.

57. Linderholm, C. (1972) *Mathematics Made Difficult*. World Publishing: New York.

58. Briggs, K. et al. (1991) Feigenvalues for Mandelsets. *Journal of Physics*. A24: 3363.

59. Moser, L. (1982) A Mathematical analysis of the game of jai alai. *American Math Monthly.* May 89(5): 292.

60. Rosenfeld, A., and Kim, C. (1982) How a digital computer can tell whether a line is straight. *American Math Monthly.* April 89(4): 230.

61. Hwang, F. (1982) New concepts in seeding knockout tournaments. *American Math Monthly.* April 89(4): 235.

62. Buchman, E. and Valentine, F. (1982) Any new Helly numbers? *American Math Monthly.* June 89(6): 370.

63. Borwein, J, Borwein, P. and Bailey, D. (1989) Ramanujan, modular equations, and approximations to pi or how to compute one billion digits of pi. *American Math Monthly.* March 96(3): 201.

64. Eidswick, J. (1986) Cubelike puzzles – what are they and how do you solve them? *American Math Monthly.* March 93(3): 157.

65. Turner, J. (1986) On caterpillars, trees, and stochastic processes. *American Math Monthly.* March 93(3): 205.

66. Allen, T. (1986) On Polya's orchard problem. *American Math Monthly.* Febr 93(2): 98.

67. McCoy, T. (1986) On $\cos F(x) = F(\sin X)$. *American Math Monthly.* Febr 93(2): 111.

68. Beck, A. (1986) The broken spiral theorem. *American Math Monthly.* April 93(4): 293.

69. Strang, G. (1982) The width of a chair. *American Math Monthly.* Oct 89(8): 529.

70. Göbel, F. (1982) How many different rinds can you peel from a sequence? *American Math Monthly.* Febr 89(2): 113.

71. Wilker, J. (1982) Rings of sets are really rings. *American Math Monthly.* March 89(3): 211.

72. Fearnley-Sander, D. (1982) Hermann Grassmann and the prehistory of universal algebra. *American Math Monthly.* March 89(3): 161.

73. Reilly, R. (1982) Mean curvature, the laplacian, and soap bubbles. *American Math Monthly.* March 89(3): 180.

74. Taussky, O. (1988) How I became a torchbearer for matrix theory. *American Math Monthly.* Nov 95(9): 801.

75. Morgan, F. (1988) Area-minimizing surfaces, faces of grassmannians, and calibrations. *American Math Monthly.* Nov 95(9): 813.

76. Brooks, R. (1988) Constructing isospectral manifolds. *American Math Monthly.* Nov 95(): 823.

77. Kalmbauer, K. (1988) On a property of $x^n e^{-x}$. *American Math Monthly.* June 95(6): 551.

78. Froemke, J. and Grossman, J. (1988) An algebraic approach to some number-theoretic problems arising from paper-folding regular polygons. *American Math Monthly.* April 95(4): 289.

79. Robinson, R. (1988) How big a slice can you make through a cube? *American Math Monthly.* April 95(4): 331.

80. Brenner, J. (1988) An elementary approach to $y'' = -y$. *American Math Monthly.* April 95(4): 344.

81. Passoja, D. and A. Lakhtakia (1992): Variations on a Persian theme. *J. Rec. Math.*, 24: 1-5. (The paper deals with variations on the Pascal triangle that are connected with the square roots of integers).

82. Mathews, J. (1990) Gear tains and continued fractions. *American Math Monthly.* June 97(6): 505-510.

83. Gallian, J. and Douglas, J. (1988) Homomorphisms from $Z_m[i]$ into $Z_n[i]$ and $Z_m[\rho]$ into $Z_n[\rho]$, where $i^2 + 1 = 0$ and $\rho^2 + \rho^1 + 1 = 0$. *American Math Monthly.* March 95(3): 247.

84. Koblitz, N. (1988) Problems that teach the obvious but difficult. *American Math Monthly.* March 95(3): 254.

85. MacKinnon, N. (1989/1990) Modelling Monopoly. *Math Spectrum.* 22(2): 39. (Describes computer simulations of the famous Monopoly game.)

86. Reid, W. (1967) Weight of an hourglass. *American Journal of Physics.* April 35(4): 351-352.

87. Letter in *New Scientist*, 6 October 1990, page 66, reporting that Beardsley's illustration to Pope's 1896 *Rape of the Lock* is reminiscent of *M*-Set.

88. M. Batty and P. Longley (1987) Urban shapes as fractals. *Area.* 19: 215- 221.

89. Rosato, A., Prinz, F., Swendsen, R. (1987) Why the Brazil nuts are on top: size segregation of particulate matter by shaking. *Phys. Rev. Let.* 58(10): 1028-1040. (Describes Monte Carlo methods).

90. Schröder, M. (1983) Where is the next Mersenne prime hiding? *Math. Intelligencer.* 5(3): 31-33.

91. Fairfield, J. (1983) Segmenting blobs into subregions. *IEEE Trans. Sys. Man and Cyber.* SMC-13: 363-367.

92. Pickover, C. (1988) Symmetry, beauty and chaos in Chebyshev's Paradise. *The Visual Computer: An International Journal of Computer Graphics*, 4: 142-147.

93. Kulpa, Z. (1987) Putting order in the impossible. *Perception.* 16: 201-214.

Where are the remaining titles for "The Top 100"? They'll appear in a forthcoming book, when I receive additional suggestions from readers.

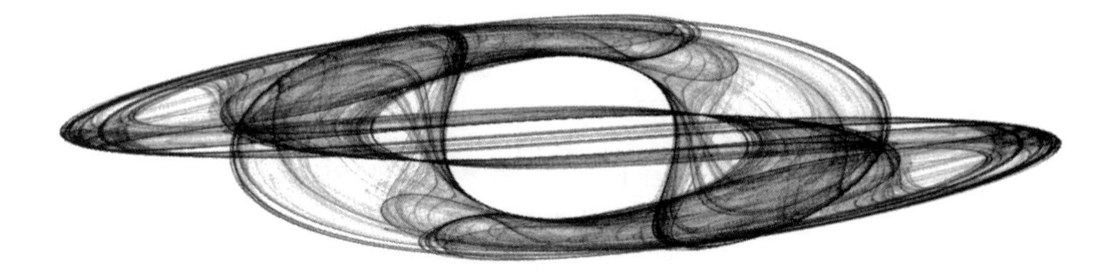

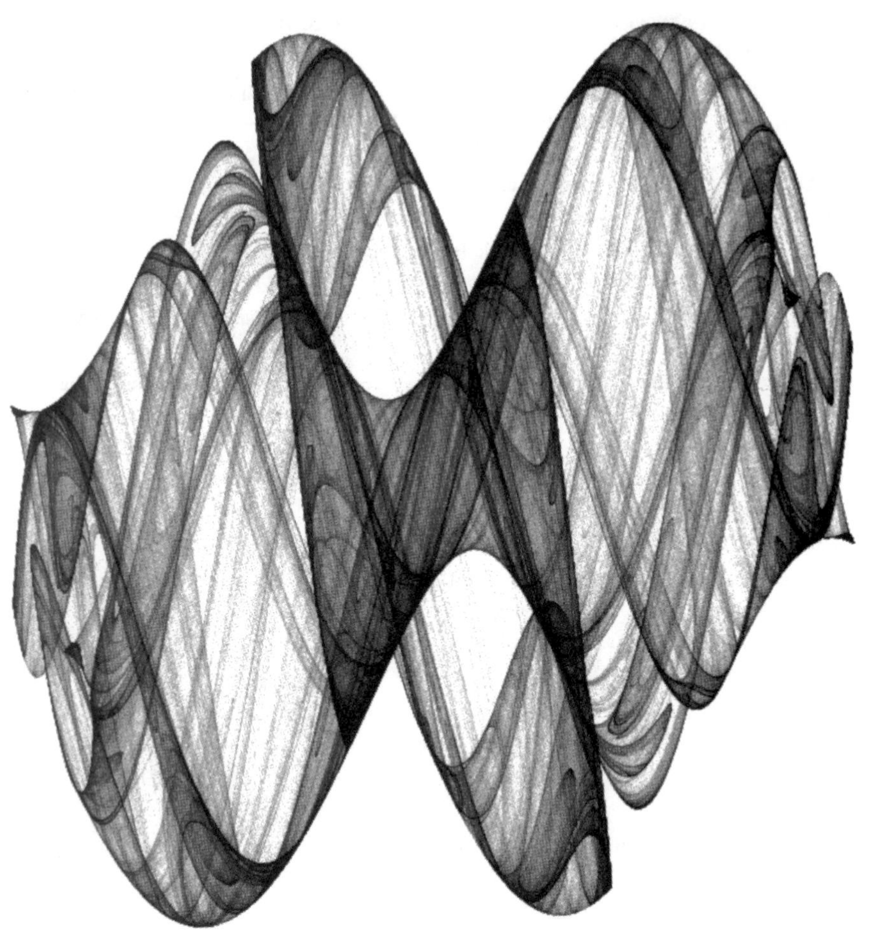

"Chaos," Cross said.
He pointed to... a flickering candle flame.
"It appears to be a very simple system...
a wick, the wax... yet it's inherently unpredictable.
The best computer in the world... could not predict
the pattern this light will take
from moment to moment.
It's the secret of the universe."

"Sounds messy," Duvall said.

"But it's what makes the universe
so full of possibilities,
gives us what free will we have,
or at least lets us think that we have it."

- *Dark Matter*, Garfield Reeves-Stevens

Appendix E

Build Your Own Globular Cluster!

"The pattern, projected in two dimensions on a computer screen, may even fool a connoisseur of globular clusters into believing you are displaying a real object."
Kik Velt, 1986 *Sky and Telescope*

In 1844, the infamous Earl of Rosse (1800-1867) turned his brand new telescope to the heavens to study the tight, compact assemblages of stars known as *globular clusters*. Rosse (pictured at left), an Irish astronomer and telescope maker, had recently resigned from his position in parliament to pursue his passion for astronomy. He used his wealth to build an immense reflecting telescope which exceeded all technological capabilities of his time.[27] As he trained his monster telescope to the heavens, he became particularly interested in one cluster called Messier 2 in the constellation Aquarius (Figure E.1). There are about 140 globular star clusters in our Milky Way galaxy. These star agglomerations are so dense that they look like a single fireball. A typical globular cluster contains several hundred thousand stars, and the stars are most closely packed in the center.[28] Towards the periphery of the cluster, the star count drops quickly. Astronomers have noticed that globular clus-

[27] Until his father's death in 1818, Rosse was known as Lord Oxmantown. He was the chancellor of the University of Dublin in 1862. In 1844 he constructed a telescope with a massive six-foot reflector to study globular clusters.

[28] The density of stars may exceed a hundred solar masses per cubic light-year.

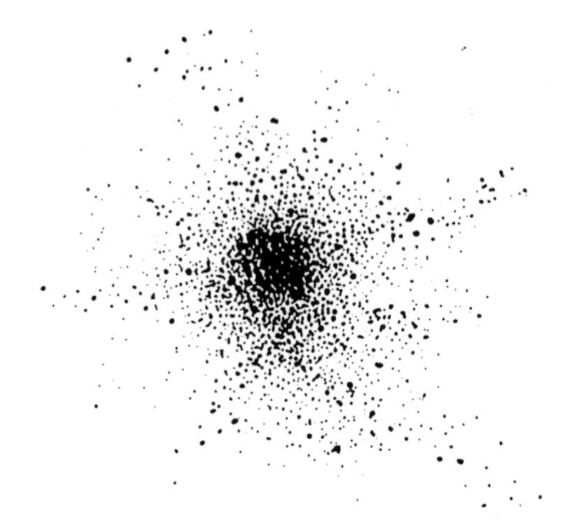

Figure E.1. *The Earl of Rosse's drawing of the globular cluster Messier 2 in Aquarius.* Some observers have claimed to seen various "star chains" in globular clusters. Are such patterns merely a trick played by their eyes attempting to find structure in a random distribution?

ters tend to arrange themselves symmetrically about the Milky Way galaxy and tend to concentrate near its center (Figure E.4). Most of the stars in globular clusters are very old, dating back to the time of the formation of the galaxy itself.

Some observers claim to see dark lanes and strings of stars in globular clusters – vague structure in the mostly homogenous spherical conglomeration of stars. Are such patterns merely a trick played by their eyes attempting to find order in a random distribution? One good way to answer this question is to use random numbers in a graphical simulation of a globular cluster and view the results.

My personal interest in globular clusters began with an amazing paper in *Sky and Telescope*, published in 1986.[29] This short article gives a great introduction to the field of globular clusters and also gives simple program code enabling readers to create their own cluster using computer graphics. In the article, Kik Velt notes that a globular cluster is characterized by only two parameters, the tidal radius and the core radius. About half the stars lie within the core, where the density is nearly uniform. Farther out the density declines dramatically and reaches zero where tidal forces from the Milky Way overcome the gravitational attraction of the cluster itself. This attraction tears away outlying stars. These so-called "tidal effects" are only important very far from the cluster's center, and we can ignore them in this chapter.

---

[29] Velt, K. (1986) Astronomical computing: making your own globular cluster. *Sky and Telescope*. April, 71(4): 398-399. (The paper is by Kik Velt, Nieuwe Bussummerweg 124, 1272Cl Huizen, Netherlands, and communicated by Roger W. Sinnott.)

Figure E.2. *Computer simulation of a globular cluster.* I computed this using an IBM RISC System/6000 which allows me to fly around the cluster in real-time. Hundreds of globular clusters surround the disk of our galaxy, concentrating near its center. They contain old stars and provide clues to the formation process of the galaxy itself. Globular clusters consist of 50,000 to a million stars gravitationally bound together. Some astrophysicists think the centers of globular clusters contain small black holes.

A good model for a globular cluster treats it as though it were a gravitationally bound cloud of gas, with a local density ρ obeying the following law:

$$\rho = \frac{1}{(1+r^2)^n} \tag{E.1}$$

The local density is a function of the distance r from the center. Notice that as r increases, ρ decreases. Another parameter which effects the density is n, the "polytrophic index" (typically $2.5 < n < 3.2$). Kik Velt uses $n = 3$ for his simulations of clusters.

Code E.1 shows you how to draw this cluster on your computer. It's programmed in the language C, and modelled after the BASIC code which Kik Velt gave in the *Sky and Telescope* article, with a few minor corrections. To get your own computer to generate globular clusters, you need a random number generator and some mathematical constraints that insure that the final distribution of stars obeys the density function defined in the previous ρ equation. You may wish to

Figure E.3. *Photo of a real globular cluster (M55).*

obtain the original *Sky and Telescope* article for details.[30] In Code E.1 the variable *numstars* is the number of stars in your simulated cluster. I've set this value to 10,000. The variable $r0$ is the core radius. Increasing the value of $r0$ leads to a bigger cluster. The variable $R1$ is the inflection point of the density curve. What happens as you change $R1$?

If you don't have access to 3-D graphics, simply project the cluster into the x-y plane by plotting only the x and y values, and omit z. The results on the computer screen are so realistic that even seasoned astronomers may be fooled into believing that the picture is that of a real object.

With computer graphics, it's not too hard to show what the sky would look like while you stand on a planet on the edge of the cluster and gaze into the heart of the cluster. You can even fly an imaginary spaceship deep into the heart, or spend time exploring and portraying the sky as seen from somewhere on the fringe of the cluster. Why not render several globular clusters in the same sky? Create a new universe. By using random numbers, you can insure that no two imaginary worlds will ever be the same. In 1844, the Earl of Rosse said, "I fear that no amount of optical power will make these objects better known to us." Could the Earl have imagined flying a spaceship on a computer graphics screen through a simulated galactic cluster, and gazing at the incredible lamp of stars all around him?

---

[30] Kik Velt uses calculus to integrate the density (ρ) equation to determine the cumulative mass, C contained within a sphere or radius r. $C = (\pi/2) \arctan r + (\pi/2)r/(1 + r^2) - \pi r/(1 + r^2)^2$.

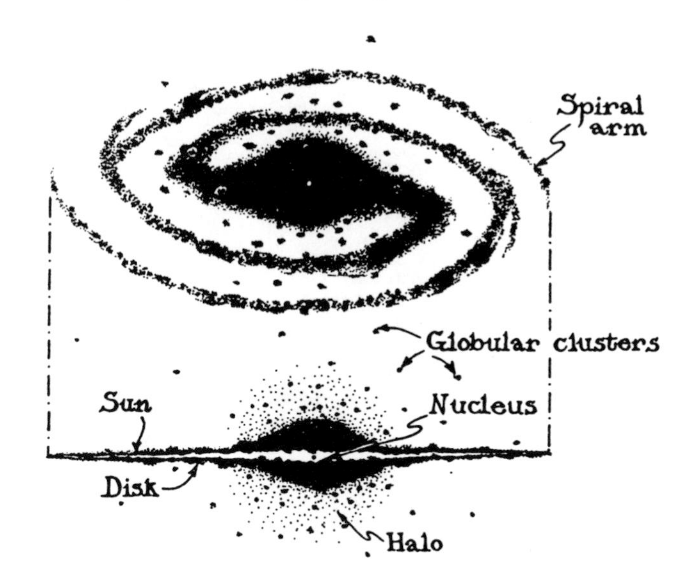

Figure E.4. *Milky Way galaxy.* The positions of globular clusters, little independent star systems, are shown in this rough illustration of our Milky Way galaxy. (From Pagels, H. (1985) *Perfect Symmetry: The Search for the Beginning of Time.* Simon and Schuster: NY. © 1985 by H. Pagels.)

E.1 Digressions

1. In this simulation, all stars have the same brightness. How realistic is this simplification? I achieve a slight gradation in brightness by plotting stars which are closer to the viewer's eye with a slightly larger radii than those far away. How realistic is this?
2. Do you see any of the hypothetical "star chains" in the random number simulation? Are the star chains that some astronomers see only the chance lineups of randomly positioned stars?
3. Write a computer game where you hide your spaceship near some star in the globular cluster. Your opponent's objective is to try to find you within some fixed amount of time.

```
ALGORITHM: How to Create a Globular Cluster

#include <math.h>
float s1,r,x,y,z,numstars,a,b,c,d,rnd,r0,r1,r2,r3,pi,c0,c1;
int i,k;
main()
{
   r0 = 20.0; r2 = r0*r0; r3 = r2*r0;
   pi = 3.14159; c0 = pi*pi*r3/4.0;
   srand(923784651); r1 = r0/sqrt(2.0);
   numstars = 10000;
   for (i = 1; i <= numstars; i++) {
      /* rnd is a random number (0 to 1) */
      rnd = ((float) rand()/32767.);
      c = c0*rnd; r = r1;
      for(k = 1; k <= 5; k++){
         /* newton-raphson iteration */
         a  = r/r0; c1 = atan(a)*0.5*r3;
         a  = 1+a*a; c1 = c1+r*0.5*r2/a;
         c1 = pi*(c1-r*r2/(a*a));
         d  = 4.0*pi*r*r/(a*a*a);
         r  = r+(c-c1)/d;
       }
       /* 3-D position */
       do {
              x = ((float) rand()/32767.) - 0.5;
              y = ((float) rand()/32767.) - 0.5;
              z = ((float) rand()/32767.) - 0.5;
              s1 = sqrt(x * x + y * y + z * z);
       } while (s1 > 0.5);
       /* point is in sphere */
       r = r/s1; x = x*r; y = y*r; z = z*r;
       DrawDotAt(x, y, z);
    } /* number of stars */
}
```

Pseudocode E.1. *How to create a globular cluster.* (The program coded here is in the style of the C language.)

Figure E.5. *The Milky Way galaxy.* Here an astronomer looks at the stellar system of the Milky Way galaxy reduced 100,000,000,000,000,000,000 times. The head of the astronomer is approximately in the position occupied by our sun. (From Gamow, see Credits.)

Appendix F

Messages for ET

"Is mankind alone in the universe? Or are there somewhere other intelligent beings looking up into their night sky from very different worlds and asking the same kind of question?" Carl Sagan and Frank Drake, 1975, *Cosmology+1*

"My heart started beating harder. I was sitting up, a question just forming in my mind... What would be moving the door? Then I saw edging around it a compact figure." So begins Whitley Strieber's controversial best-seller *Communion.* In the book, he describes how, on December 26, 1985, he was visited by a UFO or alien presence while at a secluded cabin in upstate New York. Whether or not you believe his story, the possibility of extraterrestrial visitors has fascinated humans for centuries. Modern-day scientists are also concerned with the possibility of communicating with extraterrestrials.

At the end of this chapter is a message humankind transmitted to the Great Cluster in Hercules using the huge Arecibo radio telescope located in Puerto Rico. The message was sent in a binary code consisting of 1's and 0's from the 1000-foot antenna at Arecibo. You can decode the message by breaking up the digits into 72 consecutive groups of 23 digits each, and arranging the groups one under the other, reading right to left and then top to bottom. If you color the 1 digits black, you'll be able to discern a schematic representation of a human being towards the bottom of the message. Also encoded in the message are: the chemical formulas for elements of the DNA molecule (phosphate group, the deoxyribose sugar, and the organic bases thymidine, adenine, guanine, and cytosine), the numbers 1 to 10 in binary, the atomic numbers for hydrogen, carbon, nitrogen, oxygen, and phosphorous, the number of humans on earth, and a graph showing the solar system with the earth displaced towards the figure of the human being. Both the height of the human being and the diameter of the Arecibo telescope are given in units of the wavelength that was used to transmit the message: 12.6 centimeters.

How would you code a message to be interpreted by extraterrestrials? Should we be sending messages to the stars? What information would you send?

On April 10, 1992, a University of California, Berkeley group mounted an instrument on the Arecibo dish's focal point. The new device is capable of simultaneously monitoring 4 million radio channels. For the next 5 years, in a project called SERENDIP III, the instrument will scan for extraterrestrial life by examining any curious signals coming from the heavens. It will survey a fifth of the sky every 6 months at a cost of about $70,000 per year. The search is perhaps the most powerful of the several other SETI (Search for Extraterrestrial Life) projects since 1960. (For further information, see: Holden, C. (1992) The Big Ear is Listening. *Science*. May 29, 256: 1276.)

Do you think the majority of the people on Earth would be happy to receive an intelligent signal from an advanced extraterrestrial civilization? What effect would this have on politics, religion, and philosophy?

```
0 0 0 0 0 0 1 0 1 0 1 0 1 0 0 0 0 0 0 0 0 0 0 0 0 1 0 1 0 0 0 0 0 1 0 1 0
0 0 0 0 0 0 1 0 0 1 0 0 0 1 0 0 0 1 0 0 0 1 0 0 1 0 1 1 0 0 1 0 1 0 1 0 1
0 1 0 1 0 1 0 1 0 1 0 0 1 0 0 1 0 0 0 0 0 0 0 0 0 0 0 0 0 0 0 0 0 0 0 0 0
0 0 0 0 0 0 0 0 0 0 0 0 0 0 0 0 1 1 0 0 0 0 0 0 0 0 0 0 0 0 0 0 0 0 0 0 0
1 1 0 1 0 0 0 0 0 0 0 0 0 0 0 0 0 0 0 0 0 0 0 1 1 0 1 0 0 0 0 0 0 0 0 0 0
0 0 0 0 0 0 0 0 1 0 1 0 1 0 0 0 0 0 0 0 0 0 0 0 0 0 0 0 0 0 0 1 1 1 1 1 0
0 0 0 0 0 0 0 0 0 0 0 0 0 0 0 0 0 0 0 0 0 0 0 0 0 0 0 0 0 0 0 1 1 0 0 0 0
1 1 1 0 0 0 1 1 0 0 0 0 1 1 0 0 0 1 0 0 0 0 0 0 0 0 0 0 0 0 0 1 1 0 0 1 0
0 0 0 1 1 0 1 0 0 0 1 1 0 0 0 1 1 0 0 0 0 1 1 0 1 0 1 1 1 1 1 0 1 1 1 1 1
0 1 1 1 1 1 0 1 1 1 1 1 0 0 0 0 0 0 0 0 0 0 0 0 0 0 0 0 0 0 0 0 0 0 0 0 0
0 1 0 0 0 0 0 0 0 0 0 0 0 0 0 0 0 0 0 1 0 0 0 0 0 0 0 0 0 0 0 0 0 0 0 0 0
0 0 0 0 0 0 0 0 0 0 0 1 0 0 0 0 0 0 0 0 0 0 0 0 0 0 0 0 0 1 1 1 1 1 1 0 0
0 0 0 0 0 0 0 0 0 0 0 1 1 1 1 1 0 0 0 0 0 0 0 0 0 0 0 0 0 0 0 0 0 0 0 0 0
0 0 1 1 0 0 0 0 1 1 0 0 0 0 1 1 1 0 0 0 1 1 0 0 0 1 0 0 0 0 0 0 0 1 0 0 0
0 0 0 0 0 0 1 0 0 0 0 1 1 0 1 0 0 0 0 1 1 0 0 0 1 1 1 0 0 1 1 0 1 0 1 1 1
1 1 0 1 1 1 1 1 0 1 1 1 1 1 0 1 1 1 1 1 0 0 0 0 0 0 0 0 0 0 0 0 0 0 0 0 0
0 0 0 0 0 0 0 0 0 1 0 0 0 0 0 0 1 1 0 0 0 0 0 0 0 0 0 1 0 0 0 0 0 0 0 0 0
0 0 1 1 0 0 0 0 0 0 0 0 0 0 0 0 0 0 0 1 0 0 0 0 0 1 1 0 0 0 0 0 0 0 0 0 0
1 1 1 1 1 1 0 0 0 0 0 1 1 0 0 0 0 0 0 1 1 1 1 1 0 0 0 0 0 0 0 0 0 0 1 1 0
0 0 0 0 0 0 0 0 0 0 0 0 1 0 0 0 0 0 0 0 0 1 0 0 0 0 0 0 0 0 1 0 0 0 0 0 1
0 0 0 0 0 0 1 1 0 0 0 0 0 0 0 1 0 0 0 0 0 0 0 1 1 0 0 0 0 1 1 0 0 0 0 0 0
1 0 0 0 0 0 0 0 0 0 0 1 1 0 0 0 1 0 0 0 0 1 1 0 0 0 0 0 0 0 0 0 0 0 0 0 0
0 1 1 0 0 1 1 0 0 0 0 0 0 0 0 0 0 0 0 0 1 1 0 0 0 1 0 0 0 0 1 1 0 0 0 0 0
0 0 0 0 1 1 0 0 0 0 1 1 0 0 0 0 0 0 1 0 0 0 0 0 0 0 1 0 0 0 0 0 0 1 0 0 0
0 0 0 0 0 1 0 0 0 0 0 1 0 0 0 0 0 0 0 1 1 0 0 0 0 0 0 0 0 1 0 0 0 1 0 0 0
0 0 0 0 0 1 1 0 0 0 0 0 0 0 0 1 0 0 0 1 0 0 0 0 0 0 0 0 0 1 0 0 0 0 0 0 0
1 0 0 0 0 0 1 0 0 0 0 0 0 0 1 0 0 0 0 0 0 0 1 0 0 0 0 0 0 0 1 0 0 0 0 0 0
0 0 0 0 0 0 1 1 0 0 0 0 0 0 0 0 0 1 1 0 0 0 0 0 0 0 0 1 1 0 0 0 0 0 0 0 0
0 1 0 0 0 1 1 1 0 1 0 1 1 0 0 0 0 0 0 0 0 0 0 1 0 0 0 0 0 0 0 0 1 0 0 0 0
0 0 0 0 0 0 0 0 0 0 1 0 0 0 0 0 1 1 1 1 1 0 0 0 0 0 0 0 0 0 0 0 0 1 0 0 0
0 1 0 1 1 1 0 1 0 0 1 0 1 1 0 1 1 0 0 0 0 0 0 1 0 0 1 1 1 0 0 1 0 0 1 1 1
1 1 1 1 0 1 1 1 0 0 0 0 1 1 1 0 0 0 0 0 1 1 0 1 1 1 0 0 0 0 0 0 0 0 0 1 0
1 0 0 0 0 0 1 1 1 0 1 1 0 0 1 0 0 0 0 0 0 1 0 1 0 0 0 0 0 1 1 1 1 1 1 0 0
1 0 0 0 0 0 0 1 0 1 0 0 0 0 0 1 1 0 0 0 0 0 0 1 0 0 0 0 0 1 1 0 1 1 0 0 0
0 0 0 0 0 0 0 0 0 0 0 0 0 0 0 0 0 0 0 0 0 0 0 0 0 0 0 0 0 0 0 0 1 1 1 0 0
0 0 0 1 0 0 0 0 0 0 0 0 0 0 0 0 0 0 1 1 1 0 1 0 1 0 0 0 1 0 1 0 1 0 1 0 1
0 1 0 0 1 1 1 0 0 0 0 0 0 0 0 0 1 0 1 0 1 0 1 0 0 0 0 0 0 0 0 0 0 0 0 0 0
0 0 1 0 1 0 0 0 0 0 0 0 0 0 0 0 0 0 1 1 1 1 1 0 0 0 0 0 0 0 0 0 0 0 0 0 0
0 0 0 1 1 1 1 1 1 1 1 1 0 0 0 0 0 0 0 0 0 0 0 0 1 1 1 0 0 0 0 0 0 0 1 1 1
0 0 0 0 0 0 0 0 0 1 1 0 0 0 0 0 0 0 0 0 0 0 1 1 0 0 0 0 0 0 0 1 1 0 1 0 0
0 0 0 0 0 0 0 1 0 1 1 0 0 0 0 0 1 1 0 0 1 1 0 0 0 0 0 0 0 1 1 0 0 1 1 0 0
0 0 1 0 0 0 1 0 1 0 0 0 0 0 1 0 1 0 0 0 1 0 0 0 0 1 0 0 0 1 0 0 1 0 0 0 1
0 0 1 0 0 0 1 0 0 0 0 0 0 0 0 1 0 0 0 1 0 1 0 0 0 1 0 0 0 0 0 0 0 0 0 0 0
0 1 0 0 0 0 1 0 0 0 0 1 0 0 0 0 0 0 0 0 0 0 0 0 1 0 0 0 0 0 0 0 0 0 1 0 0
0 0 0 0 0 0 0 0 0 0 0 0 1 0 0 1 0 1 0 0 0 0 0 0 0 0 0 0 0 1 1 1 1 0 0 1 1
1 1 1 0 1 0 0 1 1 1 1 0 0 0
```

from 'Jupiter'

Appendix G

Planet Music

Can you name this British composer known primarily for his orchestral suite *The Planets*, first performed in 1918? (See Appendix N, "Author's Notes" for the answer.) Below his photograph is a piece of music he titled "Jupiter." He also composed operas and other instrumental music.

In *The Planets* the themes for each planet are suggestive of the historical and astrological association of the planet. Thus, the movement for Mars is martial in nature, while the movement for Venus is romantic, and the movement for Mercury is quick-moving and sprightly. Each movement has a subtitle, for example Mars: "The Bringer of War," Venus: "The Bringer of Peace," and Jupiter: "The Bringer of Jollity."

The Jupiter piece is described in the following, from the record, "The Planets, Op. 32, the Philadelphia Orchestra, Eugene Ormandy, Conductor" (RCA Red Seal):

> Jupiter, the Bringer of Jollity is the one movement that bears clear witness to [the composer's] love of English folksong and dance; the merrymaking is resolutely English in verve, with pentatonic figurations and characteristic modal inflections. The broad singing melody that lies at the center of the movement was later set to the words of Spring-Rice, "I vow to my country"; for better or for worse, its patriotic associations are too deeply ingrained for us to be able to imagine how it sounded in its pristine state. Many have questioned the appropriateness of the tune in its context, even without its later connotations, but the fact remains that as a melody it is a splendid invention, and a moment to be awaited eagerly is its momentary return in the bass of the orchestra under gigantic catherine-wheels of sound just before the exuberant foot-stamping coda.

Appendix H

Meditations on Transcendentals

"Mathematics is a lesser activity than religion in the sense that we've agreed not to kill each other but to discuss things"

Richard Preston, 1992, *The New Yorker*

Section 8.1, "The Pi Slaves" discussed transcendental numbers such as π and e whose digits never end, nor has anyone detected an orderly pattern in their arrangement. To complete the survey in section *H.2* of the present chapter, I asked colleagues if there were any other famous transcendental numbers.

H.1 Liouville and Deckert Numbers

One famous example of a transcendental number "discovered" much later than π or e is the Liouville number named after its inventor, French mathematician J. Liouville. First discussed in 1851, you can compute this fascinating number by:

$$\sum_{k=1}^{\infty} a_k r^{-k!} \tag{H.1}$$

where $0 \le a_k < r$. The numbers a_k are integers. The resulting number is a Liouville number of base r. If the values for a_k are all 1, and $r = 10$ we get:

$$\frac{1}{10} + \frac{1}{10^{1 \times 2}} + \frac{1}{10^{1 \times 2 \times 3}} + \cdots \tag{H.2}$$

The decimal value can easily be written down! It is:

0.110001000000000000000001000 ...

which has a one in the 1st, 2nd, 6th, 24th, etc. places and zeros elsewhere. What is this number like for other values for a_k and r? Kenneth L. Deckert of Almaden, California writes: "For another interesting example, let values for a_k be the digits in the fractional part of the decimal expansion of π (pi)." Can you think of any interesting properties that the Deckert number should exhibit? Here are some related questions and curiosities.

1. Is the following formula true or false?

$$(1+\pi)\times(1-\pi)=1-\pi^2. \qquad (H.3)$$

2. Wayne Delia of Fishkill, New York asks if the number π^2 ever occurs in geometry or physics.
3. Dennis Gordon of Madison, Wisconsin remarks, "I like the equation $\pi^4+\pi^5 \sim e^6$ presented in *Mazes for the Mind*. However, I like $2^{(\ln 3/\ln 2)}=3$ somewhat better."
4. In 8.1, "The Pi Slaves" I asked if a transcendental number raised to a transcendental power produces a transcendental result. Leif Schioler of Denmark notes: e is transcendental, $\ln 2$ (natural log of 2) is also transcendental, but $e^{\ln 2}=2$ is not transcendental.

H.2 The 15 Most Famous Transcendental Numbers

After conducting a brief survey of readers, I made a list of the fifteen most famous transcendental numbers as defined in the *Glossary*. Can you list these in order of relative fame and/or usage?

1. $\pi = 3.1415\ldots$
2. $e = 2.718\ldots$
3. Euler's constant, $\gamma = 0.577215\ldots = \lim_{n\to\infty}(1+1/2+1/3+1/4+\cdots+1/n-\ln(n))$ (Not proven to be transcendental, but generally believed to be by mathematicians.)
4. Catlan's constant, $G=\sum(-1)^k/(2k+1)^2 = 1-1/9+1/25-1/49+\cdots$ (Not proven to be transcendental, but generally believed to be by mathematicians.)
5. Liouville's number (discussed in previous section).
6. Chaitin's "constant," the probability that a random algorithm halts. (Noam Elkies of Harvard notes that not only is this number transcendental but it is also incomputable.)
7. Chapernowne's number, 0.12345678910111213141516171819202122232425... This is constructed by concatenating the digits of the positive integers. (Can you see the pattern?)

8. Special values of the zeta function, such as $\zeta(3)$. (Transcendental functions can usually be expected to give transcendental results at rational points.)
9. ln(2).
10. Hilbert's number, $2^{\sqrt{2}}$. (This is called Hilbert's number because the proof of whether or not it is transcendental was one of Hilbert's famous 100 problems. In fact, according to the Gelfond-Schneider theorem, any number of the form a^b is transcendental where a and b are algebraic ($a \neq 0, a \neq 1$) and b is not a rational number. Many trigonometric or hyperbolic functions of non-zero algebraic numbers are transcendental.)
11. e^{π}
12. π^e (Not proven to be transcendental, but generally believed to be by mathematicians.)
13. Morse-Thue's number, 0.01101001 ... (See *Mazes for the Mind* for more information.)
14. i^i (Here i is the imaginary number $\sqrt{-1}$. If a is algebraic and b is algebraic but irrational then a^b is transcendental. Since i is algebraic but irrational, the theorem applies. Note also: i^i is equal to $e^{-\pi/2}$ and several other values. Consider $i^i = e^{i \log i} = e^{i \times i\pi/2}$. Since log is multivalued, there are other possible values for i^i.)
15. Feigenbaum numbers, e.g. 4.669 .... (These are related to properties of dynamical systems with period-doubling. The ratio of successive differences between period-doubling bifurcation parameters approaches the number 4.669 ..., and it has been discovered in many physical systems before they enter the chaotic regime. It has not been proven to be transcendental, but is generally believed to be.)

Keith Briggs from the Mathematics Department of the University of Melbourne in Australia computed what he believes to be the world-record for the number of digits for the Feigenbaum number:

```
4.
66920160910299067185320382046620161725818557747576863274565l
3430041343302113147371386897440239480138171659848551898151 34
```

Briggs carried out the computation using special-purpose software designed by David Bailey of NASA Ames running on an IBM RISC System/6000. The computation required a few hours of computation time. For more information, see: Briggs, K. (1991) A precise calculation of the Feigenbaum constants, *Math. Comp.* 57: 435.

Question: Is -1^{-i} a transcendental number?

In 8.1, "The Pi Slaves," I posed the question: Is there a compact formula relating e, π, i and ϕ, the golden ratio? One answer is: $e^{i\pi} + 2\phi = \sqrt{5}$.

H.3 Calculating Pi

"The digits of π beyond the first few decimal places are of no practical or scientific value. Four decimal places are sufficient for the design of the finest engines; ten decimal places are sufficient to obtain the circumference of the earth within a fraction of an inch if the earth were a smooth sphere."

Petr Beckmann, *A History of Pi*

In 1989, the Chudnovsky brothers, two Columbia University mathematicians, computed over one billion digits of π using a Cray 2 and an IBM 3090-VF computer. Their 1,011,196,691 digits would stretch 1,580 miles if printed using this typeface. The formula they used, elephantine as well as diophantine, looked like this:

$$\sum_{n=0}^{\infty} |c_1 + n| \times \frac{(6n)!(-1)^n}{(3n)!n!^3(640{,}320)^{3n}} \tag{H.4}$$

$$= \frac{(640{,}320)^{3/2}}{163 \times 8 \times 27 \times 7 \times 11 \times 19 \times 127} \times \frac{1}{\pi} \tag{H.5}$$

$$c_1 = \frac{13{,}591{,}409}{163 \times 2 \times 9 \times 7 \times 11 \times 19 \times 127} \tag{H.6}$$

In the past, in order to compute π to millions of digits, mathematicians always had to start from scratch each time. Their new equation overcomes this liability. Toward the end of the summer of 1991, the brothers had computed π to over two billion digits.

H.3.1 Idiot Savants

"Mathematical objects such as the number pi seem to exist in an external, objective reality. Numbers seem to exist apart from time or the world. Numbers might exist even if the universe did not. Pi may even exist apart from God, in the opinion of some mathematicians, for while there is reason to doubt the existence of God, by their way of thinking there is no good reason to doubt the existence of the circle."

Richard Preston, 1992, *The New Yorker*

In 1844, Johann Martin Zacharias Dase (1824-1861), a human computer, calculated π correct to 200 places in less than two months: $\pi =$

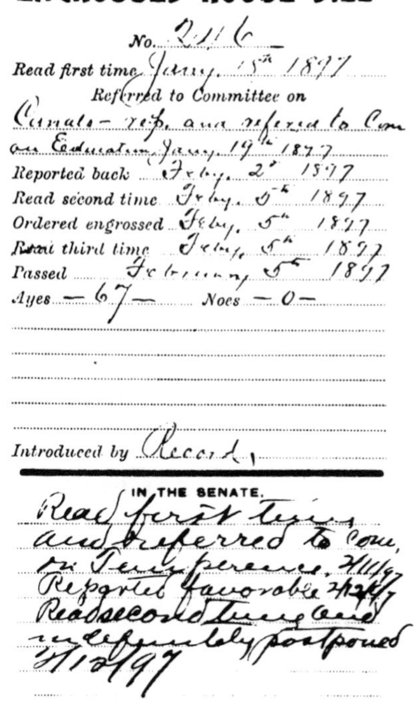

ENGROSSED HOUSE BILL

No. 246

Read first time Jany. 18th 1897

Referred to Committee on Canals — rep. and referred to Com on Education Jany. 19th 1897

Reported back Feby. 2d 1897

Read second time Feby. 5th 1897

Ordered engrossed Feby. 5th 1897

Read third time Feby. 5th 1897

Passed February 5th 1897

Ayes — 67 — Noes — 0 —

Introduced by Record

IN THE SENATE.

Read first time and referred to Com. on Temperance. 2/11/97

Reported favorable 2/12/97

Read second time and indefinitely postponed 2/12/97

Figure H.1. *Facsimile of a bill that attempted to legislate the value of Pi.* Bill No. 246, passed by The Indiana House of Representatives in 1897, legislated the value of π to 9.2376.

3.14159 26535 89793 23846 26433 83279 50288 41971 69399 37510 58209 74944 59230 78164 06286 20899 86280 34825 34211 70679 82148 08651 32832 06647 09384 46095 50582 23172 53594 08128 48111 74502 84102 70193 85211 05559 64462 29489 54930 38196.

Dase had an incredible brain. He could give the number of sheep in a flock after a single glance. He could multiply two 8-digit numbers in his head in 54 seconds, two 40-digit numbers in 40 minutes, and two 100 digit numbers in 8 hours! Dase performed such computations for weeks on end, running as an unattended supercomputer. He would break off his calculation at bedtime, store everything in memory, and resume calculation after breakfast. Occasionally, Dase had a system crash.

To compute π Johann Dase used: $\pi/4 = arctan(1/2) + arctan(1/5) + arctan(1/8) \ldots$ with a series expansion for each arctangent. Dase ran the arctangent job in his brain for almost two months.

H.3.2 Viete's Amazing Pi Formula

> *"Pi is not the solution to any equation built from a less than infinite series of whole numbers. If equations are trains threading the landscape of numbers, then no train stops at pi."* Richard Preston, 1992, *The New Yorker*

Amateur mathematician Francois Viete published the following interesting formula for π in 1593 in his *Variorum de rebus mathematicis responsorum liber VIII* (Various mathematical problems, vol. 8):

$$\pi = \frac{2}{\sqrt{1/2} \times \sqrt{1/2 + 1/2\sqrt{1/2}} \times \sqrt{1/2 + 1/2\sqrt{1/2 + 1/2\sqrt{1/2}}} \times \cdots} \tag{H.7}$$

Viete was a lawyer by profession and rose to the position of councillor of Parliament of Brittany, until forced to flee during the persecution of the Huguenots.

H.3.3 Wallis's Formula

John Wallis (1616-1703) in his *Arithmetica infinitorum* (1655) derived the famous formula which bears his name:

$$\pi = 2\,\frac{2 \times 2 \times 4 \times 4 \times 6 \times 6 \ldots}{1 \times 3 \times 3 \times 5 \times 5 \times 7 \ldots} \tag{H.8}$$

Nilakantha, an astronomer who lived on the Kerala coast of India, described the following formula in Sanskrit poetry around the year 1500:

$$\frac{\pi}{4} = \frac{1}{1} - \frac{1}{3} + \frac{1}{5} - \frac{1}{7} + \frac{1}{9} - \cdots \tag{H.9}$$

Appendix I

Lewis Carroll

I.1 Alice in Wonderland

"What do you think he's dreaming about?" asked Tweedledee.
Alice said "Nobody can guess that."
"Why about you! And if he left off dreaming about you... you'd be nowhere. Why you're only a sort of thing in his dream." Lewis Carroll

Those of you who are Lewis Carroll aficionados may be curious as to how he arrived at his title *Alice in Wonderland.* On June 10, 1864 he wrote the following letter to his friend Tom Taylor:

> I should be very glad if you could help me in fixing on a name for my fairy-tale.... The heroine spends an hour underground, and meets various birds, beasts, etc. (*no* fairies), endowed with speech. The whole thing is a dream, but *that* I don't want revealed till the end. I first thought of "Alice's Adventures Under Ground," but that was pronounced too like a lesson-book... Then I took "Alice's Golden Hour," but that I gave up, having a dark suspicion that there is already a book called "Lily's Golden Hours." Here are other names I have thought of:
>
> Alice among the
> (elves, goblins)
>
> Alice's
> (hour, doings, adventures)
> in
> (elf-land, wonderland)
>
> Of all these I at present prefer "Alice's Adventures in Wonderland." Perhaps you can suggest a better name than any of these.

I.2 Lewis Carroll Reading List

The following is a list of books by Lewis Carroll.

1. *Alice in Wonderland* (1865)
2. *Phantasmagoria and Other Poems* (1869)
3. *Through the Looking-Glass* (1871)
4. *The Hunting of the Snark* (1876)
5. *Rhyme? and Reason?* (1883)
6. *Sylvie and Bruno* (1889)
7. *Sylvie and Bruno Concluded* (1893)
8. *Three Sunsets and Other Poems* (1898)

Also of interest:

1. Gardner, M. (1960) *The Annotated Alice*. Potter: NY.
2. Crutch, D. (1979) *The Lewis Carroll Handbook*. Dawson-Archon: Folkestone, Kent.
3. Collingswood, S. (1898) *The Life and Letters of Lewis Carroll.* Unwin: London.
4. Goodacre, S. H. (1977) Alice's changes of size in Wonderland. *Jabberwocky* 6: 20-24.
5. Clark, A. (1980) *Lewis Carroll: A Biography*. Praeger: NY.

I.3 Grinning Cheshire Cat Found

A London newspaper recently reported that the inspiration for the vanishing Cheshire Cat in Lewis Carroll's *Alice in Wonderland* may have been discovered in a tiny town in England. The Lewis Carroll Society, with 350 members, said it found a carving of a cat's head in a church where Carroll's father was a rector. The riddle was solved when Chicagoan Joel Birenbaum noticed a small smiling cat which began to disappear as he knelt at the altar of St. Peter's Church in Croft, about 200 miles northeast of London. He found that, like the beloved literary feline who disappears leaving only his grin behind, all he could see of the wall carving was the cat's ear-to-ear grin.

Appendix J

For Further Reading

J.1 Chaos

"Do not follow where the path may lead. Go, instead, where there is no path and leave a trail." Anonymous

"The dreams of men belong to God." S. R. Donaldson

1. Crutchfield, J., Farmer, J., Packard, N. (1986) Chaos. *Scien. Amer.* 255: 46-57;
2. Dewdney, A. K. (1985) Computer Recreations. *Scien. Amer.* 253: 16-24.
3. Dewdney, A. K. (1987) Probing the strange attractions of chaos. *Scien. Amer.* July Issue, 108-111.
4. Devaney, R. (1986) Chaotic bursts in nonlinear dynamical systems. *Science.* 235: 342-345.
5. Dixon, R. (1992) The pentasnow gasket and its fractal dimension. In *Fivefold Symmetry*. Hargittai, I., ed. World Scientific: New Jersey.
6. Glass, L., Mackey, M. (1988) *From Clocks to Chaos: The Rhythms of Life*. Princeton Univ. Press: New Jersey.
7. Gleick, J. (1987) *Chaos: Making a New Science*. Viking: New York.
8. Heck, A. and Perdang, J. (1991) *Applying Fractals in Astronomy*. Springer: NY. (Technical reading. Topics include: The fractal structure of quantum space-time, fractal aspects of galaxy clustering, pulsating stars and fractals...)
9. Lorenz, E. (1963) Deterministic nonperiodic flow. *J. Atmos. Sci.* 20: 130.
10. Mandelbrot, B. (1983) *The Fractal Geometry of Nature*. Freeman, San Francisco.
11. Markus, M. (1990) Chaos in maps with continuous and discontinuous maxima. *Computers in Physics*. Sept/Oct 4(5): 481-493.
12. May, R. (1976) Simple mathematical models with very complicated dynamics. *Nature.* 261: 459-467.
13. Moon, F. (1987) *Chaotic Vibrations*. John Wiley and Sons: New York. (Moon gives many practical examples of chaos in real physical systems.)

14. Pickover, C. (1990) *Computers, Pattern, Chaos and Beauty.* St. Martin's Press: New York.

15. Sprott, J. C. (1993) Automatic generation of strange attractors. *Computers and Graphics.* 17(3): 325-332. (This paper introduces the Lyapunov method I used for automatically selecting visually interesting attractors.)

16. Sprott, J. C. (1993) *Strange Attractors: Creating Patterns in Chaos.* M&T Books: New York.[31]

17. Wolf, A., Swift, J., Swinney, H. and Vastano, J. (1985) Determining the Lyapunov exponents from a time series. *Physica.* 16D: 285-317.

J.2 Image Processing

1. Pavlidis, T. (1982) *Algorithms for Graphics and Image Processing.* Computer Science Press: NY.

2. Pratt, W. (1978) *Digital Image Processing.* Wiley: NY.

J.3 The Solar System

1. Boulet, D. (1991) *Methods of Orbit Determination for the Microcomputer.* Willmann-Bell, Inc. ISBN 0-943396-34-4. (Orbital motion. Broad scope.)

2. Hynes, S. (1991) *Planetary Nebulae.* Willmann-Bell, Inc. ISBN 0-943396-30-1. (Observing advice, and a catalog of 1,340 objects.)

3. Montenbruck, O., and Pfleger, T. (1991) *Astronomy on the Personal Computer.* Springer-Verlag: NY. 255 pages and floppy disk. (Time and date calculations, the effect of mutual gravitational perturbations on orbits, lunar occultations, orbits of planets, asteroids, and comets. Pascal Code.)

4. Moore, P., Hunt, G., Nicolson, I., and Cattermole, P. (1990) *The Atlas of the Solar System.* Crescent Books/Crown Publishers: NY.

---

[31] When I asked Dr. J. Clint Sprott, the guru of strange attractors, to describe himself, he told me the following. "I'm sure I have the world's largest collection of strange attractors. I estimate about a million. I've looked at maybe 5% of them. The others sit in a compact code on my disk. I'm busily doing statistical studies on them. Who am I? B.S. in physics from MIT in 1964, Ph.D. in physics from University of Wisconsin-Madison in 1969. Worked at Oak Ridge National Lab before returning to a faculty position in the Physics Dept at U.W. in 1973 where I've been ever since. My training and research has mostly been in experimental plasma physics. I have been heavily involved in physics popularization through a program called 'The Wonders of Physics' that involves public presentations of dramatic demonstrations of physical phenomena, production of videotapes, and development of physics educational software. I'm technical editor of Physics Academic Software. I've written books such as ***Introduction to Modern Electronics*** (Wiley, 1979) and ***Numerical Recipes and Examples in BASIC*** (Cambridge Univ Press, 1991). I became a chaos enthusiast in 1989."

5. Kippenhahn, R. (1990) *Bound to the Sun.* Freeman: NY. 282 pages. (Solar system, planets, comets....)

6. Meeus, J. (1991) *Astronomical Algorithms.* Willmann-Bell, Inc. (Position of solar system bodies, eclipse predictions, 3-D plots.)

7. Dairymple, G. (1991) *The Age of the Earth.* Stanford University Press: Calif. 474 pages. (Explains the evidence and logic that have led scientists to conclude that the Earth and the other parts of the solar system are four and one-half billion years old.)

8. Rukl, A. (1990) *Atlas of the Moon.* Paul Hamlyn Pub: London. (76 maps covering the Moon's earthside.)

9. Schaaf, F. (1991) *Seeing the Solar System.* Wiley: NY. 208 pages. (53 short project descriptions for ar.ateur astronomers.)

10. Shklovsky, I. (1991) *Five Billion Vodka Bottles to the Moon.* Norton: NY. (Written by one of the leading Soviet astrophysicists of this century.)

J.4 The Universe

1. Brandt, J. and Chapman, R. (1992) *Rendezvous in Space: The Science of Comets.* Freeman: NY. (The first comet book to incorporate the findings from Halley's return. Highlights comet research since 1950. How to: optimize comet watching, plot comet orbits, report a new comet. 168 illustrations.)

2. Dobson, J. (1991) *How and Why to Make a User-Friendly Telescope.* Contact: Everything in the Universe, 5248 Lawton Ave., Oakland, CA 94618.

3. Gamow, G. (1947) *One, Two, Three... Infinity.* Dover, NY.

4. Gingerich, O. (1977) *Cosmology+1.* (Readings from *Scientific American.*) Freeman: NY.

5. Maran, S. (1993) *The Astronomy Astrophysics Encyclopedia.* Van Nostrand: NY. (Contains the current collective knowledge of today's leading reseachers. 1000 pages).

6. Pagels, H. (1985) *Perfect Symmetry: The Search for the Beginning of Time.* Bantam: NY.

7. Powell, C. (1992) The golden age of cosmology. *Scientific American.* July 267(1): 17.

8. Krupp, E. C. (1991) *Beyond the Blue Horizon.* HarperCollins: NY. 387 pages. (The study of astronomical concepts and traditions across time and culture. How North American natives, Australian aborigines, Babylonians, Greeks, and others have viewed the sun, moon, planets, stars, comets, and the Milky Way.)

9. Drake, F. and Sobel, D. (1992) *Is Anyone Out There? The Search for Extraterrestrial Intelligence.* Delacorte Press: NY.

J.5 Pickover Books

1. Pickover, C. (1992) *Mazes for the Mind: Computers and the Unexpected.* St. Martin's Press, 175 Fifth Ave, New York, NY 10598 (USA). ISBN 0-312-08165-0. Phone 1-800-221-7945. (Published in Japan by Hakuyo-Sha.)
2. Pickover, C. (1991) *Computers and the Imagination.* St. Martin's Press, 175 Fifth Ave, New York, NY 10598 (USA). ISBN 0-312-06131-5. Phone 1-800-221-7945. (U.K. Publisher: Alan Sutton Publishing, Phoenixmill, Far Throupp, Stroud Glouchestershire GL5 2BU UK.)
3. Pickover, C. (1990) *Computers, Pattern, Chaos and Beauty.* St. Martin's Press, 175 Fifth Ave, New York, NY 10598 (USA). ISBN 0-312-04123-3. Phone: 1-800-221-7945. (U.K. Publisher: Alan Sutton Publishing, Phoenixmill, Far Throupp, Stroud Glouchestershire GL5 2BU UK.)
4. Hargittai, I. and Pickover, C. (1992) *Spiral Symmetry.* World Scientific Publishing, Suite 1B, 1060 Main St., River Edge, New Jersey 07661. ISBN 981-02-0615-1. Phone: 800 227-7562. (Topics: Spirals in nature, art, and mathematics. Fractal spirals, plant spirals, artist's spirals, the spiral in myth and literature.)
5. Pickover, C. (1994) *The Pattern Book: Fractals, Art, and Nature.* World Scientific Publishing, Suite 1B, 1060 Main St, River Edge, New Jersey 07661. Phone: 800 227-7562.
6. Pickover, C. (1994) *Visions of the Future: Art, Technology, and Computing in the 21st Century* (2nd Edition). St. Martin's Press, 175 Fifth Ave, New York, NY 10598 (USA). ISBN 0-312-08481-1. Phone 1-800-221-7945. (U.K. Publisher: Science Reviews Limited, PO Box 81, Northwood, Middlesex HA6 3DN England.)
7. Pickover, C. and Tewksbury, S. (1994) *Frontiers of Scientific Visualization.* Wiley: New York. (Topics: computer graphics, computer art, virtual reality, fractals, unusual graphics of genetic sequences, etc.)
8. Anthony, P. and Pickover, C. (1994) *Spider Legs*, in press. (Science-fiction novel.)
9. Pickover, C. (1994) *Visualizing Biological Information.* World Scientific Publishing, Suite 1B, 1060 Main St, River Edge, New Jersey 07661. Phone: 800 227-7562. (Topic: The creative use of computer graphics to find patterns in DNA and other biological sequences.)
10. Pickover, C (1992) *Mit den Augen des Computers.* ISBN 3-87791-323-7. Markt&Technik: Hans-Pinsel Strasse 2, D-8013.
11. Pickover, C (1992) *Computers, Fractals, Chaos* (In Japanese). Hakuyo-Sha, 3 Niban-Cho, Chiyodo-ku, Tokyo 102, Japan.

Appendix K

Other Chaotic Attractor Equations

The following papers and books all give explicit formulas or computational recipes for generating beautiful chaotic attractors. To encourage your involvement, I have listed some program outlines for generating these attractors.

1. Lauwerier, H. (1990) *Fractals*. Princeton University Press. (The section on Gumowski and Mira lists chaotic equations which produce patterns resembling feathers. Code K.2 shows you how to produce the feather pattern, a representation of a dynamical system. Simply plot a dot at positions determined by x and y through the iteration. Use double precision variables.)
2. Ikeda, K. (1979) Multiple-valued stationary state and its instability of the transmitted light by a ring cavity system. *Optical Communications*. 30: 257. (This paper describes how to generate the Ikeda attractor, a swirling intricate 2-D attractor. Code K.1 shows you how to produce the Ikeda pattern. Simply plot the position of variables j and k through the iteration. The variables *scale, xoff,* and *yoff* position and scale the image to fit on the graphics screen.)
3. Pickover, C. (1990) *Computers, Pattern, Chaos and Beauty*. St. Martin's Press: NY. (This book contains recipes for many chaotic attractors. Code K.3, from this book, gives a recipe for producing a variety of dynamical systems. The color assignment is based on the number of iterations needed to produce a particular point on the plot. If color options are not available, line 5 may be omitted. Code K.4 is a recipe for a 3-D attractor.)
4. Fay, T. (1989) The Butterfly Curve. *American Math. Monthly*. 96(5): 442 - 443. (See Code K.5.)
5. Nussbaum, R., Peitgen, O. (1984) Special and spurious solutions of $\dot{x}(t) = -\alpha f(x(t-1))$. Memoirs of the American Mathematical Society. 51: 1- 129. (This paper describes interesting 2-D attractors.)
6. Chossat, P., Golubitsky, M. (1988) Symmetry-increasing bifurcations of chaotic attractors. *Physica D*. 32: 423-426. (Beautiful kaleidoscopic attractors.)
7. Field, M., Golubitsky, M. (1990) Symmetric chaos. *Computers in Physics*. Sept/Oct 4(5): 470-479. (Beautiful kaleidoscopic attractors are generated by the Chossat-Golubitsky formula: $f(\zeta, \lambda) = (\alpha u + \beta v + \lambda)\zeta + \gamma\bar{\zeta}^{m-1}$ where

```
ALGORITHM: How to Create an Ikeda Attractor.

c1 = 0.4, c2 = 0.9,  c3 = 6.0, rho = 1.0;
for (i = 0, x = 0.1, y=0.1;i<=3000;i++) (
  temp = c1 - c3 / (1.0 + x * x + y * y);
  sin_temp = sin(temp);
  cos_temp = cos(temp);
  xt = rho + c2 * (x*cos_temp-y*sin_temp);
  y =  c2 * (x * sin_temp + y*cos_temp) ;
  x = xt;
  j = x * scale + xoff;
  k = y * scale + yoff;
)
```

Pseudocode K.1. *How to create an Ikeda attractor.* (The program coded here is in the style of the C language.)

$u = \zeta\bar{\zeta}$ and $v = (\zeta^m + \bar{\zeta}^m)/2$. ζ is complex. $\alpha, \beta, \gamma, \lambda$, and ϕ are constants. $\bar{\zeta}$ is the complex conjugate of ζ.)

```
ALGORITHM: How to Create Feather Fractals.

aa = -0.48; b = 0.93; p=9200000; c=2.0-2.0*aa;
x = 3.0; y = 0;
w = aa*x + c*(x*x)/(1. + x*x);
for(n = 0; n <= p; n++){
    PlotDotAt (x,y);
    z = x; x = b*y + w; u = x*x;
    w = aa*x + c*u/(1. + u); y = w-z;
}
```

Pseudocode K.2. *How to create a Feather Fractal.* Examine the region between -10 and 10 in the x and y directions. (The program coded here is in the style of the C language.)

```
ALGORITHM : Bifurcation Plot Generator

INPUT: Min and Max picture boundaries, beta, resolution,
iteration, X0
Typical Parameter Values: min=0, max=125, N=300, res=400,
beta=5, X0=0.9
OUTPUT: Bifurcation-Plot ( Lambda vs. X(time) )

1 do lambda = min to max by (max-min)/res;
2   x=x0;                   (* Set initial point, X0 *)
3   do i = 1 to N;          (* Iteration loop        *)
4     x = lambda*x*(1+x)**(-beta);
5     SetColor(i)           (* Set color based on in *)
6     PlotDot(lambda,x)     (* Plot dot at (lambda,x)*)
7   end;                    (* End Iteration loop    *)
8 end;                      (* End lambda loop       *)
```

Pseudocode K.3. *Bifurcation Plot Generator.*

```
ALGORITHM 3-D Strange Attractor Generator

TYPICAL PARAMETER VALUES:
xxmin=-2; xxmax=2, yymin=-2, yymax=2  (* picture boundaries  *)
pres  = 1600                          (* picture resolution  *)
iter1 = 1000; iter2 = 5000;
(* iter1*iter2 = total number of iterations *)

METHOD: A 5-parameter Dynamical System
OUTPUT: Pixel array containing the output picture intensities.
NOTES:  Try experimenting with different values of e which
can control the degree of randomness of the system.

xinc=pres/(xxmax-xxmin);        (*controls x posit.*)
yinc=pres/(yymax-yymin);        (*controls y posit.*)
a=2.24;b=.43;c=-.65;d=-2.43;e=1;
p(*,*)=0;                       (*init. p array *)
x,y,z=0;                        (*starting point*)
do j = 1 to iter1;
 do i = 1 to iter2;
  xx = sin(a*y)  -z*cos(b*x);
  yy = z*sin(c*x)-cos(d*y);
  zz = e*sin(x);
  x = xx; y=yy; z=zz;
  if xx<xxmax & xx>xxmin & yy < yymax &
   yy > yymin then do;
   xxx= (xx-xxmin)*xinc;     (* scale *)
   yyy= (yy-yymin)*yinc;     (* scale *)
   p(xxx,yyy) = p(xxx,yyy) + 1;
  end; /* then do */
 end;      (* i       *)
end;       (* j       *)
(*P now contains the intensities for each pixel in the picture*)
```

Pseudocode K.4. *3-D strange attractor generator.*

```
ALGORITHM: How to Create a Butterfly Curve

OUTPUT: Plot points at locations specified by
        variables xx and yy.
NOTE:   Assume screen goes from 0 to 100 in x and y directions.

1  pi = 3.1415;
2  DO theta = 0 to 100*pi by .010;
3   r = exp(cos(theta)) - 2*cos(4*theta) + (sin(theta/12))**5;
4   x = r * cos(theta);/* convert from polar coordinates */
5   y = r * sin(theta);
6   xx = (x * 6) + 50; /* scale factors to enlarge and
7   yy = (y * 6) + 50;    center the curve */
8   IF theta = 0 THEN MovePenTo(xx,yy);
9                ELSE DrawTo(xx,yy);
10 END;
```

Pseudocode K.5. *How to create a Butterfly Curve.*

Appendix L

Science-Fiction

"Life is a movement from the forgotten into the unexpected." Loren Eisely

L.1 Mathematics in Science-Fiction

There are many intriguing examples of science-fiction books dealing with mathematics. A number of these books have been inspired by mathematics dealing with higher spatial dimensions and strange geometrical topologies. Others deal with extraterrestrial creatures based on mathematics or concerned with mathematical issues. The following is a list of some favorites.

1. Fadiman, C. (1958) *Fantasia Mathematica.* Simon and Schuster: NY (A book of stories and diversions all drawn from the universe of mathematics).
2. Fadiman, C. (1981) *The Mathematical Magpie.* Simon and Schuster: NY. (A collection of short stories and cartoons relating to mathematics.)
3. Burger, D. (1969) *Sphereland.* Crowell. (Extends Abbott's *Flatland* to include Einsteinian curved spacetime.)
4. Feldman, A. (1963) The Mathematicians. In *50 Short Science Fiction Stories.* Asimov, I. and Conklin, G., eds. Collier: NY. (Describes a mathematical race of extraterrestrials.)
5. Zebrowski, G. (1985) "Gödel's Doom." (A change in a central mathematical theorem alters reality. This story appears in several anthologies.)
6. Wells, H. (1896) "The Plattner Story." In *The Country of the Blind and Other Stories.* T. Nelson and Sons: London. (British author Wells is considered by many as the father of science-fiction.)
7. Breuer, M. (1929) "The Captured Cross-Section." (Breuer was an American medical doctor and science-fiction short story writer.)
8. Russell, B. (1954) "The Vision of Professor Squarepunt."
9. Anthony, P. (1992) *Fractal Mode.* Ace-Putnam: NY. (Describes a Mandelbrot set universe.)
10. Anthony, P. (1991) *Virtual Mode.* Ace-Putnam: NY. (The main character, Colene, is fascinated by fractals).

11. Clarke, A. (1989) *Ghost from the Grand Banks.* Bantam: NY. (A female character becomes insane after exploring the Mandelbrot set.)

12. Anthony, P. (1976) *OX.* Avon: NY. (Cellular automata creatures).

13. Simak, C. (1981) *Project Pope.* Ballantine: NY. (Equation creatures).

14. Prachett, T. (1990) *Pyramids.* Corgi. (One of the characters, a camel, is the best mathematician in the world. Zeno makes an appearance with some dangerous paradox testing.)

15. Adams, D. (1981) *The Hitch Hiker's Guide to the Galaxy.* Crown: NY. (Marvin, the paranoid android, claims various mathematical results. Also, a godlike computer declares that the answer to the mystery of the universe is forty-two.)

16. Horgan, J. (1991) *Entoverse.* Ballantine Books: New York : (Mathematical creatures.)

17. Rucker, R. (1980) *White Light.* Ace: NY. (The protagonist goes to the afterworld, where there is a Hilbert's hotel and a transfinitely tall mountain.)

18. DeLillo, D. (1976) *Ratner's Star.* Vintage: NY. (Eccentric mathematicians decode a binary message from the stars.)

19. Heinlein, R. And He Built a Crooked House. In Fadiman, C. (1958) *Fantasia Mathematica.* Simon and Schuster: New York. (In this story, Heinlein tells the tale of a California architect who constructs a 4-dimensional house. He explains that a 4-dimensional house would have certain advantages:

> "I'm thinking about a fourth spatial dimension, like length, breadth, and thickness. For economy of materials and convenience of arrangement you couldn't beat it. To say nothing of ground space – you could put an eight-room house on the land now occupied by a one-room house."

Unfortunately, once the builder takes the new owners on a tour of the house, they can't find their way out. Windows and doors which normally face the outside of the house now face inside. Needless to say some very strange things happen to the poor people trapped in the house.)

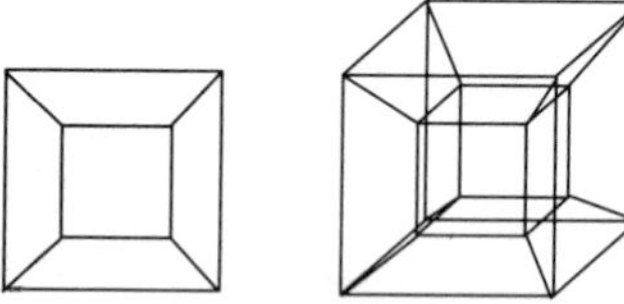

20. Rucker, R. (1985) *Mathenauts: Stories of Mathematical Wonder.* Morrow: NY.

21. Forward, R. (1985) *Rocheworld.* Baen Books: NY. (Set in the 21st century, this book describes intelligent aliens who are fond of mathematics and are more mathematically advanced than humans.)

22. Foster, A. D. (1985) *Sentence to Prism.* Ballentine: NY.

23. Zamyatin, Y. (1983) *WE.* Avon: NY.

24. Delany, S. (1963-1968) *The Fall of the Towers (Trilogy).* Bantam: NY. (Involves battles between creatures that are ratios. The battle strategy involves attempting to cause irrational resonances in an opponent's harmonic. To these creatures there is a very subtle difference between 1/2 and 2/4. The battle spreads to include creatures who are musical notes. These creatures find a difference between C-sharp and D-flat.)

25. Abbott, E. (1884) *Flatland.* Dover: NY. (Mathematician Edwin A. Abbott discusses two-dimensional worlds.)

26. Hinton, C. (1907) *An Episode of Flatland.* Swan Sonnenschein: London. (Speculative essays on multidimensional space.)

27. Dewdney, A. (1983) *Planiverse.* Posieden: NY. (Creatures in a two-dimensional universe.)

28. Odle, E. (1923) *The Clockwork Man.* Doubleday: NY. (Humans master the concept of 4-D reality.)

29. Kuttner, H. (1943) "Mimsy Were the Borogoves." In *Isaac Asimov Presents the Great Science Fiction Stories, Volume 5, 1943.* Edited by Isaac Asimov and Martin H. Greenberg. DAW Books: NY. (Humans master the concept of 4-D reality.)

30. Gardner, M. (1946) *The No-Sided Professor.* Prometheus: NY. (The Möbius strip – a surface with one side – is featured. This story appears in several anthologies.)

31. Nearing, H. (1954) "The Hermeneutical Doughnut" (The Möbius strip – a surface with one side – is featured.)

32. Sagan, C. (1985) *Contact.* Pocket: NY. (The digits of pi are a coded message from God.)

33. Rucker, R. (1983) *The Sex Sphere.* Ace: NY. (Multidimensional fantasy.)

34. Knuth, D. (1974) *Surreal Numbers.* Addison-Wesley: MA.

35. Nim, P. (1978) *Double Möbius Sphere: A Story of the Shape of the Universe.* Pocket: NY.

36. Lem, S. (1974) *Cyberiad.* Harcourt Brace Javanovich: CA. (Cybernetics. Robot-constructor fables.)

37. Vardeman, R. E. (1989) *Weapons of Chaos.* Berkeley Pub.: NY. (A good trilogy of books relating to chaos and attractor theory.)

L.2 Computers in Science-Fiction

Here are a few titles of books where computers play a central role.

1. Van Vogt, A. (1940) *Slan.* Amereon, Ltd: NY. (Electric filing cabinets.)

2. Smith, E. (1986) *Skylark of Valeron.* Berkley: NY. (A mapping navigational computer the size of a planet helps the hero find his way.)

3. Manning, L. (1932) "Master of the Brain." (A mechanical brain administers the entire world.)

4. Jones, D. (1966) *Colossus.* Putnam: NY. (A computer takes over the world by force.)

5. Herbert, F. (1966) *Destination Void.* Berkley: NY. (The computer as God.)

6. Heinlein, R. (1966) *The Moon is a Harsh Mistress.* Ace: NY. (The computer as a playful companion.)

7. Gibson, (1984) *Neuromancer.* Ace: NY. (The computer is indifferent to human kind and has unfathomable purposes.)

8. Adams, D. (1979) *Hitch-Hiker's Guide to the Galaxy.* Pan: London. (A godlike computer declares that the answer to the mystery of the universe is forty-two.)

9. Lem, S. (1974) *Cyberiad.* Harcourt Brace Jacanovich: CA. (Cybernetics, computer science, science end philosophy with a unique humorous style. Robot-constructor fables.)

10. Gibson, W. and Sterling, B. (1991) *The Difference Engine*. Bantam Books: NY. (Describes an alternate past where Babbage engines have become a powerful tool.)
11. Card, O. (1986) *Speaker for the Dead*. Tor: NY. (Sentient computer.)
12. Card, O. 1986) *Xenocide* Tor: NY. (Sentient computer.)

Here are some others: Asimov's "The Computer that Went on Strike" and "The Last Question," Amminnus' "The Thought Machine," Anthony's *Heaven Cent*, Boulle's "The Man Who Hated Machines," Cameron's *Cybernia*, Chaulker's *Quest for the Well of Souls*, Crichton's *The Terminal Man*, Deighton's *The Billion Dollar Brain*, Ellison's "I have No Mouth and I Must Scream," Farmer's *The Gods of Riverworld*, Hodder-Wiliams' *Fistful of Digits*, Clarke's *2001: A Space Odyssey*, and Kagain's *Hellspark*.

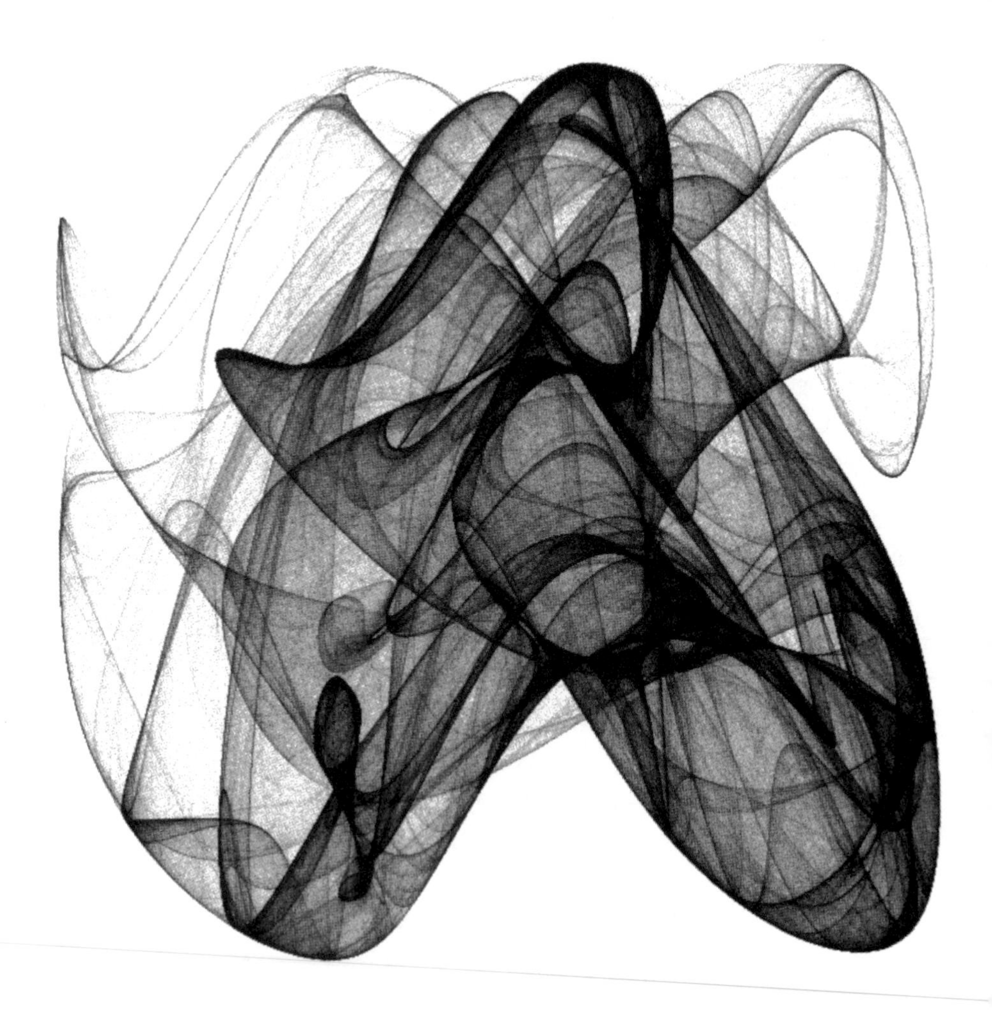

Appendix M

Unusual Resources and Related

M.1 Software

There are many educational programs available for those of you wishing to explore the solar system. Here are a few examples of "freeware" available for a nominal charge from many on-line services or individuals, and other interesting software.[32]

M.2 Latööcarfian Explorer Program

You'll certainly enjoy the *Latööcarfian Explorer Program* by John Matthews, M.D. Matthews, a computer enthusiast in southwest Ohio, has implemented the Latööcarfian algorithm on a personal computer. Named "Cliff," his program runs under System 7.0 (Apple Macintosh computer) and is optimized for Macs with floating point co-processor and color display hardware. (There's also a version for low end Macs. IBM versions will be available in the future.) "Cliff" is freeware, available for a nominal charge from many on-line services or Mac User Groups. It is also available directly from the author for $20, and includes some of his other demos and utilities on the disk. Contact: John B. Matthews, M.D., Gem City Software, 6415 Tantamount Lane, Dayton, Ohio 45449.

See Code M.1 for a BASIC program.

---

[32] There are a few places in this book where companies or products are mentioned. Although all of the products listed provide a stimulus for the imagination, they are listed for illustrative purposes only. ***The author does not endorse any particular company, technology, software, or product***, nor does he accept responsibility for the selection of any products by the reader. The opinions expressed in this book are the author's and do not represent the opinions of any organization or company.

```
20  REM   The values of a, b, c, and d may be chosen at random in a
30  REM   range:  ( -3.0 < a, b < 3.0, 0.5 < c, d < 1.5 )
80  REM   To produce the King's dream, use the following constants:
95  REM   ( a = -0.966918, b = 2.879879, c = 0.765145, d = 0.744728 ).
98  REM
100 CLS : SCREEN 9 ' Clear Screen and Establish VGA Screen Mode
110 INPUT " Enter A,B,C,D ==> ";A,B,C,D     ' Get Parameters
120 LOCATE 2,2 : PRINT "Press a Key to Stop...." ' Inform how to Stop
130 LOCATE 3,2 : PRINT "Iterations:"        ' Set up Informational Line
140 X=.1 : Y=.1                             ' Initialize X and Y
150 FOR N=1 TO 10000000#                    ' Loop 10,000,000 Times
160   XN = SIN(Y*B) + C*SIN(X*B)            ' Calculate New X
170   YN = SIN(X*A) + D*SIN(Y*A)            ' Calculate New Y
180   X  = XN : Y = YN                      ' Restore to Original Variables
190   X2 = X*25+250:Y2=Y*50+200             ' Scale to fit the Screen
200   PSET(X2,Y2),1                         ' Plot this Point
210   X$ = INKEY$ : IF X$<>"" THEN 240      ' Stop if User presses a Key
220   LOCATE 3,13 : PRINT N                 ' Display Current Counter Value
230 NEXT N                                  ' Iterate
240 LOCATE 4,1 : PRINT " Done."             ' Terminating Message
```

Pseudocode M.1. *BASIC listing for the Latööcarfian equations.* (Program courtesy of Tom Rankin.)

M.3 The Pope Examines the Sky

What would the sky have looked like when Pope Clement XI gazed up at the starry heavens over the Vatican exactly at midnight on October 10, 1721? Suppose you were alive in the future, say A.D. 4000, and leaned your head out of your New York City apartment to look at the sky? What would you see? EZCosmos (for IBM compatibles) graphically answers these questions, converting your PC into an interactive planetarium and time-machine, showing you the sky at any time and any place on Earth from 4000 BC to AD 10,000. The program recognizes over a thousand cities. Alternatively you can input any latitude or longitude. After making the necessary calculations, EZCosmos displays the heavens at the location and date provided. EZCosmos includes 10,000 objects (stars, planets, nebulas). Want to know where Alpha Centuri is? Just type its name and your computer will highlight it. To identify any spot in the heavens, simply move your cursor to the object. The computer instantly provides the object's name and other details. Contact: Future Trends Software, 1508 Osprey Dr., Suite 103, De Soto, TX 75115.

M.4 World on a Canadian Disc

You don't need to take a trip to the moon and look back at our azure planet in order to get a good view of the globe and it yearly changes. Thanks to GEOSCOPE, an "interactive global change encyclopedia," you can use a personal computer to explore everything from ozone concentration to the evolution of verdant forests in Zaire. The Canadian Space Agency and other firms have placed more than 150 data sets and satellite images on a PC diskette. The most enjoyable part of GEOSCOPE is its "philosophy workroom," where students can ponder the erudite words and images of Carl Sagan and Galileo. For more information, contact the Canadian Space Agency, FAX 514 496-4220. Also see: *Science* May 22, 1992, 256(5060): 1137.

M.5 Others

1. *Planet*, by Peter Cramer, Quentin Herr, Michael Collette, and Steve Jogan, of the Physics Courseware Project at Case Western Reserve University. *Planet* simulates the motion of a planet around a star, using an analytical solution rather than a numerical one. It is designed for use in a standard introductory undergraduate physics course, and allows students to visualize the changes in a planet's velocity during its orbit. (Contact: Prof. Peter Cramer, Dept. Physics, University Circle, Case Western Reserve University, Cleveland, OH 44106. The program runs on MS-DOS and Macintosh.)
2. *Dance of the Planets*, by Arc Science Simulations, P.O. Box 1955 N. Loveland, CO 80539. (IBM Compatible, 650K, EGA/VGA, HD. Math co-processor recommended.) Through orbital simulation, the sky and the solar system can come to life on your computer. You can find orbital resonances and chaos, lost comets, and errant asteroids. Has over 6300 solar system objects in a beautiful starry sky of 11,000 stars and non-stellar objects.
3. *Pickover Software Sampler*, Bourbaki, PO. Box 2867, Boise, Idaho 83701. (Software for drawing and exploring many of the images in *Computers, Pattern, Chaos and Beauty*, St. Martin's Press: NY).
4. *Distance Sun Windows*, by Virtual Reality Laboratories, 2341 Ganador Ct., San Luis Obispo, CA 93401. (All-purpose planetarium program for the IBM PC.)
5. *Arizona Database*, by Arizona Database, 3135 S. 48th St., Suite 3, Tempe, AZ 85282. (Integrated catalogue of 43,228 deep-space objects on an IBM PC. Permits interesting astronomical searches. For example, you can locate all globular clusters of magnitude 9 to 13 that lie within a specified rectangle of the sky.)
6. *Deep Space 3D*, by David Chandler, PO Box 309, La Verne, CA 91750. (Calculates and displays comets on 3-D stereo star charts. IBM PC.)
7. *Fractal Discovery Laboratory*. Designed for a science museum or school setting. "Entertaining for a four-year-old, and fascinating for the mathematician." Earl Glynn, Glynn Function Study Center, 10808 West 105th Street, Overland Park, KS 66214-3057.
8. *CAL*, a fractal plotter for IBM compatible PCs which allows the user to create images from one of 50 formulas supplied, or to enter their own equations. Contact: Tim Harris, 5 Burnham Park Road, Peverell, Plymouth England.

M.6 Newsletters and Magazines

1. *bOING-bOING Magazine*. Unusual newsletter covering topics such as: cyberspace, data encryption, chaos, artificial life, cryonics, nanotechnology, self-organization, game theory, evolution, brain toys, programming, alien life, psychedelia. Very weird short stories. Fringe culture. Cyberpunk. A "perpetual novelty brain jack designed to demolish the gray shield of consensus reality." Sample, $3.95; four issues, $14. 4500 Forman Ave #2, Toluca Lake, CA 91602

2. *QUANTUM*. Beautiful, well-illustrated, glossy, student magazine of math and science. Highly recommended. Conveys the breadth and wonder of math and physics. Contact: Springer-Verlag New York Inc., Attn: Journal Promotion Department, 175 Fifth Ave, New York, NY 10010.

3. *AMYGDALA*, a fascinating newsletter on fractals. Write to AMYGDALA, Box 219, San Cristobal, New Mexico 87564 for more information.

4. *The Best of Journal of Chaos and Graphics*, three volumes in one wild and informal paperback book. Topics: fractals, computer art, devil's curves, music, and much more. Published by: Media Magic, PO Box 507, Nicasio, CA 94946 USA (Toll free: 1-800-882-8284. $10.00, plus $2.00 postage and handling, US orders; $4.00 postage and handling non-US orders.)

5. *Powell's Technical Bookstore Newsletter*. 33 NW Park Ave., Portland, Oregon 97209. (Computing, electronics, engineering, and science books.)

6. *YLEM – Artists using science and technology*. This newsletter is published by an organization of artists who work with video, ionized gases, computers, lasers, holograms, robotics, and other nontraditional media. It also includes artists who use traditional media but who are inspired by images of electromagnetic phenomena, biological self-replication, and fractals. Contact: YLEM, Box 749, Orinda, CA 94563.

7. *Recreational and Educational Computing Newsletter*. Dr. Michael Ecker, 909 Violet Terrace, Clarks Summit, PA 18411. (Devoted to the playful interaction of computers and "mathemagic" – from digital delights to strange attractors. Puzzles, program teasers.)

8. *MEDIA MAGIC: Computers in the Arts and Sciences*, PO Box 507, Nicasio, California 94946. This fine company distributes books, videos, prints, and calendars. Catalog available.

9. *ART MATRIX*, creator of beautiful postcards and videotapes of exciting mathematical shapes. Write to ART MATRIX, PO Box 880, Ithaca, New York 14851 for more information.

10. *Small Computers in the Arts Newsletter*. An organization that supports artists working with small computers. 5132 Hazel Ave., Philadelphia, PA 19143.

11. *Zometool*, a construction kit that turns abstract geometrical concepts into working models. The ultimate tool for understanding the structure of space. BioCrystal, P.O. Box 7053, Boulder CO 80306-7053.

12. *Dinosaur Society*, a nonprofit alliance of scientists, artists, educators, and writers worldwide. They publish a monthly newspaper for children, *Dino Times*, and have a children's dinosaur club to give children accurate up-to-date images of dinosaurs, dinosaur scientists, and the scientific process. *The Dinosaur Report* covers the same subjects but is for adults. (Did you know that a new dinosaur species is discovered, on average, every seven weeks?) Contact: The Dinosaur Society, PO Box 2098, New Bedford, MA 02162.

13. *Maze Magazine*, mazes for children and adults. STILES, Dept. GM1, 17 North, Manchester, MA 01944.
14. *Skullduggery,* an amazing catalog of prehistoric skulls. (I have the saber tooth tiger and dire wolf skulls in my office.) Contact: Skullduggery, 624 South B Street, Tustin, CA 92680.
15. *Computer-aided Hypervisualization Books*, information on visualization techniques helping researchers graph beyond 3 dimensions. CAH Research Center, PO Box 1003, Kent-Ridge Post Office, Kent-Ridge Crescent, Singapore 0511.
16. *International Association for Astronomical Arts*, for anyone interested in astronomical "space" art. Contact: IAAA, 4160 Willows Road, Alpine CA 91901. In Europe, contact: IAAA, 99 Southam Road, Hall Green, Birmingham B28 0AB England. (See Appendix O, "Credits and Additional References" for more information.)
17. *HyperSpace*, a fascinating journal on all subjects relating to higher dimensional geometries, geometry and art, and unusual patterns. The journal has articles in English and Japanese. Contact: Japan Institute of Hyperspace Science, c/o K. Miyazaki, Graduate School of Human and Environmental Studies, Kyoto University, Sakyo-ky, Kyoto 606 Japan.
18. *Extropy*, unusual journal devoted to nanotechnology, life extension, artificial life, digital economies, and related. Contact: Extropy, PO Box 57306, Los Angeles CA 90057-0306.

M.7 Dreams

"We are such stuff as dreams are made of." Shakespeare

1. *Dream Network*, quarterly magazine serving a community of laypeople and professionals involved in studying dreams. Includes book reviews and letters to the editor. How to start your own dream group. Tibetan dream yoga. Human/insect relations in dreams. Contact: 1337 Powerhouse Lane, Suite 32, Moab, Utah 84532.
2. *NightLight*, quarterly publication on lucid dreams. The goal is to advance research on consciousness and to apply the results to help humanity. Contact: Lucidity Institute, 2555 Park Boulevard, #2, Palo Alto, CA 94306.
3. *Dream Catchers*, artistic basket-like items to catch nightmares, based on ancient Indian legends. Contact: Poteet, 216 M Passo del Pueblo Norte, Suite 205, Taos, NM 87571.

M.8 Fractals in the UK

1. *FRAC'Cetera* is both a user's group and an electronic newsletter. The former deals with users of the fractal explorer program *Fractint*. The latter is a frequently updated compendium of information about publications, videos, software, digests, journals, etc. which deal with fractals. Contact: FRAC'Cetera, Le Mont Ardaine, Rue des Ardaines, St. Peters, Guernsey, C.I., United Kingdom.

2. *Fractal Report*, a great newsletter on fractals. Published by J. de Rivaz, Reeves Telecommunications Lab. West Towan House, Porthtowan, Cornwall TR4 8AX, United Kingdom.
3. *Strange Attractions*. A store devoted to chaos and fractals (fractal art work, cards, shirts, puzzles, and books). For more information, contact: *Strange Attractions*, 204 Kensington Park Road, London W11 1NR England.

M.9 Astronomy Society Catalog

Explore the cosmos with slides, videos, books, software, and gifts. The non-profit *Astronomical Society of the Pacific* has served as a bridge between astronomers and the public since 1889. With members in 50 states and over 75 countries, the society is an important resource for astronomy teachers, students, and enthusiasts. Their catalog sells magnificent color posters of Jupiter and Saturn (with their satellites), various nebulae, galaxies, other planets and moons. Also listed are giant moon maps, comet posters, spectra, and astronomical image processing software with 19 sample images of planets and stars (MS-DOS). Also: interactive planetarium software for the Mac, Amiga, and IBM compatibles, and "Greetings from Outer Space" tourist cards from the cosmos. Members of the society receive a magazine, monthly sky calendars, star maps, etc. Contact: Astronomical Society of the Pacific, 390 Ashton Avenue, San Francisco, CA 94112.

M.10 The Aurora Watcher's Handbook

Auroras are beautiful, luminous, atmospheric phenomena occurring in the vicinity of the earth's northern or southern magnetic pole. They're visible from time to time at night and popularly called the Northern (or Southern Lights). The name "aurora" comes from the Roman goddess of the dawn, represented as rising with rosy fingers from the saffron-colored bed of Tithonous.

The *Aurora Watcher's Handbook* tells you all you would want to know about auroras – what causes them, where they can be seen, and how to capture them on film. The book is illustrated with cartoons, color plates, and scientific drawings. Various legends and myths concerning auroras are discussed, as well as geophysical explanations of aurora structure. (Davis, N. (1992) *The Aurora Watcher's Handbook*. University of Alaska Press, 230 p., paper, $20.00.)

M.11 Lawyers in Space!

In *Space Policy: An Introduction*, Professor Nathan Goldman discusses space law, and the roles of government and private industry in developing the resources of outer space. He also examines the civilian and military implications that space development has for our global society. Interestingly, the author is a professor of space law and a lawyer specializing in space law and high technologies. (Iowa State University Press, 1992, 321 p., hardcover, $37.95.)

M.12 Who Were The Celestial Police?

Are there any planets in the Solar System which we have not yet discovered? Could they contain life?

As early as 1772 astronomers thought that there was an undiscovered planet near Jupiter. So enthralled were they with the idea that, in 1800, a team of astronomers began a systematic telescopic search for a planet between Mars and Jupiter. The team of astronomers called themselves the "celestial police," and their leader was the wizened J. H. Schröter.

The celestial police had a big surprise in store for them. Instead of finding a hidden planet, they stumbled upon many very tiny planets which we now call asteroids. In 1801, the celestial police discovered the first asteroid. They named it Ceres.[33] Three more asteroids, Pallas, Juno, and Vesta, were discovered before the end of 1807.

Most of these minor planets or planetoids lie in the region between orbits of Mars and Jupiter. It's hard to believe that by 1990 over 3000 asteroids have been named. All are small, and only Ceres has a diameter greater than 900 km. NASA's Near Earth Asteroid Rendezvous (NEAR) mission will send a small spacecraft on a year-long inspection of the asteroid Nereus from as close as a mile away. The spacecraft is planned to be launched in 1998, and two years later it will orbit Nereus.

In 1992, proud NASA scientists unveiled a color image of Gaspra – the most detailed photo ever taken of an asteroid. The photo reveals more than 600 meteorite impact craters on the 19-kilometer-long rock, some as small as 100 meters in diameter.

[33] Technically speaking, the discover of Ceres, Piazzi, was not an official member of the "police" at this time, although he joined later!

Appendix N

Author's Notes

"These Notes as I see them relate not to lectures but to feeling. I'm sure my readers differ from me on many things, but I hope that we share the essence of wonder and longing for what we may never quite understand." Piers Anthony, 1991, *Virtual Mode*

Piers Anthony, one of science fiction and fantasy's most prolific talents, established the interesting practice of placing an "Author's Note" section at the end of some of his books. In these notes, he gives that slice of his life occurring during the writing of his novel, complete with discussions of social issues and unfinished thoughts. In keeping with Piers, I include this section to give you a slice of some of the mail I have received while writing *Chaos in Wonderland.* Also included are miscellaneous recent thoughts, answers to some of the questions posed earlier in this book, and some reader responses to my previous books.

The Gleichniszahlen-Reihe Monster World-Record

In *Computers and The Imagination* I offered an award to the reader who computed the largest "Gleichniszahlen-Reihe Monster" number sequence. The "Gleichniszahlen-Reihe Monster" refers to a number sequence with some rather strange and compelling properties. Because the sequence never seems to contain a number greater than 3, you don't need large computers to begin exploring.

Consider the integer sequence $u_{r,n}$, where r is the row number, and n the column number:

```
1
1 1
2 1
1 2 1 1
1 1 1 2 2 1
  ...
```

You probably can't guess the numerical entries for the next row. However, the answer is actually simple, when viewed in hindsight. To appreciate the answer, it helps to speak the entries in each row out loud. Note that row Two has two "ones," thereby giving the sequence 2 1 for the third row. Row Three has one "two" and one "one." Row Four has one "one," one "two," and two "ones." From this, an entire sequence $u_{r,n}$ can be generated. This interesting sequence was described in a German article, where M. Hilgemeier called the sequence "Die Gleichniszahlen-Reihe," which translates into English as "the likeness sequence." The sequence, also extensively studied by John H. Conway, grows rather rapidly. For example, row 16 is:

13211321322113311213211331121113122112132113121113222
1123113112221131112311332111213211322211312113211

The largest number *u* can contain is 3. (Can "3 3 3" ever occur in the sequence?)

So far, the world-record holder for this sequence is Charles Ashbacher of Cedar Rapids, Iowa. On May 29, 1992 he sent me a diskette containing nearly 894,816 digits for row 50, which he computed using a FORTRAN program he had written. He also computed the sequence for row 53, which contained nearly 1,982,718 digits. The number would not fit on a diskette. Ashbacher estimates that row 53, if printed on paper, would require about 417 pages. In August 1992, Ashbacher computed row 56. The number of digits is in the range 4,391,696 to 4,391,703. The size of the data file containing the number is roughly 5205 KB. The computation required 9 minutes using a VAX 4000. About 1 minute of this time was spent simply dumping the contents of the array to a file. Ashbacher discovered that the number of digits in a likeness sequence for row 77 would break the one billion mark, requiring 1.2 GB of memory. For more observations on this sequence, see *Computers and the Imagination*.

Planet Music

The composer of the planet music in Appendix G, "Planet Music" is Gustav Holst (1874-1934).

Life Inside Planets

There is growing popular and scientific interest in exobiology, the study of potential life on other planets (See Figure N.1.) The bold title of *USA TODAY*'s front-page cover story (July 1, 1992) screamed out at readers:

ET May Live Deep Inside Other Planets

This popular newspaper indicated to its 6.6 million readers that microscopic organisms drawing their energy solely from subterranean chemical processes – not sunlight – thrive six to seven miles deep in the Earth's crust. In the article, Professor Thomas Gold of Cornell University suggests that similar bacteria may be living inside other solid planets. "We regard our surface life as being the marvel of creation," said Gold. But "it may be the other is very common and that we are the oddity." Bizarre worms living without the sunlight critical to surface life forms and at much higher temperatures have already been discovered. These worms eat bacteria that thrive on the subterranean chemicals. The studies are reviewed in the July 1st (1992) issue of the *Proceedings of the National Academy of Sciences*.

Hippias' Quadratrix

Hippias of Elis, who came to Athens in the second half of the 5th century B.C., studied the behavoir of the interesting formula:

$$2r = \pi\rho(\sin\theta)/\theta \qquad (N.1)$$

in polar coordinates. ($\rho = \sqrt{x^2 + y^2}$, $\theta = arctan(y/x)$.) After some manipulation, we have

$$y = x\cot(\pi x/2a) \qquad (N.2)$$

THE NATION'S NEWSPAPER — 50 CENTS

USA TODAY

No. 1 IN THE USA . . . 6.6 MILLION READERS A DAY

WEDNESDAY

'NEW YORKER' SHUDDERS IN ITS BOOKS

TINA BROWN TO EDIT FAVE LITERARY MAG, 1D

▶ ON TIME WARNER, 4B

★★★ 'LEAGUE' NOT QUITE A HOME RUN 1D

WOMEN CARRY EDDIE'S ★★ 'BOOMERANG,' Review, 7D

ET may live deep inside other planets

By Tim Friend
USA TODAY

People seeking extraterrestrial life may just have been looking in the wrong place.

New studies on Earth indicate microscopic organisms that draw their energy solely from subterranean chemical processes — not sunlight — thrive 6 to 7 miles deep in the planet's crust.

Similar bacteria may be living inside other solid planets because the interiors are likely to be similar, says Thomas Gold, professor of geochemistry and planetology at Cornell University in Ithaca, N.Y.

"We regard our surface life as being the marvel of creation," says Gold. But "it may be the other is very common and that we are the oddity."

Bizarre life forms such as 10-foot worms — first found around cracks in the ocean floor — show life can exist:

▶ Without the sunlight critical to surface life forms.

▶ At much higher temperatures than on the surface.

It was then found the worms eat bacteria that thrive on raw materials from Earth's core.

Gold says studies now show these bacteria exist throughout the crust; similar organisms could exist on any planet with a hot core.

How can we find out? By obtaining samples from young craters on the moon and from landslides on Mars where subsurface organisms would have been exposed, he says.

The studies are reviewed in today's *Proceedings of the National Academy of Sciences.*

Figure N.1. *Front page of USA today.* There is growing popular and scientific interest in exobiology, the study of potential life on other planets. Like the lifeforms in *Chaos in Wonderland*, some scientists believe that primitive life has evolved *inside* planets. (Figure © 1992, USA TODAY. Reprinted with permission.)

Amazingly, Hippias and another Greek mathematician Dinostratus (350 B.C.) only saw the tip of the iceberg for this curve – they only knew about the shape of the curve on the interval $-a < x < a$. (It looks like the top of a bald man's head on this interval.) Later, in the 17th century, the full behavior of this curve became known. What amazing things happen to the curve at higher and lower values of x? Try plotting the curve yourself.

Hermann Schubert, Sirus, Microbes, and Pi

In 1889, Hamburg mathematics professor Hermann Schubert described how there is no practical or scientific value to knowing π to more than a few decimal places:

> "Conceive a sphere constructed with the earth at its center, and imagine its surface to pass through Sirius, which is 8.8 light years distant from the earth [8.8 years × 186,000 miles per hour]. Then imagine this enormous sphere to be so packed with microbes that in every cubic millimeter millions of millions of these dimunivitve animalcula are present. Now conceive these microbes to be unpacked and so distributed singly along a straight line that every two microbes are as far distant from each other as Sirus from us, 8.8 light years. Conceive the long line thus fixed by all the microbes as the diameter of a circle, and imagine its circumference to be calculated by multiplying it diameter by π to 100 decimal places. Then, in the case of a circle of this enormous magnitude even, the circumference so calculated would not vary from the real circumference by a millionth part of a millimeter. This example will suffice to show that the calculation of π to 100 or 500 decimal places is wholly useless."

The Safford Number: 365,365,365,365,365,365

What is special about the huge number 365,365,365,365,365,365? The story begins with the calculating prodigy Truman Henry Safford (1836-1901) of Royalton, Vermont. At the age of 10, Reverend H. W. Adams asked him to square, in his head, the number 365,365,365,365,365,365. Dr. Adams reported:

> He flew around the room like a top, pulled his pantaloons over the tops of his boots, bit his hands, rolled his eyes in their sockets, sometimes smiling and talking, and then seeming to be in agony, until in not more than minute said he, 133,491,850,208,566,925,016,658,299,941,583,255!

Truman Safford graduated from Harvard, became an astronomer, and soon lost the amazing computing powers he had in his youth.

Juggler Number Contest

In *Computers and the Imagination* an award of 50 dollars was offered by the publisher for the printout of the largest Juggler number computed by a reader. Juggler numbers are defined by:

$$j(n) = \begin{cases} \left[n^{(1/2)} \right] & \text{if } n \text{ even,} \\ \left[n^{(3/2)} \right] & \text{if } n \text{ odd} \end{cases} \qquad (N.3)$$

where n is any initial positive integer. The bracket signs indicate that non-integer values are to be truncated to the maximum integer equal to or smaller than the number enclosed (i.e., $4.1 \rightarrow 4$). This means that you start with any integer and raise it to either of the two choices of powers (1/2 or 3/2) depending on whether or not it is even or odd, and repeat the operations over and over again. This sequence is produced by iterative rules: apply the rule to the current number in the sequence and you get the next one.

On June 27, 1992 Harry J. Smith of Saratoga, California beat his own previous world-record by computing a new behemoth Juggler number – the 972,463-digit giant for the starting number 48,443. The sequence starts:

$J(0) = 48443$

$J(1) = 10662183$

$J(2) = 34815273349$

and reaches its peak at $J(60)$ which has 972,463 digits. Alas, as hypothesized for all Juggler numbers, Smith's mighty sequence meets its demise at $J(157)$ at which point the Juggler sequence has decayed back to 1. Figure N.2 shows the first and last few digits of the world-record Juggler number. Harry Smith writes:

> "The research and development I had to do to compute this number on a PC was quite interesting. I knew from my earlier work that starting from $J(0) = 48443$ produced a very large number because my program ran out of memory while computing this sequence. The numbers were getting too large to be held in a Turbo Pascal array. In Turbo Pascal, arrays are limited to 65536 bytes."

```
            Juggler Sequence starting with x(0) = 48443 (JugglerW)
x(60) =
178,34693,40516,49917,39321,74350,73083,71062,13275,79274,14151,7
6887,82954,14812,50298,05610,76020,34505,51264,49900,10719,04853,
90310,21396,75508,10814,41772,60113,44775,35851,44106,14254,47343
,29065,03549,24185,18165,08550,87105,97822,64083,57062,42262,4211
8,63867,46675,77908,72179,15357,38866,65697,19465,03570,16212,831
20,68520,18217,86265,15264,93384,84982,67426,92997,04404,84059,94
577,89621,98169,31367,60594,53211,49999,14288,40254,38452,48998,5
4732,33764,13482,39256,44505,66046,79224,67233,44199,56666,04687,
97225,42220,93042,21133,87644,49994,78297,43490,56759,87604,05085
,16915,95365,86640,63886,83434,92515,75338,99544,38983,17613,5178
4,89462,22073,36967,82512,01460,08052,18188,16186,40903,06484,975
49,25819,71162,07888,14175,75562,72167,25955,46592,95394,62593,19
549,85547,25497,15871,19485,86867,48522,90296,44985,77269,33417,8
5271,82129,23073,99091,44207,47943,12895,84273,05331,71002,10397,
78189,72246,09111,53388,10432,42277,31753,86323,62891,04104,82346
,25247,28177,12890,88135,16277,81850,49785,59807,93733,26996,9637
2,00379,07521,62166,93983,82280,63468,67875,51526,87973,11399,154
26,41318,52992,53279,17880,91302,30737,89121,81453,83843,93363,28
. . .

3054,64916,53409,99122,04266,11068,37494,58203,38383,91667,01270,
90245,16774,91254,85875,83321,07524,63341,62997,27587,43246,36090
,86908,07906,48760,83639,79434,83214,54730,10046,26774,41910,1147
7,71631,62081,38266,68124,89652,97888,26641,37770,65739,22410,628
68,19971,75486,18890,27995,34284,74783,81466,88480,20255,11417,65
772,82789,25495,60918,02526,01861,80407,06024,65260,75907,49299,9
8770,77890,19223,35637,02348,87323,20656,41785,72010,15755,19890,
97368,16891,95998,29631,45662,24863,12821,60797,77535,36580,86405
,49823,46336,46158,52885,93462,46963,96383,81475,59736,45355,4652
3,25180,88022,97322,31773,85043,07122,33971,15052,29164,12195,619
68,54966,66390,40830,13981,08379,93386,55111,25953,88112,95619,65
136,87619,77749,80745,60913,22959,60952,79950,99540,94187,38167,8
2502,94368,21497,65352,91149,82350,00104,31981,81940,62657,61079,
48295,86572,73665,16676,40536,04549,68638,37618,29914,96520,36611
,67707,48815,29144,93609,11599,18713,47901,93158,68830,61937,3214
4,16963,63148,53625,33412,31137,66030,66596,63587,56580,78446,044
91,80197,14096,07110,58765,77388,60097,36411,10776,61475,38592,32
553,56732,39236,79619,78297,22571,46322,25797,32284,21657,36230,7
1628,78952 (972463)
```

Figure N.2. *The world's largest Juggler number.* On June 27, 1992 Harry J. Smith computed the world-record holder for Juggler numbers. Shown here are the first and last few digits for $J(60)$ which has 972,463 digits. Printing the entire number would require the same number of pages as in *Chaos in Wonderland*.

The software package he finally used to break the world-record (and which performed the multiple precision integer arithmetic required) was custom built in the object-oriented programming language Turbo C++ for Windows Version 3.0, by Borland International. His computer is an IBM AT compatible 33 Mhz 486 with 16 megabytes of RAM and a 330 megabyte hard disk drive. He found that DOS did not provide the necessary memory.

Once Smith conquered the memory problem, he found that it would take a week just to compute the next number in the sequence! To overcome the time limitation, Smith next used Fast Fourier Transform techniques to speed up the large number arithmetic, and a variant of Newton's method to quickly compute the square roots. As a result, the computation of the world record Juggler number required 28 hours. Smith is happy to answer questions regarding his software and methodology. Contact him at: Harry J. Smith, 19628 Via Monte Dr., Saratoga, CA 95070.

Figure N.3. *Halftoning pattern based on random maze.* Fritz Lott (Golden Valley, MN) computed a portrait of me using a maze pattern described and illustrated in *Mazes for the Mind*. To create the maze pattern, randomly orient a square tile with a single diagonal line through it. Continue until an entire checkerboard array is filled with tiles. Lott's halftoning effect requires additional processing once the maze pattern is created. (Image © 1993, W.R. Beamish Co.)

Roger Caws (West Sussex, UK) has proposed and studied a "reverse" Juggler sequence. He writes, "It is possible to map all possible reverse Juggler sequences by making $j(1) = 1$ and then producing all possible next values considering the following: $j(n + 1)$ is the set of odd integers from $j(n)^{2/3} \ldots < (j(n) + 1)^{2/3}$ + set of even integers from $j(n)^2 \ldots < (j(n) + 1)^2$. "

Factorial Bases

For those of you not familiar with numbers represented in bases other than 10 (which is the standard way of representing numbers) consider how to represent any number in base 2. Numbers in base 2 are called binary numbers. The presence of a "1" in a digit position of a number base 2 indicates that a corresponding power of 2 is used in determine the value of the binary number. A 0 in the number indicates that a corresponding power of 2 is absent from the binary number. For example, the binary number 1111 represents $(1 \times 2^3) + (1 \times 2^2) + (1 \times 2^1) + (1 \times 2^0) = 15$. The binary number 1000 represents $1 \times 2^3 = 8$. Here are the first eight numbers represented in binary notation:

0, 1, 10, 11, 100, 101, 110, 111,...

It turns out that any number can be written in the form $c_n b^n + c_{n-1} b^{n-1} + \cdots c_2 b^2 + c_1 b^1 + c_0 b^0$, where b is a *base* of computation and c is some positive integer less than the base.

With this introduction to bases, let's turn to the strange concept of "factorial base" brought to my attention by Julian Rosenman, M.D. Dr. Rosenman explains that it is possible to write any number in the form: $\ldots c_1 3! + c_2 2! + c_3 + c_4 1/2! + c_5 1/3! + c_6 1/4! + \cdots$ The values for the coefficients vary, and have certain restrictions. For example, c_1 must be less than 4. (4 cannot be used because it would apply to another coefficient c_0, i.e. $4 \times 3! = 4!$ which would be written as $c_0 \times 4!$) Likewise, c_2 must be less than 3 and so on. Here are some examples. $12 = 2 \times 3! = 200$. Also $23 = 3 \times 3! + 2 \times 2! + 1 = 321$ Fractions follow the same rules: $1/5 = 1/3! + 0/4! + 4/5! = 0.0104$. Interestingly, all fractions terminate and do not give rise to repeating decimals. This is not too hard to prove. (Try writing out 1/7. The answer requires terms out to $c/7!$). Again, you cannot have 3/3! because this would be written as 1/2!

With this background, next consider numbers such as $e = 1 + 1 + 1/2! + 1/3! + 1/4! \ldots$ In the factorial base, $e - 2 = 0.1111111 \ldots$, a repeating decimal! It does not terminate and hence cannot be a fraction. $e - 2$ also has a non-random occurrence of digits in this "base," but in base ten it is totally irregular with no apparent pattern.

Fractal Poetry

These days fractal designs are everywhere: on shirts, mugs, puzzles, and comic books. Fractal theory is even beginning to used in poetry. For example, Rodrigo Siqueira (a student of electronic engineering, data compression, and differential cryptanalysis at Universidade de Sao Paulo in Sao Paulo, Brazil) creates poems where words are arranged on fractals called Cantor dusts. As background, a symmetrical Cantor set can be constructed by taking an interval of length 1 and removing its middle third (but leaving the end points of this middle third). This leaves two smaller intervals, each one-third as long. In the symmetrical case, the middle thirds of these smaller segments are removed and the process is repeated. The symmetrical Cantor set has a "measure zero," which means that a randomly thrown dart would be very unlikely to hit a member. At the same time it has so many members that it is in fact uncountable, just like the set of all of the real numbers between 0 and 1. Here is one poem he "exhibited" at a Fractal Exposition in Sao Paulo.

```
                     THE CANTOR DUST

...in...the...structure...of...dynamical...systems.....
.lies..a..new..vision...        ...of.order.&.chaos......
.complex...  .creation..        ...order...    ..chaos....
.can  .you.  see.  .the.        .flow  in..   the  path..
.f r  .a c.  t u.  .r e.        .i n   t o    .m i  n d.
|| || || ||  || || || ||        || || || ||   || || || ||
```

Figure N.4. *Maze W-69, by Sampei Seki (Osaka, Japan).* In my book *Mazes for the Mind* there was a discussion on unusual 3-D mazes constructed from stairs. The staircase maze here is courtesy of the journal *HyperSpace* (see M.6, "Newsletters and Magazines"). Can you travel from A to B or from C to D with the condition that the total number of steps is 69. You must *add* the number of stairs when you travel up and *subtract* the number of stairs when going down.

A Nonlinear Recurrence Yielding Binary Digits

The enigmatic properties of the sequence

$$1, 2, 3, 4, 6, 9, 13, 19, 27, 38, 54, 77 \ldots$$

defined by the recurrence

$$u_1 = 1, u_{n+1} = \lfloor \sqrt{2}\,(u_n + 1/2) \rfloor, n \geq 1, \qquad (N.4)$$

where $\lfloor x \rfloor$ denotes the floor of x, the largest integer not larger than x, was discussed in the June 1991 issue of *Mathematics Magazine* (Rabinowitz and Gilbert, 64(3)). They note the unusual property that $u_{2n+1} - 2u_{2n-1}$ is just the nth digit in the binary expansion of $\sqrt{2}$.

Sequence Music

Nich Mucherino (Bridgeport, CT) writes that he was delighted by the "Drums of Ulupu" in *Mazes for the Mind.* He says, "You are right. Recursive sequences seem to have unexpected coherence to the ear when played on instruments, even when they do not make obvious sense as a sequence to the eye." He reports on his own experiments with rhythms based on computer solutions to the Tower of Hanoi puzzle. The rhythm is created by assigning an instrument to each peg of the puzzle. Each time a disk is lifted off a peg, or placed on a peg, a beat is played by the corresponding instrument. Surprisingly, when he hooked up a computer (running the algorithm to solve the puzzle) to an electronic music synthesizer, the results had a strange African, perhaps Salsa feel to them.

Mal Lichtenstein (San Diego, CA) listened for a few minutes to an 81-element ANA sequence and Morse-Thue sequence described in *Mazes for the Mind.* They sounded very similar to him. He notes that the ratios A/N and 0/1 approach one in both sequences. He also notes that at most there are two of the same elements in succession for both sequences.

Human Life-span Perception Plots

Jeremy Weinstein (Walnut Creek, CA) writes

"*Mazes for the Mind* has inspired me to plot a human life-span in terms of its 'perceived' length. In other words, the first year is one, the second year is one half (since to the two-year-old his second year is half of his life), the third year is another third, and so on quasi-logarithmically until the 70th year is 1/70th. Using this method, an 82-year-old's life is not half over at 41 but rather at 7."

Chess Knight Magic Square

In *Mazes for the Mind* I discussed the magic square, often defined as a matrix divided into N^2 cells in which the integers from 1 to N^2 are placed in such a manner that the sums of the rows, columns, and both diagonals are identical. My favorite magic square, invented by 18th century mathematician Leonhard Euler is shown in the following diagram:

```
1  48 31 50 33 16 63 18
30 51 46  3 62 19 14 35
47 2  49 32 15 34 17 64      Chess Knight
52 29  4 45 20 61 36 13      Magic Square
5  44 25 56  9 40 21 60
28 53  8 41 24 57 12 37
43 6  55 26 39 10 59 22
54 27 42  7 58 23 38 11
```

Each horizontal or vertical row totals 260. Stopping halfway on each gives 130. Even more exciting is that a chess Knight, starting its L-shaped moves from the upper left box (marked "1") can hit all 64 boxes in numerical order. (You can locate the "2" to find the knight's first move.) Can you trace out the path of the knight through the board?

Ronald Brown (Pennsylvania) points out that my example is not a true magic square because the sum of the *diagonal* numbers is not 260. Over the years, Ron Brown has collected several knight's "near" magic squares that have identical row and column sums but whose diagonals lead to different sums. All the near magic squares he has found so far have a row and column sum of 260. He asks: Does there exist a knight's "true" magic square, i.e. a knight's tour that is also a true magic square with identical row, column, and diagonal sums? Does there exist a near magic tour that has row and column sums that are different than 260?

Circular Chess

Mazes for the Mind discussed a variation of chess played on a circular playing board. Recently a new product called *Centre Chess* came to market. It follows almost the same format as the traditional chess game, except for two "passing zone" wedges and "no man's zones." As a result, the rook moves in an arc and the bishop in an S-shaped path. Contact: Amerigames International, 15 Barlow Ave., Glen Cove NY 11542.

Another recent chess product includes Treyshah chess which permits three players. The board has hexagonal spaces so playing pieces can move in six directions instead of four. Contact: Shah Games, Box 701431, Salt Lake Citv, UT 84170-1431.

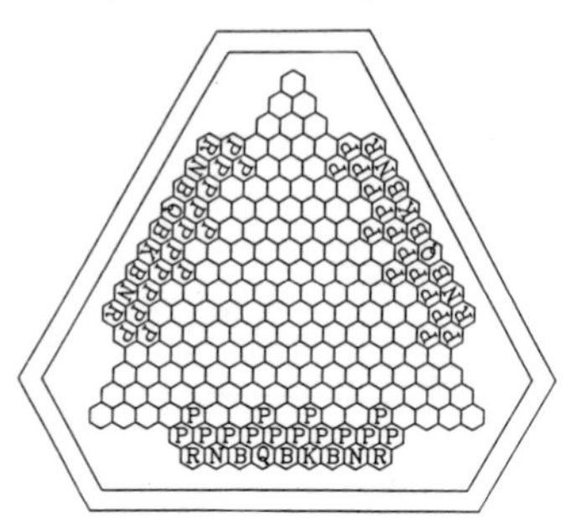

Desktop Aquarium

Mazes for the Mind discussed electronic ant farms. On a related topic, various electronic aquaria have recently been turned into products. For example, with El-Fish software, you can create an electronic fish tank complete with fish, aquatic plants, and coral. An artificial life program allows the fish to breed and mutate to form interesting new species. Contact: Maxis, 2 Theatre Square, Suite 230, Orinda, CA 94563-3346. (VGA monitor, 386 computer.)

Thomas Edison and the Personal Computer

In *Computers and the Imagination* I discussed the scientific and social impact of a personal computer placed in the year 1900 with manuals and a built-in reliability for one year. Charles Ashbacher (Cedar Rapids, IA) notes, "Everyone forgot that the greatest inventor of all time, Thomas Edison, was alive and at his peak in that year. Given the tremendous capability of the man, he would have learned of the PC and most likely would have been exposed to the physical nature of the machine. He would have been able to deduce much of the mechanism and we would have had much of our modern electronics decades later."

Jurassic Park

In *Mazes for the Mind*, I mentioned that the first published genetic sequence of a dinosaur occurred in Michael Crichton's novel *Jurassic Park* (1990). In his book, Crichton lists some 1000 bases starting with:

```
GCGTTGCTGG CGTTTTTCCA TAGGCTCCGC CCCCCTGACG
```

In my book I asked: "Do you think Crichton chose this sequence randomly, or is there some particular reason he selected this sequence of bases for use in his novel?"

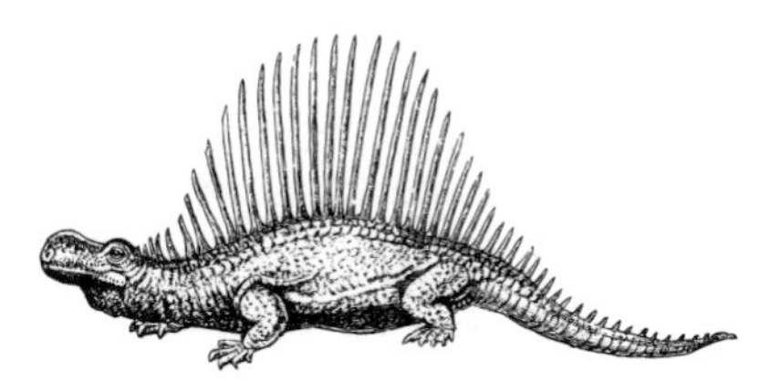

Mark Boguski, M.D., Ph.D. (of the National Center for Biotechnology Information in Bethesda, Maryland) has published a fascinating scientific paper which answers this question. Mark's first guess was that the sequence was either random or a chicken gene artfully mutated to take into account 270 million years of evolution. In his paper, he writes: "You can imagine my disappointment when this exotic sequence turns out to consist of pieces of lowly old (bacterial) pBR322, a man-made cloning vector and one of the most common pieces of DNA in the world." Mark used a Macintosh computer to search through genetic data bases to find the match. For more information: Boguski, M. (1992) A molecular biologist visits *Jurassic Park. BioTechniques.* 12(5): 668-669.

Unusual References

The entire June 1992 issue of the journal *IEEE Engineering in Medicine and Biology* was devoted to fractals and scaling theory (Vol 11, No 2). Topics in the issue included: fractal mechanisms in the electrophysiology of the heart, fractal time in protein switching, methods for estimating the fractal dimension from self-affine signals, fractal physiology, and others. • Neil Sloane's book *A Handbook of Integer Sequences* (Academic Press, 1978) is an amazing resource of unusual number sequences. A sequel is rumored to be in the works. • Mann, S. (1992) Chaos theory and strategic thought. *Parameters.* Autumn, pgs 54-68 (Mann writes, "Our views of the world are grounded in metaphors. The metaphors our culture prefers are Newtonian, yet the world is dynamical and nonlinear. The results is misperception and failure, particularly in national security affairs.") • Collins, J. et al. (1992) A random number generator based on the logit transform of the logistic variable. *Computers in Physics.* Nov/Dec 6(6): 630-632. (They discuss how to create random numbers using the logistic equation.) • Collins, J. and Stewart, I. (1993) Hexapodal gaits and coupled nonlinear oscillator models. *Biological Cybernetics.* 68: 287-298. (They discuss the use of nonlinear oscillators in modelling the walking of insects.) • Stone, D. (1969) Jumping Chinese. *Geotimes.* Oct., 14: 8. (He discusses the catastrophic (geophysical) consequences of all "seven hundred million Chinese" people simultaneously jumping off a two meter high stand. The various kinds of earthquake damage are discussed.) • Damme, R. (1989) On the dimension of a part of the Mandelbrot set. *J. Phys. A: Math Gen.* 22:5249-5258. • Pickover, C. (1993) Recursive worlds. *Dr. Dobb's Software Journal*, Sept., 18(9): 18-26. • Pickover, C.

(1994) The fractal golden curlicue is cool. *J. Recr. Math.*, in press. (On the thermodynamics of fractal curlicues.)

Gleichniszahleninventar Sequence

Roger Hargrave (West Sussux, UK) was inspired by the Gleichniszahlen-Reihe sequence discussed in *Computers and the Imagination* to extend the idea to a variation in which row $r+1$ takes into account *all* occurrences of each character in a single row r. For example, the sequence starting with 123 is: 123, 111213, 411213, 14311213,... He named this the *Gleichniszahleninventar Sequence* since Inventar is the German word for inventory. Oddly he finds that all his sequences finally oscillate between 23322114 and 32232114.

1597 Problem

In *Mazes for the Mind* I discussed integer solutions to $x=\sqrt{(1597y^2+1)}$. Paul Tourigny found the following amazing solution to a related problem I posed: $x=\sqrt{(1597y^2-1)}$. His solution is: $x=50976049658416210793518 2$ and $y=1275597675372598479252 5$. He believes this to be the smallest integer solution. Tourigny also found the following solution to one of the Brahmagupta problems posed in *Mazes for the Mind*: $66249^2-53\times 9100^2=1$.

Ten Formulas That Changed the Face of the World

In *Mazes for the Mind* I listed the "Ten Mathematical Formulas That Changed the Face of the World." Charles Ashbacher, book editor for the *Journal of Recreational Mathematics*, wrote to me with "significant disagreements with the list." For the record, Charles' Top Ten, with some of his explanations, are:

1. $1-2=-1$ (The positive integers are intuitively obvious. This formula establishes the existence of negative integers, the first "non-intuitive" set of numbers imagined by humans.)
2. $\sqrt{2}\neq m/n$ (This formula established the existence of irrational numbers and was the first instance where it was proven that some things will never be known.)
3. $a0b=a\times base\times base+0\times base+b$ (This formula establishes the concept of positional notation and the use of zero as a place-holder. This eliminated enormously cumbersome systems such as Roman numerals and greatly sped up all manner of computation. It also allowed arithmetic to be mechanized.)
4. $F=ma$
5. $E=mc^2$
6. $V=IR$
7. $\lambda=h/mv$
8. $F=(Gm_1m_2)/r^2$
9. $C=2\pi r$
10. $e^{\ln x}=x$

Errors

Frequently I receive letters with suggestions regarding my previous books. For example, Robert Stong (Charlottesville, VA) points out that at better value for Figure 9.15 in *Computers, Pattern, Chaos and Beauty* is $\mu=(-1.75,0)$. In Figure 9.22, the picture boundaries for λ are actually for z. In *Computers and the Imagination*, page 203, the Lute of Pythagoras ratios should be $CE/EG=EG/GI$. The initial isosceles triangle for the Lute should be a "Golden Triangle," i. e. one where $BC/AC=BA/AC=\phi$. Mal Lichtenstein (San Diego, CA) and others point out that the equation for the Dudley triangle should have:

mod $m + n + 1$. Michael Gordy (Somerville, MA) notes that Pseudocode 24.3 should use $w \rightarrow w/2 + z/4$. In the original version of *Computers and the Imagination*, the figures on 356 and 357 should be swapped. The figure for scorpion geometry is missing a Farsi character at the right vertex. In Chapter 29, the sequence is for row 16 not 15. Harry J. Smith (Saratoga, CA) and David Edelheit (Oyster Bay Cove, NY) report a few round-off errors for the Robbins numbers in *Mazes for the Mind*. The listed Robbins numbers were one too small for $n = 5, 7, 12, 15, 18, 19, 20, 24$, and 25. The error was probably due to the fact that even though all Robbins numbers are integers, some of the intermediate results in the algorithm I used are not. If intermediate results are stored as integers, some errors may occur. Smith and Edelheit found that $R(32)$ is Robbinmorphic since it ends in 32. The equation Smith suggests to compute the Robbins numbers is: $R(n) = R(n-1) \times (2n) \times (2n+1) \times \ldots (3n-2)/((n) \times (n+1) \times \ldots (2n-2))$. This equation can be easily implemented with an algorithm that has all integer intermediate results.

Penrose maze by Sampei Seki.
What's the shortest path from monkey A to B to C back to A?

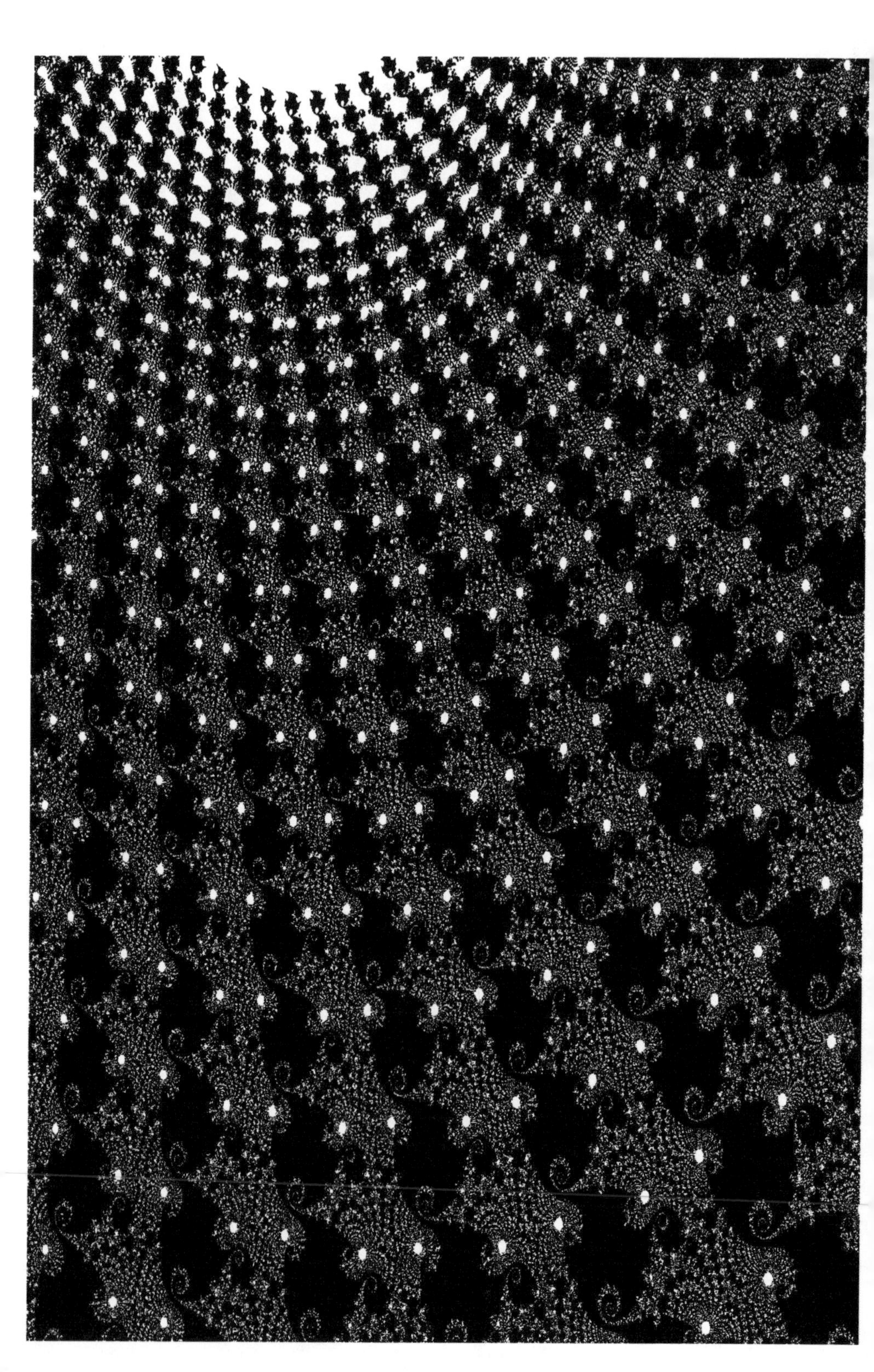

Appendix O

Credits and Additional References

Several wonderful artists have allowed me to reprint their astronomical art in this book. Dr. Paul Hartal (pictured at left) is an artist whose ideas, like his artwork, seems to span space and time. Not only is he a painter and poet, and an inventor of "lyrical conceptualism" (a new element in the periodic table of art), but Dr. Hartal is also the director of the Center for Art, Science, and Technology, in Montreal. The center is a cultural network of scientists and artists dedicated to the cause of "integrative research" and global cooperation. A man of many odysseys, Paul Hartal was born in Hungary and now lives in Canada. A pioneer of space exploration, he is an active member of the Swiss-based Orbiting Unification Ring Satellite project. His prize-winning artwork has been exhibited internationally. (See Figure 34.1, and the figures on pages 99, 102, 115, 131, 137, 151, 157, 168, 172, 176, and 200 for examples of his work.) Hartal can be reached at the Center for Art, Science, and Technology, P.O. Box 1012, St. Laurent, Montreal (Quebec) Canada H4L 4W3.

Arthur Gilbert of Derbyshire (U.K.) is a member of the International Association for the Astronomical Arts (IAAA) and has had his astronomical art exhibited internationally. (See page 1 and 5 for examples.) The IAAA is the first and only guild of astronomical artists. Originally conceived by a small core of professionals, they have grown to include student members and associate members including agents, collectors, planetariums, and writers. Their goal is to nurture the field of space art, providing a forum for the exchange of ideas and information.

Beth Avery of Portola Valley, California (pictured at top on the next page) is interested in astronomy, cosmology, and psychology. Her work has been exhibited around the world and in numerous magazines. She is a member of the Association of Science Fiction and Fantasy Artists (ASFA) as well as the IAAA. She writes, "The relationship of science and art to the human spirit is somewhat like food and water to our bodies; each is different but essential for a healthy life.... Art, in the highest form reaches into the depths of human consciousness to a realm where words fail." (See Color plate 15 and the figure on page 101.)

Gerardo Amor and Alisa Franklin (below) use a style of painting requiring spray paint. The method was created about eight years ago in Mexico City. Gerardo and Alisa spend their time "painting and dreaming." When I last talked to Alisa in Westchester, New York, she was about to spend several weeks exploring India. (Color plates 1 and 19 are examples of their work. See also the figures facing the Preface and on the Preface page, facing Part 1, and on pages 116, 128, and 156.)

The Zebrowski quotes scattered throughout the book come from: Zebrowski, G. (1992) "Life in Gödel's Universe: Maps All the Way." *OMNI*. April. 14(7): 53. The Packard quotations come from an interview in the January 1992 issue of *OMNI* (pg 85).

I thank R. Stong, K. McCarty, A. Laktakia, M. Frame, J. Steinbach, J. Warnock, H. Roth, and E. Poole for useful comments.

The astronomical photos of Jupiter, Ganymede, and related are courtesy of NASA, the Jet Propulsion Laboratory, and the California Institute of Technology. The Callisto photo is courtesy of the Voyager Experiment Team Leader, Dr. Bradford A. Smith, and the National Space Science Data Center. The Latööcarfian creatures' images come from: Huber, R. (1980) *Treasurer of Fantastic and Mythological Creatures*. Dover: NY. These renderings are based on photographs of soapstones carvings taken by Dennis Richard and published under the title *The Fantastic Art of Clark Ashton Smith* (Mirage Press, 1973). The coordinates for the atoms of aluminum gallium arsenide come from Tom Theis.

The idea that the Latööcarfian's aluminum gallium arsenide heads emit light as they think is not entirely fanciful. Nick Holonyak Jr. and his colleagues at the University of Illinois at Urbana-Champaign have made oxides of gallium arsenide to create laser diodes which emit, channel, and guide light beams. For further information, see Pennisi, E. (1992) Steamed, gallium arsenide guides light. *Science News*. April 4, 141(14): 214.

The phantom limb pain discussed at the end of Part II, "The Dream-Weavers of Ganymede" has been discussed in articles such as: Melzak, R. (1992) Phantom limbs. *Scientific American*. April, 266(4): 120-126. Phantom limb sensations have been recognized for some time. In the short story "The Case of George Dedlow," published in 1866, the protagonist loses an arm to amputation during the Civil War and later wakes up in a hospital not knowing both his legs have been amputated. Here is an excerpt from the story:

> I was suddenly aware of a sharp cramp in my left leg. I tried to get at it ... with my single arm, but, finding myself too weak, hailed an attendant. "Just rub my left calf if you please."
>
> "Calf? ... You ain't got none, pardner. It's took off."

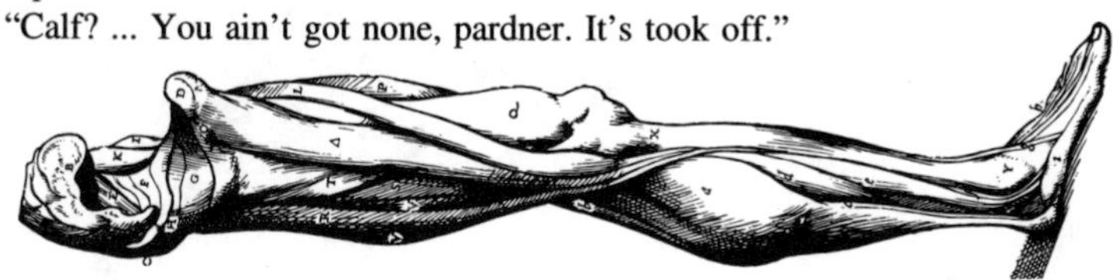

The intriguing idea of microchip ants with moving zinc oxide legs is almost a reality today. In 1991 Johannes Smits, a materials scientist at Boston University, developed a chip-etching process that neatly produces zinc-oxide beams which are thinner than a human hair and capable of bending 90 degrees in response to an electrical stimulus. This work is described in: Freedman, D. (1992) High hopes. *Discover*. May. 13(5): 20. Other work in slightly larger robot insects is being conducted by Rodney Brooks at MIT. In fact, all kinds of advanced, insect-like robots are being developed in Rodney Brooks' lab at the Massachusetts Institute of Technology (see Figure O.1).

The *Astrolabium Galileo Galilei* time piece described in Part II, "The Dream-Weavers of Ganymede" is available on earth (for navigation on Earth, not Ganymede) from Ulysse Nardin, 3 rue de Jardin, CH-2400 Le Locle, Switzerland. The price is $49,000 and up.

Tiny spiral antennas were mentioned in Part II, "The Dream-Weavers of Ganymede." Today, spiral antennas smaller than a human hair have been constructed at the National Institute of Standards and Technology (NIST). They are as small as a grain of sand and can be used to detect heat from missiles, buildings, and other objects.

There really is a substance called "armalcolite," the lunar material that was brought back to earth by astronauts Armstrong, Aldrin, and Collins and named in their honor. Armalcolite is an orthorhombic titnate of magnesium and iron first found in lunar rocks and analyzed in 1969. Six research groups independently discovered armalcolite (FeO - 5MgO - STi_2O_5) in their examination of different lunar samples collected from Mare Tranquillitatis. The first terrestrial occurrence of armalcolite was reported from DuToitspan in South Africa.

The idea of the jagged Koch snowflake sword carried by Prohaptor warriors is my own, but it was inspired by recent research in fractal drums. In 1991, Bernard Sapoval and his colleagues at the Ecole Polytechnique in Paris found that fractally shaped drum heads are very quiet when struck. Instead of being round like an ordinary drum head, these heads resemble a jagged snowflake. Sapoval cut his fractal shape out of a piece of metal and

stretched a thin membrane over it to make a drum. When a drummer bangs on an ordinary drum, the vibration spreads out to affect the entire drum head. With fractal drums, some vibrational modes are trapped within a branch of the fractal pattern. Faye Flam in the December 13th, 1991 issue of *Science* (vol. 254, p. 1593) notes: "If fractals are better than other shapes at damping vibrations, as Sapoval's results suggest, they might also be more robust. And that special sturdiness could explain why in nature, the rule is survival of the

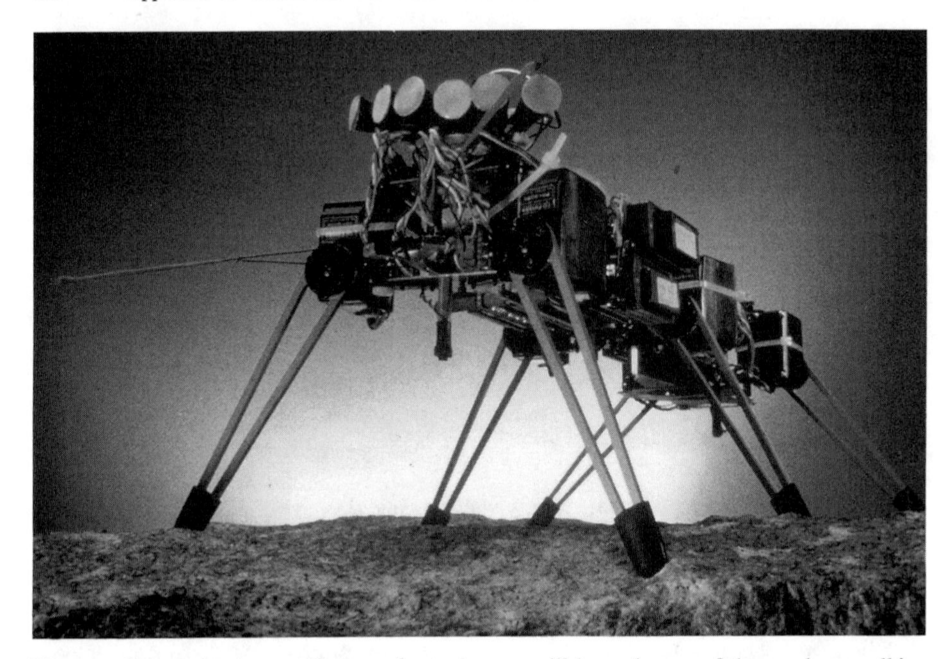

Figure O.1. *Robot insects.* Perhaps future houses will have dozens of these robots walking around – for both amusement and housecleaning. (Photo courtesy of R. Brooks, Artificial Intelligence Laboratory, MIT.)

fractal." Fractal shapes often occur in violent situations where powerful, turbulent forces need to be damped: the surf-pounded coastline, the blood vessels of the heart (a very violent pump), and the wind- and rain-buffeted mountain.

Today, a small number of computer chips are made using gallium arsenide, in comparison to more common "CMOS" and "bipoloar" chips which rely on different substances. The advantage of gallium arsenide is that it is faster than traditional semiconductors. However, one disadvantage is that gallium arsenide circuits require more power than CMOS technology. Another current disadvantage is that the gallium arsenide manufacturing process requires more steps than those for silicon bipolar or CMOS technology. Therefore, gallium arsenide chips are more expensive.

The information on calcite spines of the pipe-world worm *Aysheaia* (spines which are also used in Thorn's weapons) comes from: Pool, R. (1990) Materials tips from sea urchins. *Science.* Nov. 250: 629. Some of the descriptions of the physical characteristics and mating behavior of the Navanax people were stimulated by an article on sea slugs: Berreby, D. (1992) Sex and the single hermaphrodite. *Discover* June. 88-93. The quote from zoologist Janet Leonard comes from this article.

Several quotations on civilization and history come from: Durant, W., and Durant, A. (1968) *The Lessons of History.* Simon and Schuster: NY.

The idea of oscillating reactions forming a chemical computer is stimulated by: Browne, M. (1992) Chemists' new tools: molecular see-saws. (*NY Times*, Tuesday, April 28, C1). In papers recently published in the Proceedings of the National Academy of Sciences, Dr. John Ross of Stanford University and colleagues at the Max Planck Institute in Göttingen, Germany, offer rough blueprints for a hypothetical chemical computer.

The figures for the cosmological models resembling filaments and a folded cloth come from the fascinating paper: Klein, C. (1982) Woven heaven, tangled earth: a weaver's paradigm of the mesoamerican cosmos. In *Ethnoastronomy and Archaeosatronomy in the American Tropics*. Aveni, A. and Urton, G., eds. (Annals of the New York Academy of Sciences, Vol. 385, New York, NY). The artist is Henry F. Klein.

Several illustrations, including the fruit with the complicated network of tunnels, come from: Gamow, G. (1947) *One, Two, Three... Infinity.* Dover, NY. The frontispiece for Appendix H, "Meditations on Transcendentals" and the picture of Galileo come from: Dantzig, T, (1939) *Number: The Language of Science.* Macmillan: NY. Some of the unusual π trivia comes from Beckmann, P. (1971) *The History of Pi.* St. Martin's Press: NY.

The mole people were modelled after terrestrial mole-rats. To learn more about mole-rats, see: Braude, S. and Lacey, E. (1992) The underground society: the secret, communal life of the naked mole-rat. *The Sciences.* May/June 23-28. The introductory drawing of the airplanes traveling to the planets comes from Brubacher, A. (1928) *The Volume Library.* Educators Association: NY. The John Celestian poetry comes from: Celestian, J. F. (1973) *Words From A Graveyard* Exposition Press: NY.

Some of the facts relating to entomophagy discussed in Chapter 35, "The Glass Girls of Ganymede" come from: Rennie, J. (1992) Entomophagy. *Scientific American*, August 267(2): 20.

The idea for the inorganic double helices in Chapter 20, "Ganymedean Blood and Biology" comes from: Soghomonian, et al. (1993) An inorganic double helix: hydrothermal synthesis, structure, and magnetism of chiral $[(CH_3)_2NH_2]K_4$ $[V_{10}O_{10}(H_2O)_2(OH)_4(PO_4)_7]$ • $4H_2O$. *Science.* March 12, 259: 1596.

If you wish to learn more about electrorheological fluids, such as those making up the blood of Ganymedean wildlife, see: Ruthen, R. (1992) Fickle fluids. *Scientific American.* July. pg. 111.

The idea of the Glass Girls's stomachs serving as both a digestive and reproductive organ was inspired by the reproductive strategies of certain frogs. See for example: Duellman, W. (1992) Reproductive strategies of frogs. *Scientific American.* July. pgs. 80-87. Duellman also discusses transparent glass frogs.

The detailed, annotated maps of Ganymede are from: Moore, P., Hunt, G., Nicolson, I., and Cattermole, P. (1990) *The Atlas of the Solar System.* Crescent Books/Crown Publishers: NY. Some of the drawings of strange creatures come from: Haeckel, E. (1974) *Art Forms in Nature.* Dover: NY. The figure beneath the Henri Poincare quote is a collage by David Singer. The figure facing Appendix C is a collage by Jim Harter.

Clay Fried, a New York artist, drew the fractal sword and the figures on the first pages of Chapters 23, 27, 28, 30, 34, 36, and 38.

Glossary

This is an informal and brief reminder of the meanings of terms.

Amplexus. Typical mating embrace of amphibians.

Armalcolite. The lunar material that was brought back to earth by astronauts Armstrong, Aldrin, and Collins and named in their honor. See Appendix O, "Credits and Additional References" for chemical structure and more information.

Attractor. *Predictable attractors* correspond to the behavior to which a system settles down or is "attracted" (for example, a point or a looping closed cycle). The structure of these attractors is simple and well understood. A *strange attractor* is represented by an unpredictable trajectory where a minute difference in starting positions of 2 initially adjacent points leads to totally uncorrelated positions later in time or in the mathematical iteration. The structure of these attractors is very complicated and often not well understood.

Bathybius. A gelatinous deposit.

Bifurcation. Any value of a parameter at which the number and/or stability of steady states and cycles changes is called a bifurcation point, and the system is said to undergo a bifurcation.

Bifurcation equation. The nonlinear equation $x_{n+1} = kx_n(1 - x_n)$ is called the bifurcation or logistic equation, and it has been used in ecology as a model for predicting population growth.

Bioluminescence. Emission of light from living organisms.

Bouillabaisse. A dish composed of fish stewed in water.

Chaos. Irregular behavior displaying sensitive dependence on initial conditions. Chaos has been referred to by some physicists as the seemingly paradoxical combination of randomness and structure in certain nonperiodic solutions of dynamical systems. Chaotic behavior can sometimes be defined by a simple formula. Some researchers believe that chaos theory offers a mathematical framework for understanding much of the noise and turbulence that is seen in experimental science.

Chaotic trajectory. A chaotic trajectory exhibits three features. 1) The motion stays within a bounded region – it does not get larger and larger without limit. 2) The trajectory never settles into a periodic pattern. 3) The motion exhibits a sensitivity to initial conditions. See also *Chaos*.

Chebyshev polynomial. Curves named after the Russian mathematician Pafrutii L. Chebyshev (1821-1894). Chebyshev polynomials of degree n are usually denoted by $T_n(x)$ (the notation comes from the

French spelling, Tchebychef). $T_n(x)$ are given by the formula $T_n(x) = \cos(n \cos^{-1}x)$.

Complex number. A number containing a real and imaginary part, and of the form $a+bi$ where $i=\sqrt{-1}$.

Converge. To draw near to. A variable is sometimes said to converge to its limit.

Cycle. The cycle describes predictable periodic motions, like circular orbits. In phase plane portraits, the behavior often appears as smooth closed curves.

Diuturnal. Long-lasting.

Dynamical systems. Models containing the rules describing the way a given quantity undergoes a change through time or iteration steps. For example, the motion of planets about the sun can be modelled as a dynamical system in which the planets move according to Newton's laws. A discrete dynamical system can be represented mathematically as $x_{t+1}=f(x_t)$. A continuous dynamical system can be expressed as $dx/dt=f(x,t)$.

Electrorheological fluids. Liquids which can be transformed rapidly from a liquid to a solid and back again by varying an electric field. All electrorheological fluids are suspensions of particles, for example polymer or graphite particles in silicone or mineral oil.

Feedback. The return to the input of a part of the output of a system.

Fibonacci numbers. The sequence of numbers (1, 1, 2, 3, 5, 8, ...) is called the *Fibonacci sequence* after the wealthy Italian merchant Leonardo Fibonacci of Pisa, and it plays important roles in mathematics and nature. These numbers are such that, after the first two, every number in the sequence equals the sum of the two previous numbers $F_n = F_{n-1} + F_{n-2}$.

Fixed point. A point which is invariant under the mapping (i.e., $x_t = x_{t+1}$ for discrete systems, or $x=f(x)$ for continuous systems). A particular kind of fixed point is a *center*. For a center, nearby trajectories neither approach nor diverge from the fixed point. In contrast to the center, for a *hyperbolic fixed point*, some nearby trajectories approach and some diverge from the fixed point. A *saddle point* is an example of a hyperbolic fixed point. An *unstable fixed point* (or repulsive fixed point or repelling fixed point) x of a function occurs when $f'(x)>0$. A *stable fixed point* (or attractive fixed point) x of a function occurs when $f'(x)<0$. For cases where $f'(x)=0$ higher derivatives need to be considered.

Fractals. Objects (or sets of points, or curves, or patterns) which exhibit increasing detail ("bumpiness") with increasing magnification. Many interesting fractals, like coastlines, are self-similar, showing similar features as they are enlarged. B. Mandelbrot informally defines fractals as "shapes that are equally complex in their details as in their overall form. That is, if a piece of a fractal is suitably magnified to become of the same size as the whole, it should look like the whole, either exactly, or perhaps only after slight limited deformation."

Fritinancy. Twittering.

Gasket. A piece of material from which sections have been removed. *Mathematical gaskets*, such as Sierpiñski gaskets, can be generated by removing sections of a region according to some rule. Usually the process of removal leaves pieces which are similar to the initial region, thus the gasket may be defined recursively. The Sierpiñski gasket looks like a triangular net and is named after the prolific Polish mathemetician Waclaw Sierpiñski (1882-1969).

Henon map. The Henon map defines the point (x_{n+1}, y_{n+1}) by the equations $x_{n+1} = 1.4 + 0.3y_n - x_n^2$, $y_{n+1} = x_n$. Note that there are various expressions for the Henon map, including $x_{n+1} = 1 + y_n - \alpha x_n^2$, $y_{n+1} = \beta x_n$.

Henotheism. Belief in one god without asserting that he or she is the only God.

This is considered as a stage or religious belief between polytheism and monotheism.

Heterochromic. Multi-colored.

Heterotrophic. Organisms which require an external supply of energy contained in organic compounds.

Iteration. Repetition of an operation or set of operations. In mathematics, composing a function with itself, such as in $f(f(x))$, can represent an iteration. The computational process of determining x_{i+1} given x_i is called an iteration.

Hermaphrodite. An animal with both male and female characteristics.

Hilbert curve. Fractal curve named after German mathematician, David Hilbert (1862-1942).

Hippophagous. Eaters of horseflesh.

Ichthyophagous. Eaters of fish.

Julia set. Set of all points which do not converge to a fixed point or finite attracting orbit under repeated applications of the map. Most Julia sets are fractals, displaying an endless cascade of repeated detail. An alternate definition: repeated applications of a function f determine a trajectory of successive locations x, $f(x)$, $f(f(x))$, $f(f(f(x)))$, ... visited by a starting point x in the complex plane. Depending on the starting point, this results in two types of trajectories, those which go to infinity and those which remain bounded by a fixed radius. The Julia set of the function f is the boundary curve which separates these regions.

Koch curve. A snowflake shaped fractal object. It can be created by repeating a given operation over and over again, and was first proposed in 1904 by Swedish mathematician Helge von Koch.

Lachrymose. Causing tears.

Limit. In general, the ultimate value towards which a variable tends.

Logistic equation. The nonlinear equation $x_{n+1} = kx_n(1 - x_n)$ is called the logistic equation, and it has been used in ecology as a model for predicting population growth.

Lumbricoid. Worm-like.

Lyapunov exponent. A quantity, sometimes represented by the Greek letter Λ, used to characterize the divergence of trajectories in a chaotic flow. For a 1-D formula, such as the logistic equation, $\Lambda = \lim_{N \to \infty} 1/N \sum_{n=1}^{N} \ln |dx_{n+1}/ dx_n|$.

Lorenz attractor. Squashed pretzel-like shape discovered in 1962 by MIT meteorologist E.N. Lorenz.

Mandelbrot set. A bushy, intricate object created using mathematical feedback loops. For each complex number μ let $f_\mu(x)$ denote the polynomial $x^2 + \mu$. The Mandelbrot set is defined as the set of values of μ for which successive iterates of 0 under f_μ do not converge to infinity. An alternate definition: the set of complex numbers μ for which the *Julia set* of the iterated mapping $z \to z^2 + \mu$ separates disjoint regions of attraction. When μ lies outside this set, the corresponding Julia set is fragmented. The term "Mandelbrot Set" is originally associated with this quadratic formula, although the same construction gives rise to a (generalized) Mandelbrot Set for any iterated function with a complex parameter.

Oxide. A compound of oxygen and another element.

Peano curve. Fractal curve named after Italian geometer Giuseppe Peano (1858-1932).

Pentelic. From Mount Pentelicus, near Athens. Usually applied to famous white marble found there.

Piezoelectricity. Electric polarization in a substance resulting from the application of mechanical stress to certain crystals. Also, these materials can expand or contract when exposed to a voltage.

Polynomial. An algebraic expression of the form $a_0x^n + a_1x^{n-1} + \cdots a_{n-1}x + a_n$ where n is the degree of the expression and $a_0 \neq 0$.

Rational number. A number which can be expressed as a ratio of two integers.

Recursive. An object is said to be recursive if it partially consists of or is defined in terms of itself. A *recursive operation* invokes itself as an intermediate operation.

Sierpiñski gasket. See *gasket*.

Saracenic. Pertaining to the Saracens, nomadic people of the deserts between Syria and Arabia.

Steady state. Also called equilibrium point or *fixed point*. A set of values of the variables of a system for which the system does not change as time proceeds.

Strange attractor. See *attractor*.

Terraform. To make earth-like.

Teuthology. The study of squids.

Trajectory. A sequence of points in which each point produces its successor according to some mathematical function.

Transcendental numbers. Numbers like $e = 2.718...$, $\pi = 3.1415...$, and $2^{\sqrt{3}}$. These numbers are not the solutions of polynomial expressions having rational coefficients. The digits of π and e never end, nor has anyone detected an orderly pattern in their arrangement.

Transformation. The operation of changing (as by rotation or mapping) one configuration or expression into another in accordance with a mathematical rule.

Triglyph. Ornament.

Villein. A free common villager or peasant in a feudal class system.

Watusi. A popular dance of the 1960's.

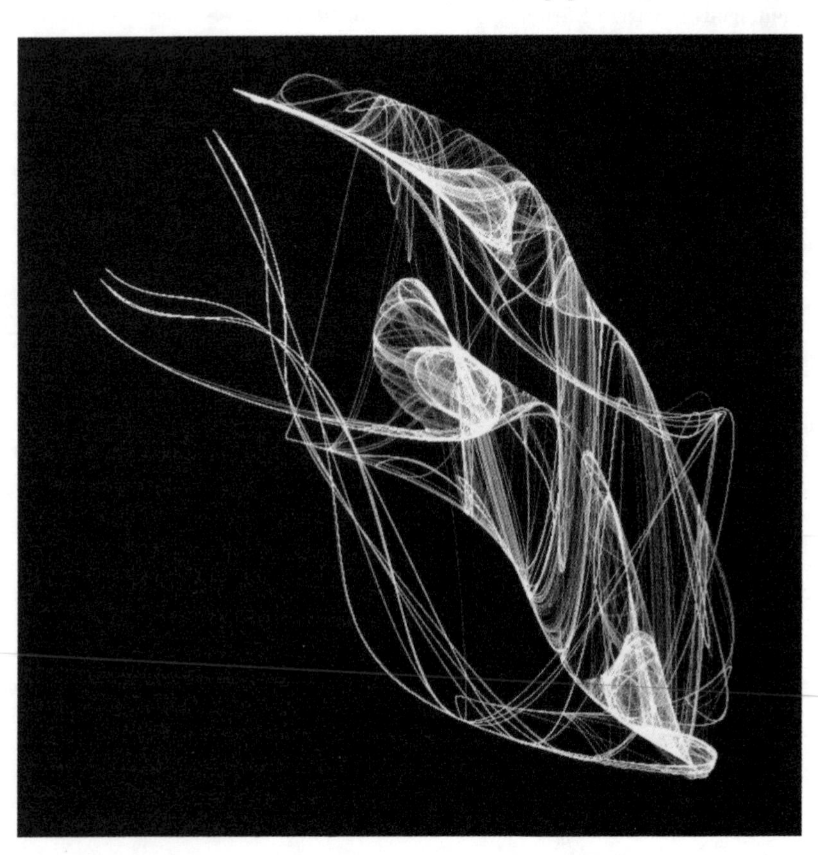

Index

D

E

F

G

H

I

J

K

L

M

U

V

W

Z

About the Author

Clifford A. Pickover received his Ph.D. from Yale University's Department of Molecular Biophysics and Biochemistry. He graduated first in his class from Franklin and Marshall College, after completing the four-year undergraduate program in three years. He is author of the popular books *Mazes for the Mind* (1992), *Computers and the Imagination* (1991) and *Computers, Pattern, Chaos, and Beauty* (1990), all published by St. Martin's Press – as well as the author of over 200 articles concerning topics in science, art, and mathematics. He is also coauthor, with Piers Anthony, of the highly-acclaimed science-fiction novel, *Spider Legs*. Pickover is currently an associate editor for the scientific journals *Computers and Graphics* and *Computers in Physics*, and is an editorial board member for *Speculations in Science and Technology*, *Idealistic Studies*, *Leonardo*, *YLEM*, and the *Poster Journal of Science.* He has been a guest editor for several scientific journals. Editor of *The Pattern Book: Fractals, Art, and Nature* (World Scientific, 1994), *Visions of the Future: Art, Technology, and Computing in the 21st Century* (St. Martin's Press, 1994), and *Visualizing Biological Information* (World Scientific, 1994), and coeditor of the books *Spiral Symmetry* (World Scientific, 1992) and *Frontiers in Scientific Visualization* (Wiley, 1994), Dr. Pickover's primary interest is in scientific visualization.

In 1990, he received first prize in the Institute of Physics' "Beauty of Physics Photographic Competition." His computer graphics have been featured on the cover of many popular magazines, and his research has recently received considerable attention by the press – including *CNN*'s "Science and Technology Week," *Science News*, *The Washington Post*, *Wired*, and *The Christian Science Monitor* – and also in international exhibitions and museums. *OMNI* magazine recently described him as "Van Leeuwenhoek's twentieth century equivalent." The July 1989 issue of *Scientific American* featured his graphic work, calling it "strange and beautiful, stunningly realistic." Pickover has received U.S. Patent 5,095,302 for a 3-D computer mouse. His hobbies include: tropical fish, prehistoric skull collecting, piano playing, science-fiction writing, African mask collecting, computer art, and electronic music synthesis. He can be reached at P.O. Box 549, Millwood, New York 10546-0549 USA.

The opinions and ideas expressed in this book are the author's and do not represent any organization or company.